T0139854

Springer Series in Translational Stroke Research

Series editor

John Zhang
Loma Linda, USA

The Springer Series in Translational Stroke Research will cover basic, translational and clinical research, emphasizing novel approaches to help translate scientific discoveries from basic stroke research into the development of new strategies for the prevention, assessment, treatment, and enhancement of central nervous system repair after stroke and other forms of neurotrauma.

More information about this series at http://www.springer.com/series/10064

Weijian Jiang • Wengui Yu
Yan Qu • Zhongsong Shi
Benyan Luo • John H. Zhang
Editors

Cerebral Ischemic Reperfusion Injuries (CIRI)

Bench Research and Clinical Implications

 Springer

Editors
Weijian Jiang
Department of Vascular Neurosurgery
PLA Rocket Force General Hospital
Beijing, China

Wengui Yu
Department of Neurology
University of California Irvine
Irvine, CA, USA

Yan Qu
Department of Neurosurgery
Tangdu Hospital of Air Force
Medical University
Xi'an, China

Zhongsong Shi
Department of Neurosurgery
Sun Yat-sen Memorial Hospital
Sun Yat-sen University
Guangzhou, China

Benyan Luo
Department of Neurology
The First Affiliated Hospital School of
Medicine, Zhejiang University
Hangzhou, Zhejiang, China

John H. Zhang
Department of Anesthesiology
Loma Linda University
Loma Linda, CA, USA

ISSN 2363-958X   ISSN 2363-9598 (electronic)
Springer Series in Translational Stroke Research
ISBN 978-3-030-07958-1   ISBN 978-3-319-90194-7 (eBook)
https://doi.org/10.1007/978-3-319-90194-7

Printed on acid-free paper

This Springer imprint is published by the registered company Springer International Publishing AG
part of Springer Nature.
The registered company address is: Gewerbestrasse 11, 6330 Cham, Switzerland

Preface

Introduction

On October 2, 2016, the Pangu Stroke Conference was held in Xinqiao Hotel, Hangzhou, China. About 80 prominent clinical and basic stroke researchers from the United States and China came together to discuss the most controversial issues in ischemic stroke management. The key issues under debate were the recanalization-induced cerebral ischemia and reperfusion injury (CIRI) (Chaps. 1, 2, 4–8), the safety and efficiency of late recanalization for chronic cerebral arterial obstruction (Chaps. 13–15), perspectives of neuroprotective therapies immediately after recanalization, characteristics of posterior circulation stroke, new stroke imaging technique applications, stroke and diabetes (Chaps. 10 and 11), cerebral vascular autoregulation and systemic immune response after stroke, as well as the role that cerebral venous system played in stroke.

Recanalization-Related Injury

Stroke therapy now encounters a new era highlighted with the application of endovascular therapy. However, over-reperfusion and recanalization-related brain injury have also emerged as new enigmas.

Professor Weijian Jiang from the PLA Rocket Force General Hospital presented a clinical case of a patient with acute ischemic stroke, who had a stroke with middle cerebral artery occlusion (MCAO) 4 h previously and received intravenous thrombolysis followed by thrombectomy. Although this patient fitted well in the time window of intravascular therapy and received each therapy following the guideline, with successful recanalization confirmed by later-on CT perfusion imaging, the patient, however, developed fatal hemisphere edema and died of circulation failure. Secondary injury following recanalization shown as hyper-reperfusion syndrome was regarded as the cause of death. Professor Jiang therefore proposed the following questions about the secondary injury induced by recanalization: In which settings will the recanalization-related injury occur?

What can we do to predict or prevent the recanalization-related injury? In which way can we differentiate it from primary brain injury caused by cerebral ischemia (Chap. 3)?

Therapies involving strict blood pressure control alongside recanalization therapy, targeted blood–brain barrier protection, the proper time point to administer anticoagulant as well as antiplatelet, and the protection of venous system during recanalization propelled the meeting to a heated debate.

Professor Wu-wei Feng from South Carolina, USA, reminded the audiences about the deviation of race, age, and stroke onset for individual stroke patients, and he argued that the treatment should be individualized. Professor Feng proposed two promising ways for reducing recanalization-related injury. One is by reducing the alteplase dosage to 0.6 mg/kg from original 0.9 mg/kg, which was proved in ENCHANTED trial, showing that low-dose alteplase was non-inferior in the ordinal analysis of modified Rankin scale scores, and causing significantly fewer symptomatic intracerebral hemorrhages. Another is to minimize the re-occlusion of the artery in thrombolytic patients according to the data of a preliminary clinical trial conducted by Professor Meng Zhang from Daping Hospital of the Third Military Medical University, in Chongqing, China. Early initiation of tirofiban reduced the risk of re-occlusion after intravenous t-PA and avoided higher risk of hemorrhagic complication. The most impressive results were that this early tirofiban therapy also significantly improved 3 months' outcomes in acute ischemic stroke patients receiving alteplase.

Professor Min Lou from the Second Affiliated Hospital of Zhejiang University, Hangzhou, China, presented the data from nearly 1000 thrombolytic patients. She called for attention to endothelial injury after thrombectomy and also preferred the early initiation of antiplatelet therapy. As one of the most experienced thrombolytic stroke specialists in China, she emphasized that the key point in thrombolysis and intravascular therapy is "fast" (Chap. 4).

Professor Yi Yang and Li-ping Liu, both from China, lectured on futile recanalization which has been defined as unfavorable 3 months neurological outcome despite successful recanalization. Various reasons for the futile recanalization were considered, including re-occlusion of the artery, hypo-perfusion due to iatrogenic reasons, microcirculation deficiency, and impaired cerebrovascular reserve capacity.

Professor Yi Yang also underlined the importance of cerebral vascular autoregulation, which is defined as the mechanism by which the constant cerebral blood flow is maintained despite changes of arterial blood pressure. Professor Yang pointed out that the progress of new imaging constructive technique and the calculation of supercomputer would surely help provide new perspective in this field.

Safety and Efficacy of Late Recanalization

Although early recanalization has now proved effective for patients arriving within the therapeutic time window, the number of patients, however, is limited to only about 7% for American stroke patients. Therefore, the question raised is whether late recanalization would help to improve patients' outcome.

Professor Weijian Jiang demonstrated another case of late recanalization from a patient with symptomatic chronic middle cerebral artery (MCA) obstruction. Anticoagulation combined with dual antiplatelet therapy was administrated upon admission. A large intracerebral hematoma with air-fluid level in the hematoma cavity was found 24 h at plain CT and micro-invasive surgery was applied to remove the hematoma, and final improvement of neurologic function was observed in the patient. Cerebral amyloid angiopathy and anticoagulation-related hemorrhage were concern for this patient, referring to the clinical and imaging presentation (Chap. 3).

Professor Wengui Yu from the University of Texas Southwestern Medical Center, USA, argued that the blood–brain barrier disruption as well as blood pressure fluctuation should be taken into consideration. During ischemic stroke, the regional perfusion was inevitably restrained. As such, a relative extremely hypo-perfusion area may be induced when the systemic blood pressure was lowered to a "normal" range. The abrupt reperfusion following recanalization would therefore bring about high risk of over-perfusion and hemorrhage. Monitoring the perfusion pressure in the ischemic area may help to adjust the systemic blood pressure (Chap. 16).

Professor Li-qun Jiao from Xuan Wu Hospital, Capital Medical University, China, presented his experience based on more than 3000 internal carotid stenosis and occlusion cases. In his opinion, early recanalization of acute stroke and late recanalization of chronic occlusion were two quite different settings. They have varied clinical presentations, perfusion status, and thereby different risks of hemorrhagic or thromboembolic incidence. Professor Li-Qun Jiao regarded the intervention of late recanalization as a secondary prevention therapy for stroke and argued that the mental status evaluation of these patients with chronic artery occlusion should be highlighted, as these patients usually presented with cognitive impairment, epilepsy, and other psychiatric symptoms. Professor Li-qun Jiao finally displayed a clinical case of late recanalization with significant hyper-perfusion using transcranial Doppler (TCD) monitoring. This phenomenon sustained for 5 days after intervention and gradually faded away without any complication. It would be a quite interesting issue for exploring the mechanism of this late recanalization-related hyper-perfusion.

The Application of New Imaging Technique and Strategies

The new imaging techniques including high-resolution magnetic resonance imaging (HR-MRI), computed tomography perfusion (CTP) or magnetic resonance perfusion (MRP), and sensitive weighted imaging (SWI) are applied in clinical practice and open a new page of individualized medical management.

Professor Bing Zhang from Drum Tower Hospital of Nanjing University Medical School and Professor Min Lou from the Second Affiliated Hospital of Zhejiang University gave lectures about the new imaging techniques and their implication for clinical practice. They first highlighted the importance of high-resolution magnetic resonance imaging (HR-MRI). The digital subtracted angiography (DSA) was previously assumed as the "gold standard" for evaluating vascular abnormalities. Actually, some recent studies have revealed that a proportion of patients with negative results in DSA are in fact, have various vascular abnormalities, including dissection, small plaque, and neovascularization. Additionally, DSA cannot evaluate the vessel wall while HR-MRI can, which is more meaningful in deciding therapy and evaluating surgery risk. For example, eccentric stenosis with irregular enhancement of vessel wall usually suggests unstable plaque, while longitudinal and general stenosis with uniform enhancement often indicates vasculitis. HR-MRI is also now applied in evaluating collateral and dynamic change of vessel after treatment. For the vessels obstructed chronically, the HR-MRI would reveal neovascularization into the vessel wall and a direct arterial-venous fistula, which both fade after successful recanalization.

Another new technique of CT perfusion or MR perfusion imaging has now become more prevalent in clinical practice, not only in selecting patients for receiving recanalization, but also in quantifying the function of blood–brain barrier. Blood–brain barrier is now considered critical for predicting hyper-perfusion and hemorrhagic transform.

The sensitivity weighted imaging (SWI) is valuable in studying cerebral venous changes and iron deposition. However, the pathologic mechanisms underlying the engorgement of cerebral veins are still under debate. Some researchers point out that the visibility of veins may not indicate enlargement of the venous lumen. Instead, it may be related to increased oxygen deprivation fraction or flow stasis.

One more imaging technique called diffusion tensor imaging (DTI) is now found valuable to predict motor dysfunction in stroke patients and is also important in selecting appropriate patient for intravascular therapy. Professor Min Lou proposed that it may be more feasible to develop new analytic software for current imaging. A most intriguing concept is to deliver microcapsule to the target area by intravascular device and monitor the medicine release with specific contrast.

Apparently, the flourishing new techniques of imaging are indispensable assistance for promoting our understanding of neurology.

The Characteristics of Posterior Circulation Stroke

Professor Yi-long Wang from Beijing Tiantan Hospital, Capital Medical University, and Dr. Shen-qiang Yan from the Second Affiliated Hospital of Zhejiang University presented their recent studies on posterior circulation stroke. A few novel opinions were proposed. Firstly, the different flow rate and intra-lumen pressure determine different models of hemodynamics when stroke occurs. It is well acknowledged that the posterior circulation bears more abundant collateral network, and that more plasma flows surrounding a clot when a thrombosis occurs in basilar artery. Therefore, the time window for the posterior circulation stroke is supposed to be extended. Moreover, the current scale for neurological deficiency is mostly suitable for anterior circulation stroke rather than posterior circulation stroke, especially for the widely used NIH stroke scale. Thirdly, the risk of hemorrhagic transformation varied between anterior and posterior circulation stroke in the setting of recanalization.

Dr. Shen-qiang Yan raised distinct features of top of the basilar syndrome from other basilar artery obstruction. Nowadays, although intravascular therapy is strongly recommended for large artery occlusion, whether it is the same for the top of the basilar syndrome is still controversial. According to the analysis of the corhort of about 1000 cases acute stroke patients receiving recanalization therapy, Dr. Yan highlighted the significant favorable efficacy of intravanous thrombolysis for the patients with top of basilar syndrome than other types of posterior circulation stroke. The explanation was that the embolus in top of the basilar syndrome is usually cardiac, and the visualization of both posterior cerebral arteries suggests residue flow around the clot. In this context, t-PA are more likely to reach around and lysis the clot in the top of basilar artery.

Professor Xun-ming Ji presented the clinical trial of Basilar Artery Occlusion Chinese Endovascular Trial, and in his opinion, another distinct feature of the posterior circulation stroke is the relative narrow space in the posterior fossa. It means that, when the brain stem gets swollen due to the posterior circulation infarction, the other tissues including veins and easily compressed and lead to a series of specific pathological changes.

Professor Wengui Yu from the University of Texas Southwestern Medical Center then presented three cases of posterior circulation infarction receiving late recanalization. One of them was a teenage boy who has been suffering from "locked-in" syndrome for more than 16 years. After late recanalization, these patients demonstrated mild or significant improvement of neurological function without hemorrhagic complications. Professor Wengui Yu hypothesized that posterior circulation may have lower risk of hemorrhagic complication due to abundant collaterals. Professor Wengui Yu also suggested that large-scale clinical trial for posterior circulation stroke should be considered (Chap. 16).

The Immune Regulation and Stroke

In the past, stroke was considered only as a disease of cerebral vessels that, like water pipes, were obstructed or ruptured. It was not until about a decade ago that people began to recognize that stroke and the immune system interacted with each other.

Professor Heng Zhao from Stanford University and Professor Jun-wei Hao from the General Hospital of Tianjin Medical University elaborated on the topic of immune system and stroke. Although the brain was thought to be isolated from the peripheral immune system due to the existence of blood–brain barrier (BBB), the ictus of stroke would not only lead to the disruption of BBB, but also activate the systemic immune response, including the brain cells. Therefore, the stroke patients are at high risk of infections, especially stroke-associated pneumonia (SAP). There is multi-centered large-scale clinical trial about the prophylactic use of antibiotics in SAP, yet revealed negative result. It may indicate that the main cause of SAP is not the exogenic pathogens but "the Trojan's Horse," the immune response inside the brain (Chap. 9).

Professors Zhao and Hao also suggested that atherosclerosis is closely related to the chronic inflammation of arteries and that oxidative stress is a critical part of the whole chain. Besides the ox-LDL and foam cells we are familiar with, the endothelial cells and the platelet are also considered to play special role in atherosclerosis and thromboemboli. In another aspect, cognitive impairment and dementia are spotted as having vascular reasons. Those who presented with severe leukoaraiosis also showed higher incidence of stroke in the follow-up observations. Although immune therapy for stroke is a newly sprouted direction, noteworthy, a few clinical trials conducted by Professor Fu-dong Shi in General Hospital of Tianjin Medical University showed promising results for immune regulatory medicine in stroke patients.

Professor Jun Chen from the University of Pittsburgh School of Medicine and Professor Xing-shun Xu from the Second Affiliated Hospital of Suzhou University emphasized the BBB as a therapeutic target for stroke. The study conducted by Professor Chen demonstrated the disruption of BBB and the expression of matrix metalloproteinases-9 (MMP-9) in a vivid and elaborate way. It is highly noticeable the role endothelial cells play in this course and the hemorrhage thereafter. Although several explorations of hypothermia treatment showed frustrating results, sub-hypothermia or regional hypothermia therapy is proposed for stroke patients.

The Cerebral Venous System and Stroke

Professor Xun-ming Ji from Xuanwu Hospital of Capital Medical University and Professor Yan Qu from Tangdu Hospital of Fourth Military Medical University put forward the interwoven between the cerebral venous system and stroke. The most important point is that the cerebral venous system has been neglected for a long period. Decades ago, the cerebral venous system, like arterial system, was already found composed of different grade of collaterals, including venous circle in the base of brain. There is growing evidence that the venous sinus anomaly is closely related to unfavorable outcome of stroke patients, involving malignant edema, futile recanalization, and hemorrhage. Professor Ji proposed that the increased venous pressure could produce hyper-intracranial pressure and venous infarction as well as hemorrhage, which formed a vicious circle. The device detecting fractional flow reserve in venous sinus is now being applied in those who are suspected of having venous hypertension in Xuanwu Hospital, and so is the noninvasive monitoring with those patients, said Professor Ji. Evidence is emerging that migraine is closely associated with venous outflow deficiency. Professor Wengui Yu also mentioned that the venous early opacification was related to high risk of hemorrhagic transform and may relate to failure of arterial autoregulation.

In contrast with the clinical calling for cerebral venous study, basic studies using venous sinus animal model are still inadequate and at the beginning stage. Alongside the more improvement of specific imaging technique for cerebral venous evaluation, we are able to aquire more information and thus explore more hidden treasure in this field.

Professor John. H. Zhang made a brief but thought-provoking statement in this part of discussion that focused on the physiologic structural and functional characteristics for cerebral view and called out for more insight into the cerebral venous system.

Hangzhou, China   Lusha Tong
Loma Linda, CA, USA   John H. Zhang

Contents

About the Editors

Weijian Jiang is the Chairman of New Era Stroke Care and Research Institute of PLA Rocket Force General Hospital, Beijing, China. Dr. Jiang is an established clinical researcher in endovascular therapy for cerebrovascular disease and has experience with over 20,000 interventional radiology cases. Dr. Jiang has published four books in stroke-related issues.

Wengui Yu is Professor of Clinical Neurology and Vice Chair for Hospital Affairs at the Department of Neurology and Director, Comprehensive Stroke & Cerebrovascular Center, University of California, Irvine. Dr. Yu is an established expert in neuro-ICU and stroke management.

Yan Qu is the Professor and Director of the Department of Neurosurgery at Tangdu Hospital of Air Force Medical University, Xi'an, Shaanxi, China. Dr. Qu's neurosurgical expertise includes cerebrovascular diseases and brain tumors and has published more than 50 articles in stroke research.

Zhongsong Shi is Professor and Deputy Director of the Department of Neurosurgery at Sun Yat-sen Memorial Hospital, Sun Yat-sen University, Guangzhou, China. Dr. Shi is an academic physician with neurosurgical and endovascular treatment experience for cerebrovascular disease and has published more than 30 papers in stroke research in internationally recognized journals.

Benyan Luo is a Professor and Chair of Department of Neurology at the First Affiliated Hospital of Zhejiang University. She has made substantial contribution in cerebrovascular diseases, neurodegenerative diseases, and neuropsychology.

John H. Zhang is Professor of Anesthesiology, Neurosurgery, Neurology, and Physiology and Director, Center for Neuroscience Research at Loma Linda University School of Medicine, Loma Linda, CA, USA. Dr. Zhang is an internationally recognized expert in stroke research.

Chapter 1
Gelatinase-Mediated Impairment of Microvascular Beds in Cerebral Ischemia and Reperfusion Injury

Shanyan Chen, Hailong Song, Jiankun Cui, Joel I. Shenker, Yujie Chen, Grace Y. Sun, Hua Feng, and Zezong Gu

Abstract Stroke is one of the leading causes of death, and acute ischemic stroke (AIS) is the most common form. Tissue plasminogen activator (tPA) is the only FDA-approved drug for recanalization in AIS with narrow therapeutic window. In this chapter, we will discuss the activation of gelatinases (MMP-2/9), one of the major mediators in cerebral ischemia and reperfusion injury (CIRI) with exogenous tPA in AIS. First, we briefly overview the structure of microvascular beds and the homeostasis of neurovascular unit associated with the extracellular matrix (ECM). Then we review the gelatinase-mediated degradation of ECM and the impairment of microvascular beds in AIS. Moreover, we discuss the self-perpetu-

S. Chen · H. Song
Interdisciplinary Neuroscience Program, University of Missouri School of Medicine, Columbia, MO, USA

Department of Pathology and Anatomical Sciences, University of Missouri School of Medicine, Columbia, MO, USA

J. Cui · Z. Gu (✉)
Department of Pathology and Anatomical Sciences, University of Missouri School of Medicine, Columbia, MO, USA

Harry S. Truman Memorial Veterans' Hospital Research Service, Columbia, MO, USA
e-mail: guze@health.missouri.edu

J. I. Shenker
Department of Neurology, University of Missouri School of Medicine, Columbia, MO, USA

Y. Chen · H. Feng
Department of Neurosurgery, Southwest Hospital, Third Military Medical University, Chongqing, China

G. Y. Sun
Department of Biochemistry, University of Missouri School of Medicine, Columbia, MO, USA

Department of Pathology and Anatomical Sciences, University of Missouri School of Medicine, Columbia, MO, USA

© Springer International Publishing AG, part of Springer Nature 2018
W. Jiang et al. (eds.), *Cerebral Ischemic Reperfusion Injuries (CIRI)*, Springer Series in Translational Stroke Research, https://doi.org/10.1007/978-3-319-90194-7_1

ating loop of gelatinase activation in CIRI with exogenous tPA. At last, we literature available approaches showing protective functions through blocking the
vicious circle of gelatinase activation, which may hold great promise in combined,
treatment with tPA in AIS.

Keywords Gelatinases · Microvascular beds · Cerebral ischemia and reperfusion
injury · Extracellular matrix · Mechanism-based inhibitor · tPA

1   Introduction

Stroke is a challenging medical problem world-wide as the second leading cause of
death globally [1, 2]. On the average, someone in the United States has a stroke
every 40 s, and someone dies of stroke every 4 min [3]. Stroke burden in Asia is
particularly high. Asians account for 60% of people having stroke worldwide annu-
ally [4–8]. Moreover, stroke prevalence and burden in China are increasing over the
last three decades [9].

Acute ischemic stroke or AIS, normally caused by thromboembolic obstruction
of the cerebral arteries, is a devastating neurological disease process accounting for
approximately 87% of all stroke cases, but with limited availability of effective
treatments [10]. Acute therapies for ischemic stroke are mainly based on the
"recanalization hypothesis" [11]. In the United States, recombinant human tissue-
type plasminogen activator (tPA) is the only Food and Drug Administration (FDA)-
approved pharmacological treatment for patients with AIS. Thrombolysis with tPA
can be presented within 4.5 h after stroke onset [12, 13]. However, due to risks of
hemorrhage in stroke patients, tPA must be given within a limited time window and
subject to stringent patient eligibility criteria, so only 5.8% of patients can get such
therapy [14]. Interestingly, administration of tPA intravenously shows minimal
toxicity in studies with non-ischemic mice [15] or patients in early stage after AIS
[14]. In contrast, injection of tPA into cerebral ventricles in non-ischemic mice
increased permeability of the blood-brain barrier (BBB) [15, 16]. Therefore, it is
reasonable to consider that tPA plays its fibrinolytic role as the beneficial action
inside the blood vessel. If so, then perhaps exogenous tPA may exert its deleterious
effects in brain parenchyma through impaired microvasculature [17–19].
Consequently, pre-recanalization conditions of the cerebral microvasculature may
affect outcomes of recanalization in ischemic stroke.

Matrix metalloproteinases (MMPs) are a family of 23 zinc-dependent endopeptidases in humans that serve as physiological mediators of matrix degradation
or turnover for tissue homeostasis and remodeling [20, 21]. MMPs are expressed
into the extracellular matrix (ECM) as inactive zymogens and become proteolytically active by a mechanism called 'cysteine switch' [22]. A subclass of MMPs,
gelatinases (MMP-2/-9), also call gelatinases for proteolysis of gelatin and have
been well studied as potential markers or therapeutic target in both experimental
cerebral ischemia paradigms, as well as in patients with AIS, especially in the
disruption of BBB [23–26].

In this chapter, we review the role of gelatinases in proteolytic modulation of the microvascular beds within the neurovascular unit (NVU), a conceptually functional structure in the central nervous system (CNS) of neurons, glial cells, endothelial cells and pericytes along with their tightly associated components of the ECM [27, 28]. Understanding of how aberrant activity of gelatinases contributing to the disturbances of cell-matrix homeostasis occur in the microvascular beds with respect to cerebral ischemia and reperfusion injury (CIRI) may hold significant promise for the outcome of recanalization in ischemic stroke.

2 Microvascular Beds and the Extracellular Matrix (ECM)

The BBB functions to restrict influx of specific compounds from blood to the brain parenchyma. The anatomical substrate of the BBB is the cerebral microvascular beds comprising of endothelial cells, pericytes and astrocyte end feet as well as the ECM components [29, 30]. The microvascular bed together with neurons and glial cells constitute the NVU, the basic functional unit in the CNS contributing to the dynamic regulation of microvascular permeability [30, 31].

In the brain, endothelial cells are connected to the vascular beds via paracellular tight junction proteins, such as claudins and zona occludin (ZO), which limit paracellular diffusion [32, 33]. Recent studies indicate the intricate signaling pathway regulating activity of these proteins [34]. Proinflammatory cytokines such as TNFa was shown to alter metabolism of occludin and other tight junction proteins, and leading to changes in cell permeability [35]. Pericytes are specialized cells, with features similar to vascular smooth muscle cells. They are found around the capillary endothelium on the ablumenal side, and can be interspersed at different locations along the length of the capillary. Pericytes constrict around capillaries in response to different stimuli and maintain proper capillary lumen diameter, and thereby may play an important role in regulating blood flow through capillaries [36, 37]. Analysis of pericyte-deficient mice (e.g., *Pdgfb* mutants) reveals that pericyte deficiency leads to increased BBB permeability, which probably occurs by endothelial transcytosis [38]. Paradoxically, recent reports show variations in capillary capacity to constrict in order to regulate regional blood flow occur due to the heterogeneity of pericyte forms at the pre-capillary arteriole and post-capillary venule [39]. Astrocytic end feet cover endothelial cells and pericytes. The gap occurring between the end feet allows basement membranes to contact with microglial and neuropil components [40, 41]. In this way, astrocytes maintain barrier properties of endothelial cells while still allowing some contacts at specific sites [42].

Importantly, the dynamic structure of the microvascular beds is highly regulated by interactions between different cells and ECM components [42]. The main physical function of the ECM is to provide support to the components of the NVU via linkage with different cell adhesion molecules and therefore maintaining intercellular communication. The basal lamina is a specialized layer of the ECM on the ablumenal side of endothelial cells. ECM receptors, such as integrins and dystroglycan, mediate the connections between cellular and matrix components [42]. The major protein components of the ECM in the CNS include: structural elements (type IV collagens and elastin), specialized proteins (laminins, entactin/nidogen, fibronectin, and vitronectin), and proteoglycans [heparan sulfate proteoglycans (HSPG), perlecan, and agrin] [42–49].

Laminins, major components of the ECM, are glycoproteins with the isoforms containing three chains: alpha, beta, and gamma chains [50, 51]. These trimeric proteins form a cross-like structure that can bind to other cell membrane and ECM molecules. Laminins is known to play an important role in axonal guidance during neural development [52]. Defects in laminin are associated with developmental anomolies [53] and increased susceptibility to stroke leading to neurovascular impairment [54, 55]. At the microvascular beds, laminins are secreted by endothelial cells and astrocytes located in the basal lamina, and this secretion is regulated by pericytes [38, 42, 56–58]. On the other hand, laminins also regulate pericyte differentiation and maintains BBB integrity and function [59–61] [see more details in Chap. 8 on BBB and Stroke by Yao].

Degradation of ECM components may prohibit cells from interacting with ECM, and this can cause a form of cell detachment known as anoikis, in which cells lose the interactions with their matrix [55, 62, 63]. In focal ischemic mice, ECM components such as laminin, dystroglycan and α6 integrin are degraded and these changes are correlated with vascular breakdown [64]. Acute disruption of astrocytic laminin may lead to impaired function of vascular smooth muscle cells and cerebral hemorrhage [65]. Cerebral ischemia is also associated with a decrease in the laminin-binding proteins α6β4 and dystroglycan, and this is coincident with astrocyte swelling [42, 66–68]. Moreover, laminin antibodies can disrupt cell-laminin interactions and cause neuronal cell death [55, 69].

3 Gelatinases Mediate Impairment of Microvascular Beds in CIRI

Substantial studies reveal that activation of gelatinases lead to proteolysis of ECM and subsequently impact the integrity of microvascular beds after cerebral ischemia [54, 55, 70, 71]. Gelatinases are named due to their ability to degrade gelatin (the denatured form of collagen), and are comprised mainly of MMP-2 (gelatinase A) and MMP-9 (gelatinase B). Gelatinases are widely studied in the pathogenesis of CNS diseases, and activation of gelatinases under different pathologic conditions has been shown to cause detrimental outcomes, including BBB disruption [72, 73], hemorrhage, neuronal apoptosis [69, 70], brain damage in AIS [63] and traumatic brain injury [74]. MMP-9 activity is significantly elevated in humans AIS [70, 75, 76], and its activation can be considered an independent predictor of hemorrhagic transformation in AIS [77]. In addition, cerebral infarction is reduced in mice deficient in MMP-9 [78] or after treatment with MMP inhibitors [54, 55]. Studies by our group as well as others have related aberrant MMP-9 proteolytic activity with increased permeability of the BBB, leading to brain edema and hemorrhage, and directly contributes to neuronal apoptosis and brain damage after acute cerebral ischemia [24, 63, 79, 80]. Different cell types have been shown to be important sources of MMP-9 in CNS, including endothelial cells, neutrophils, microglia/macrophage and neurons [24, 81]. The roles of pericytes in CIRI is actively investigated. Activation of MMP-9 at pericyte somata is demonstrated inducing proteolytic degradation of the BBB rapidly [82]. Detachment of pericytes from the basal lamina and migration of pericytes toward the hypoperfusion lesion may be associated with the MMPs expression of pericytes [81]. Gelatinase-induced disruption of microvascular beds may lead to multiple damages: (1) Gelatinases could degrade tight junction proteins such as occludin, claudins and ZO-1, which leads to disruption of tight junctions between adjacent endothelial cells [83, 84]. Early inhibition of MMP activity in the early stage of ischemia may promote expression of tight junction proteins and angiogenesis during recovery [85]; (2) Gelatinases could degrade ECM and cause dysfunction of endothelial cells and loss of microvascular integrity [70, 86]. Integrin–matrix interactions in the cerebral vasculature may explain the mechanism of how degradation of laminin in ECM mediates the loss of endothelial cells after ischemia [67, 87]; (3) Gelatinase disruption of the ECM may lead to contraction of pericytes and loss of pericyte-endothelial interaction [54, 88, 89]. The increased pericyte tone and death of pericytes will restrict cerebral blood flow in the ischemic area [90, 91] and result in persistent hypoperfusion even with successful recanalization by tPA [92–95]; and (4) Pericytes regulate BBB integrity with an important inhibitory mechanism to reduce the number of caveolae and rate of transcytosis [38, 96, 97]. Gelatinase-induced migration of pericytes away from the brain microvessels may relieve the inhibitory signaling of transcytosis and thus increase the trans-endothelial permeability [89, 98].

4 Activation of Gelatinases and Extravasation of tPA into Brain Parenchyma in CIRI-Induced Microvascular Bed Damage

The disruption of microvascular beds in AIS follows a biphasic time course, and may correlate with the biphasic increase of gelatinase level during CIRI [99, 100]. This biphasic feature may be explained by the self-perpetuating loop of gelatinase activation in microvascular bed damage. Initially, gelatinases are activated within 2 h after onset of cerebral ischemia and result in disruption of the microvascular beds [101]. Following the gelatinase-mediated damage of microvascular beds, peripheral immune cells, such as neutrophils and macrophages, and small molecules, such as proinflammatory cytokines, free radicals and nitric oxide may cross through the impaired microvascular bed, subsequently generate more gelatinases and exacerbate further microvascular bed dysfunction [63, 70, 102].

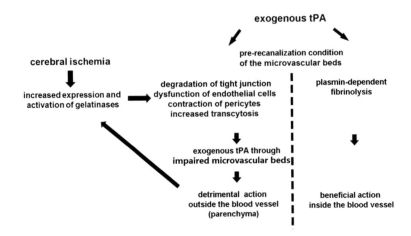

Scheme 1.1 Role of gelatinases in BBB after cerebral ischemia and after tPA treatment

Exogenous tPA can be a robust driver of the self-perpetuating loop of gelatinase activation (Scheme 1.1). The stronger benefits and less side effects of tPA treatment are only seen in patients treated in early time windows [14]. With prolonged ischemia, the risk for exogenous tPA to injure the CNS will increase. In a 10-year long clinical study, among over 50,000 tPA-treated patients, more rapid tPA therapy was associated with reduced mortality, with fewer symptomatic intracranial hemorrhages [14]. Whether exogenous tPA can reach the brain parenchyma is dependent on the state of microvascular beds before recanalization [103]. It is reasonable to consider that in the early stage of ischemia, exogenous tPA mainly promotes its intravascular fibrinolysis function and has only limited capacity to impact the healthy or less damaged microvascular beds [104]. However, with the progression of cerebral ischemia, the gelatinase-mediated degradation of the ECM may lead to disruption of the microvascular bed and exposure of the brain parenchyma to the exogenous tPA. Consequently, the exogenous tPA amplifies MMP-9 levels in brain parenchyma and exaggerate the gelatinase-mediated impairment of the NVU including damage to the microvascular beds. It has been demonstrated that MMP-9 levels are increased after tPA administration in transient focal cerebral ischemic rats [105, 106] and decreased in tPA knockout mice. Broad-spectrum MMP inhibitors decrease plasma MMP-9 levels and prevent brain hemorrhage after thrombolytic treatment of tPA [106, 107]. In clinical study, active forms of MMP-9 were found sorely in patients treated with tPA and correlated with degraded laminin [108]. Signaling mechanisms of tPA induced gelatinases may involve oxidative stress-mediated NF-kB transcription and low-density lipoprotein receptor-related protein (LRP) [107, 109]. This self-perpetuating loop of gelatinase activation mediated by exogenous tPA may also explain why higher level of plasma MMP-9 may predict worse intracranial hemorrhage after thrombolysis in human stroke [110]. Moreover, the finding that mechanical recanalization with endovascular procedures have relative longer therapeutic window (6–12 h) than with exogenous tPA (4.5 h) may be due to the lack of the toxicity of exogenous tPA [111–114].

5   Early Inhibition of Gelatinases May Escalate Neuroprotection and Extend the tPA Therapeutic Window

Although the nonfibrinolytic effects of tPA outside the blood vessel may limit its use as thrombolytic therapy in ischemic stroke, recanalization with tPA has been demonstrated in actual practice. A potential application to extend the use of tPA is offering an improved condition of the microvascular bed before recanalization by blocking the vicious circle of gelatinase, which starts as early as 1–2 h after onset of cerebral ischemia [54, 55, 101].

Many pharmacological MMP inhibitors, including batimastat (BB-94), GM 6001, doxycycline, and minocycline, have shown protective functions in AIS models [23, 115–117]. Broad-spectrum MMP inhibitors combined with tPA treatment have produced promising results to ameliorate MMP-mediated brain damage in animal models of stroke [23, 84, 118]. Minocycline has been demonstrated decreasing the levels of gelatinases in many models of AIS. In rat with embolic ischemia, combining minocycline with tPA administration decreased plasma MMP-9 levels, reduced infarction, and ameliorated brain hemorrhage [108]. Combination therapy with minocycline and candesartan also improves long-term recovery in rat model of ischemia. Two pilot clinical trials have shown minocycline to be potentially effective in AIS and in combination with tPA [76, 119], while another clinical trial showed intravenous minocycline was safe but not efficacious [120]. Large efficacy clinical trials are needed to confirm the potential treatment effect for minocycline in AIS therapy. SB-3CT, the first mechanism-based MMP inhibitor selective for gelatinases, can restore laminin degradation, antagonize neuronal apoptosis, reduce brain infarction and improve behavioral function in mouse model of AIS [54, 55]. Moreover, SB-3CT protected the laminin of pericytes and restored the contraction of pericytes [54]. SB-3CT also has the unique pharmacokinetic benefit of crossing the BBB, therefore it could reach the brain parenchyma at therapeutic concentrations as early as 10 min after intraperitoneal injection [74, 121].

Other treatments such as therapeutic hypothermia, MMP-9 gene silencing, and tissue inhibitor of matrix metalloproteinases 1, could also decrease activity of gelatinases and provide protective function in AIS [26]. The benefit of statins in decrease the risk of stroke in patients with vascular disease [122] may partially due to the reduction of gelatinase-mediated BBB permeability [123, 124]. Moreover, a recent study suggested that inactivation of actin depolymerizing factor may prevent early BBB disruption, reduce the vulnerability of the BBB to gelatinase-mediated damage, and therefore block the self-perpetuating loop of gelatinase activation in AIS [102]. In another recent study, pretreatment with rivaroxaban, a novel oral anticoagulant inhibiting factor Xa, was shown to reduce MMP-9 expression and improve the dissociations between astrocytes and pericytes after tPA administration [125].

These therapeutic features of gelatinases inhibition may prevent early microvascular bed disruption within the time window of exogenous tPA in AIS. Improved pre-recanalization conditions of the microvascular bed may prevent adverse consequences among patients treated with intravenous tPA and extend the time window for possible treatment. In the future, combination therapies of tPA with early inhibition of gelatinases may be of great potential value for AIS therapy.

References

1. Lozano R, Naghavi M, Foreman K, Lim S, Shibuya K, Aboyans V, et al. Global and regional mortality from 235 causes of death for 20 age groups in 1990 and 2010: a systematic analysis for the global burden of disease study 2010. Lancet. 2012;380:2095–128.
2. Feigin VL, Forouzanfar MH, Krishnamurthi R, Mensah GA, Connor M, Bennett DA, et al. Global and regional burden of stroke during 1990–2010: findings from the global burden of disease study 2010. Lancet. 2014;383:245–54.
3. Mozaffarian D, Benjamin EJ, Go AS, Arnett DK, Blaha MJ, Cushman M, et al. Heart disease and stroke statistics—2015 update: a report from the American Heart Association. Circulation. 2015;131:e29–322.
4. Lloyd-Jones D, Adams R, Carnethon M, De Simone G, Ferguson TB, Flegal K, et al. Heart disease and stroke statistics—2009 update: a report from the American Heart Association Statistics Committee and Stroke Statistics Subcommittee. Circulation. 2009;119:480–6.
5. Liu L, Wang D, Wong KS, Wang Y. Stroke and stroke care in china: huge burden, significant workload, and a national priority. Stroke. 2011;42:3651–4.
6. He J, Gu D, Wu X, Reynolds K, Duan X, Yao C, et al. Major causes of death among men and women in china. N Engl J Med. 2005;353:1124–34.
7. Zhou M, Wang H, Zhu J, Chen W, Wang L, Liu S, et al. Cause-specific mortality for 240 causes in china during 1990–2013: a systematic subnational analysis for the global burden of disease study 2013. Lancet. 2016;387:251–72.
8. Toyoda K, Koga M, Hayakawa M, Yamagami H. Acute reperfusion therapy and stroke care in Asia after successful endovascular trials. Stroke. 2015;46:1474–81.
9. Wang W, Jiang B, Sun H, Ru X, Sun D, Wang L, et al. Prevalence, incidence and mortality of stroke in china: results from a nationwide population-based survey of 480,687 adults. Circulation. 2017;135(8):759–71.
10. van der Worp HB, van Gijn J. Clinical practice. Acute ischemic stroke. N Engl J Med. 2007;357:572–9.
11. Rha JH, Saver JL. The impact of recanalization on ischemic stroke outcome: a meta-analysis. Stroke. 2007;38:967–73.
12. Del Zoppo GJ, Saver JL, Jauch EC, Adams HP Jr, American Heart Association Stroke Council. Expansion of the time window for treatment of acute ischemic stroke with intravenous tissue plasminogen activator: a science advisory from the American Heart Association/ American Stroke Association. Stroke. 2009;40:2945–8.
13. Jauch EC, Saver JL, Adams HP Jr, Bruno A, Connors JJ, Demaerschalk BM, et al. Guidelines for the early management of patients with acute ischemic stroke: a guideline for healthcare professionals from the American Heart Association/American Stroke Association. Stroke. 2013;44:870–947.
14. Saver JL, Fonarow GC, Smith EE, Reeves MJ, Grau-Sepulveda MV, Pan W, et al. Time to treatment with intravenous tissue plasminogen activator and outcome from acute ischemic stroke. JAMA. 2013;309:2480–8.
15. Su EJ, Fredriksson L, Geyer M, Folestad E, Cale J, Andrae J, et al. Activation of pdgf-cc by tissue plasminogen activator impairs blood-brain barrier integrity during ischemic stroke. Nat Med. 2008;14:731–7.
16. Wang L, Fan W, Cai P, Fan M, Zhu X, Dai Y, et al. Recombinant adamts13 reduces tissue plasminogen activator-induced hemorrhage after stroke in mice. Ann Neurol. 2013;73:189–98.

17. Chevilley A, Lesept F, Lenoir S, Ali C, Parcq J, Vivien D. Impacts of tissue-type plasminogen activator (tpa) on neuronal survival. Front Cell Neurosci. 2015;9:415.
18. Lo EH, Broderick JP, Moskowitz MA. Tpa and proteolysis in the neurovascular unit. Stroke. 2004;35:354–6.
19. Wang X, Rosell A, Lo EH. Targeting extracellular matrix proteolysis for hemorrhagic complications of tpa stroke therapy. CNS Neurol Disord Drug Targets. 2008;7:235–42.
20. Yong VW. Metalloproteinases: mediators of pathology and regeneration in the cns. Nat Rev Neurosci. 2005;6:931–44.
21. Rosenberg GA. Matrix metalloproteinases and their multiple roles in neurodegenerative diseases. Lancet Neurol. 2009;8:205–16.
22. Van Wart HE, Birkedal-Hansen H. The cysteine switch: a principle of regulation of metalloproteinase activity with potential applicability to the entire matrix metalloproteinase gene family. Proc Natl Acad Sci U S A. 1990;87:5578–82.
23. Asahi M, Asahi K, Jung JC, del Zoppo GJ, Fini ME, Lo EH. Role for matrix metalloproteinase 9 after focal cerebral ischemia: effects of gene knockout and enzyme inhibition with bb-94. J Cereb Blood Flow Metab. 2000;20:1681–9.
24. Turner RJ, Sharp FR. Implications of mmp9 for blood brain barrier disruption and hemorrhagic transformation following ischemic stroke. Front Cell Neurosci. 2016;10:56.
25. Dong X, Song YN, Liu WG, Guo XL. Mmp-9, a potential target for cerebral ischemic treatment. Curr Neuropharmacol. 2009;7:269–75.
26. Chaturvedi M, Kaczmarek L. Mmp-9 inhibition: a therapeutic strategy in ischemic stroke. Mol Neurobiol. 2014;49:563–73.
27. Sweeney MD, Ayyadurai S, Zlokovic BV. Pericytes of the neurovascular unit: key functions and signaling pathways. Nat Neurosci. 2016;19:771–83.
28. Dirnagl U. Pathobiology of injury after stroke: the neurovascular unit and beyond. Ann N Y Acad Sci. 2012;1268:21–5.
29. Ballabh P, Braun A, Nedergaard M. The blood-brain barrier: an overview: structure, regulation, and clinical implications. Neurobiol Dis. 2004;16:1–13.
30. Hawkins BT, Davis TP. The blood-brain barrier/neurovascular unit in health and disease. Pharmacol Rev. 2005;57:173–85.
31. Zlokovic BV. Remodeling after stroke. Nat Med. 2006;12:390–1.
32. Kniesel U, Wolburg H. Tight junctions of the blood-brain barrier. Cell Mol Neurobiol. 2000;20:57–76.
33. Stamatovic SM, Keep RF, Andjelkovic AV. Brain endothelial cell-cell junctions: how to "open" the blood brain barrier. Curr Neuropharmacol. 2008;6:179–92.
34. Dorfel MJ, Huber O. Modulation of tight junction structure and function by kinases and phosphatases targeting occludin. J Biomed Biotechnol. 2012;2012:807356.
35. Ni Y, Sun GY, Lee JC. TNFα alters occludin and cerebral endothelial permeability: role of p38MAPK. PLoS One. 2017;12(2):e0170346.
36. Winkler EA, Bell RD, Zlokovic BV. Central nervous system pericytes in health and disease. Nat Neurosci. 2011;14:1398–405.
37. Cai W, Liu H, Zhao J, Chen LY, Chen J, Lu Z, et al. Pericytes in brain injury and repair after ischemic stroke. Transl Stroke Res. 2017;8(2):107–21.
38. Armulik A, Genove G, Mae M, Nisancioglu MH, Wallgard E, Niaudet C, et al. Pericytes regulate the blood-brain barrier. Nature. 2010;468:557–61.
39. Attwell D, Mishra A, Hall CN, O'Farrell FM, Dalkara T. What is a pericyte? J Cereb Blood Flow Metab. 2016;36:451–5.
40. Lassmann H, Zimprich F, Vass K, Hickey WF. Microglial cells are a component of the perivascular glia limitans. J Neurosci Res. 1991;28:236–43.
41. Mathiisen TM, Lehre KP, Danbolt NC, Ottersen OP. The perivascular astroglial sheath provides a complete covering of the brain microvessels: an electron microscopic 3d reconstruction. Glia. 2010;58:1094–103.

42. Baeten KM, Akassoglou K. Extracellular matrix and matrix receptors in blood-brain barrier formation and stroke. Dev Neurobiol. 2011;71:1018–39.
43. Barber AJ, Lieth E. Agrin accumulates in the brain microvascular basal lamina during development of the blood-brain barrier. Dev Dyn. 1997;208:62–74.
44. Lukes A, Mun-Bryce S, Lukes M, Rosenberg GA. Extracellular matrix degradation by metalloproteinases and central nervous system diseases. Mol Neurobiol. 1999;19:267–84.
45. Fukuda S, Fini CA, Mabuchi T, Koziol JA, Eggleston LL Jr, del Zoppo GJ. Focal cerebral ischemia induces active proteases that degrade microvascular matrix. Stroke. 2004;35:998–1004.
46. Yurchenco PD, Amenta PS, Patton BL. Basement membrane assembly, stability and activities observed through a developmental lens. Matrix Biol. 2004;22:521–38.
47. Agrawal S, Anderson P, Durbeej M, van Rooijen N, Ivars F, Opdenakker G, et al. Dystroglycan is selectively cleaved at the parenchymal basement membrane at sites of leukocyte extravasation in experimental autoimmune encephalomyelitis. J Exp Med. 2006;203:1007–19.
48. Baumann E, Preston E, Slinn J, Stanimirovic D. Post-ischemic hypothermia attenuates loss of the vascular basement membrane proteins, agrin and sparc, and the blood-brain barrier disruption after global cerebral ischemia. Brain Res. 2009;1269:185–97.
49. Cardoso FL, Brites D, Brito MA. Looking at the blood-brain barrier: molecular anatomy and possible investigation approaches. Brain Res Rev. 2010;64:328–63.
50. Miner JH, Li C, Mudd JL, Go G, Sutherland AE. Compositional and structural requirements for laminin and basement membranes during mouse embryo implantation and gastrulation. Development. 2004;131:2247–56.
51. Hallmann R, Horn N, Selg M, Wendler O, Pausch F, Sorokin LM. Expression and function of laminins in the embryonic and mature vasculature. Physiol Rev. 2005;85:979–1000.
52. Chen ZL, Haegeli V, Yu H, Strickland S. Cortical deficiency of laminin gamma1 impairs the akt/gsk-3beta signaling pathway and leads to defects in neurite outgrowth and neuronal migration. Dev Biol. 2009;327:158–68.
53. Coles EG, Gammill LS, Miner JH, Bronner-Fraser M. Abnormalities in neural crest cell migration in laminin alpha5 mutant mice. Dev Biol. 2006;289:218–28.
54. Cui J, Chen S, Zhang C, Meng F, Wu W, Hu R, et al. Inhibition of mmp-9 by a selective gelatinase inhibitor protects neurovasculature from embolic focal cerebral ischemia. Mol Neurodegener. 2012;7:21.
55. Gu Z, Cui J, Brown S, Fridman R, Mobashery S, Strongin AY, et al. A highly specific inhibitor of matrix metalloproteinase-9 rescues laminin from proteolysis and neurons from apoptosis in transient focal cerebral ischemia. J Neurosci. 2005;25:6401–8.
56. Sorokin L, Girg W, Gopfert T, Hallmann R, Deutzmann R. Expression of novel 400-kda laminin chains by mouse and bovine endothelial cells. Eur J Biochem. 1994;223:603–10.
57. Sixt M, Engelhardt B, Pausch F, Hallmann R, Wendler O, Sorokin LM. Endothelial cell laminin isoforms, laminins 8 and 10, play decisive roles in t cell recruitment across the blood-brain barrier in experimental autoimmune encephalomyelitis. J Cell Biol. 2001;153:933–46.
58. Tilling T, Engelbertz C, Decker S, Korte D, Huwel S, Galla HJ. Expression and adhesive properties of basement membrane proteins in cerebral capillary endothelial cell cultures. Cell Tissue Res. 2002;310:19–29.
59. Menezes MJ, McClenahan FK, Leiton CV, Aranmolate A, Shan X, Colognato H. The extracellular matrix protein laminin alpha2 regulates the maturation and function of the blood-brain barrier. J Neurosci. 2014;34:15260–80.
60. Yao Y, Chen ZL, Norris EH, Strickland S. Astrocytic laminin regulates pericyte differentiation and maintains blood brain barrier integrity. Nat Commun. 2014;5:3413.
61. Gautam J, Zhang X, Yao Y. The role of pericytic laminin in blood brain barrier integrity maintenance. Sci Rep. 2016;6:36450.
62. Frisch SM, Francis H. Disruption of epithelial cell-matrix interactions induces apoptosis. J Cell Biol. 1994;124:619–26.
63. Gu Z, Kaul M, Yan B, Kridel SJ, Cui J, Strongin A, et al. S-nitrosylation of matrix metalloproteinases: signaling pathway to neuronal cell death. Science. 2002;297:1186–90.

64. Li L, Liu F, Welser-Alves JV, McCullough LD, Milner R. Upregulation of fibronectin and the alpha5beta1 and alphavbeta3 integrins on blood vessels within the cerebral ischemic penumbra. Exp Neurol. 2012;233:283–91.
65. Chen ZL, Yao Y, Norris EH, Kruyer A, Jno-Charles O, Akhmerov A, et al. Ablation of astrocytic laminin impairs vascular smooth muscle cell function and leads to hemorrhagic stroke. J Cell Biol. 2013;202:381–95.
66. Wagner S, Tagaya M, Koziol JA, Quaranta V, del Zoppo GJ. Rapid disruption of an astrocyte interaction with the extracellular matrix mediated by integrin alpha 6 beta 4 during focal cerebral ischemia/reperfusion. Stroke. 1997;28:858–65.
67. Tagaya M, Haring HP, Stuiver I, Wagner S, Abumiya T, Lucero J, et al. Rapid loss of microvascular integrin expression during focal brain ischemia reflects neuron injury. J Cereb Blood Flow Metab. 2001;21:835–46.
68. Milner R, Hung S, Wang X, Spatz M, del Zoppo GJ. The rapid decrease in astrocyte-associated dystroglycan expression by focal cerebral ischemia is protease-dependent. J Cereb Blood Flow Metab. 2008;28:812–23.
69. Chen ZL, Strickland S. Neuronal death in the hippocampus is promoted by plasmin-catalyzed degradation of laminin. Cell. 1997;91:917–25.
70. Lakhan SE, Kirchgessner A, Tepper D, Leonard A. Matrix metalloproteinases and blood-brain barrier disruption in acute ischemic stroke. Front Neurol. 2013;4:32.
71. del Zoppo GJ. The neurovascular unit, matrix proteases, and innate inflammation. Ann N Y Acad Sci. 2010;1207:46–9.
72. Kawakita K, Kawai N, Kuroda Y, Yasashita S, Nagao S. Expression of matrix metalloproteinase-9 in thrombin-induced brain edema formation in rats. J Stroke Cerebrovasc Dis. 2006;15:88–95.
73. Li L, Tao Y, Tang J, Chen Q, Yang Y, Feng Z, et al. A cannabinoid receptor 2 agonist prevents thrombin-induced blood-brain barrier damage via the inhibition of microglial activation and matrix metalloproteinase expression in rats. Transl Stroke Res. 2015;6:467–77.
74. Hadass O, Tomlinson BN, Gooyit M, Chen S, Purdy JJ, Walker JM, et al. Selective inhibition of matrix metalloproteinase-9 attenuates secondary damage resulting from severe traumatic brain injury. PLoS One. 2013;8:e76904.
75. Horstmann S, Kalb P, Koziol J, Gardner H, Wagner S. Profiles of matrix metalloproteinases, their inhibitors, and laminin in stroke patients: influence of different therapies. Stroke. 2003;34:2165–70.
76. Switzer JA, Hess DC, Ergul A, Waller JL, Machado LS, Portik-Dobos V, et al. Matrix metalloproteinase-9 in an exploratory trial of intravenous minocycline for acute ischemic stroke. Stroke. 2011;42:2633–5.
77. Castellanos M, Leira R, Serena J, Pumar JM, Lizasoain I, Castillo J, et al. Plasma metalloproteinase-9 concentration predicts hemorrhagic transformation in acute ischemic stroke. Stroke. 2003;34:40–6.
78. Asahi M, Wang X, Mori T, Sumii T, Jung JC, Moskowitz MA, et al. Effects of matrix metalloproteinase-9 gene knock-out on the proteolysis of blood-brain barrier and white matter components after cerebral ischemia. J Neurosci. 2001;21:7724–32.
79. Wang J, Tsirka SE. Neuroprotection by inhibition of matrix metalloproteinases in a mouse model of intracerebral haemorrhage. Brain. 2005;128:1622–33.
80. Alluri H, Wilson RL, Anasooya Shaji C, Wiggins-Dohlvik K, Patel S, Liu Y, et al. Melatonin preserves blood-brain barrier integrity and permeability via matrix metalloproteinase-9 inhibition. PLoS One. 2016;11:e0154427.
81. Hu X, De Silva TM, Chen J, Faraci FM. Cerebral vascular disease and neurovascular injury in ischemic stroke. Circ Res. 2017;120:449–71.
82. Underly RG, Levy M, Hartmann DA, Grant RI, Watson AN, Shih AY. Pericytes as inducers of rapid, matrix metalloproteinase-9-dependent capillary damage during ischemia. J Neurosci. 2017;37:129–40.

83. Yang Y, Estrada EY, Thompson JF, Liu W, Rosenberg GA. Matrix metalloproteinase-mediated disruption of tight junction proteins in cerebral vessels is reversed by synthetic matrix metalloproteinase inhibitor in focal ischemia in rat. J Cereb Blood Flow Metab. 2007;27:697–709.
84. Mishiro K, Ishiguro M, Suzuki Y, Tsuruma K, Shimazawa M, Hara H. A broad-spectrum matrix metalloproteinase inhibitor prevents hemorrhagic complications induced by tissue plasminogen activator in mice. Neuroscience. 2012;205:39–48.
85. Yang Y, Thompson JF, Taheri S, Salayandia VM, McAvoy TA, Hill JW, et al. Early inhibition of mmp activity in ischemic rat brain promotes expression of tight junction proteins and angiogenesis during recovery. J Cereb Blood Flow Metab. 2013;33:1104–14.
86. del Zoppo GJ, von Kummer R, Hamann GF. Ischaemic damage of brain microvessels: inherent risks for thrombolytic treatment in stroke. J Neurol Neurosurg Psychiatry. 1998;65:1–9.
87. Del Zoppo GJ, Milner R, Mabuchi T, Hung S, Wang X, Koziol JA. Vascular matrix adhesion and the blood-brain barrier. Biochem Soc Trans. 2006;34:1261–6.
88. Dore-Duffy P, Owen C, Balabanov R, Murphy S, Beaumont T, Rafols JA. Pericyte migration from the vascular wall in response to traumatic brain injury. Microvasc Res. 2000;60:55–69.
89. Liu S, Agalliu D, Yu C, Fisher M. The role of pericytes in blood-brain barrier function and stroke. Curr Pharm Des. 2012;18:3653–62.
90. Fernandez-Klett F, Priller J. Diverse functions of pericytes in cerebral blood flow regulation and ischemia. J Cereb Blood Flow Metab. 2015;35:883–7.
91. Hall CN, Reynell C, Gesslein B, Hamilton NB, Mishra A, Sutherland BA, et al. Capillary pericytes regulate cerebral blood flow in health and disease. Nature. 2014;508:55–60.
92. Dawson DA, Ruetzler CA, Hallenbeck JM. Temporal impairment of microcirculatory perfusion following focal cerebral ischemia in the spontaneously hypertensive rat. Brain Res. 1997;749:200–8.
93. Yemisci M, Gursoy-Ozdemir Y, Vural A, Can A, Topalkara K, Dalkara T. Pericyte contraction induced by oxidative-nitrative stress impairs capillary reflow despite successful opening of an occluded cerebral artery. Nat Med. 2009;15:1031–7.
94. Nedelmann M, Ritschel N, Doenges S, Langheinrich AC, Acker T, Reuter P, et al. Combined contrast-enhanced ultrasound and rt-pa treatment is safe and improves impaired microcirculation after reperfusion of middle cerebral artery occlusion. J Cereb Blood Flow Metab. 2010;30:1712–20.
95. An H, Ford AL, Vo K, Eldeniz C, Ponisio R, Zhu H, et al. Early changes of tissue perfusion after tissue plasminogen activator in hyperacute ischemic stroke. Stroke. 2011;42:65–72.
96. Al Ahmad A, Gassmann M, Ogunshola OO. Maintaining blood-brain barrier integrity: pericytes perform better than astrocytes during prolonged oxygen deprivation. J Cell Physiol. 2009;218:612–22.
97. Daneman R, Zhou L, Agalliu D, Cahoy JD, Kaushal A, Barres BA. The mouse blood-brain barrier transcriptome: a new resource for understanding the development and function of brain endothelial cells. PLoS One. 2010;5:e13741.
98. Takata F, Dohgu S, Matsumoto J, Takahashi H, Machida T, Wakigawa T, et al. Brain pericytes among cells constituting the blood-brain barrier are highly sensitive to tumor necrosis factor-alpha, releasing matrix metalloproteinase-9 and migrating in vitro. J Neuroinflammation. 2011;8:106.
99. Rosenberg GA, Estrada EY, Dencoff JE. Matrix metalloproteinases and timps are associated with blood-brain barrier opening after reperfusion in rat brain. Stroke. 1998;29:2189–95.
100. Jin R, Yang G, Li G. Molecular insights and therapeutic targets for blood-brain barrier disruption in ischemic stroke: critical role of matrix metalloproteinases and tissue-type plasminogen activator. Neurobiol Dis. 2010;38:376–85.
101. Heo JH, Lucero J, Abumiya T, Koziol JA, Copeland BR, del Zoppo GJ. Matrix metalloproteinases increase very early during experimental focal cerebral ischemia. J Cereb Blood Flow Metab. 1999;19:624–33.

102. Shi Y, Zhang L, Pu H, Mao L, Hu X, Jiang X, et al. Rapid endothelial cytoskeletal reorganization enables early blood-brain barrier disruption and long-term ischaemic reperfusion brain injury. Nat Commun. 2016;7:10523.
103. Benchenane K, Berezowski V, Fernandez-Monreal M, Brillault J, Valable S, Dehouck MP, et al. Oxygen glucose deprivation switches the transport of tpa across the blood-brain barrier from an lrp-dependent to an increased lrp-independent process. Stroke. 2005;36:1065–70.
104. Niego B, Medcalf RL. Plasmin-dependent modulation of the blood-brain barrier: a major consideration during tpa-induced thrombolysis? J Cereb Blood Flow Metab. 2014;34:1283–96.
105. Aoki T, Sumii T, Mori T, Wang X, Lo EH. Blood-brain barrier disruption and matrix metalloproteinase-9 expression during reperfusion injury: mechanical versus embolic focal ischemia in spontaneously hypertensive rats. Stroke. 2002;33:2711–7.
106. Tsuji K, Aoki T, Tejima E, Arai K, Lee SR, Atochin DN, et al. Tissue plasminogen activator promotes matrix metalloproteinase-9 upregulation after focal cerebral ischemia. Stroke. 2005;36:1954–9.
107. Wang X, Lee SR, Arai K, Lee SR, Tsuji K, Rebeck GW, et al. Lipoprotein receptor-mediated induction of matrix metalloproteinase by tissue plasminogen activator. Nat Med. 2003;9:1313–7.
108. Murata Y, Rosell A, Scannevin RH, Rhodes KJ, Wang X, Lo EH. Extension of the thrombolytic time window with minocycline in experimental stroke. Stroke. 2008;39:3372–7.
109. Yepes M, Sandkvist M, Moore EG, Bugge TH, Strickland DK, Lawrence DA. Tissue-type plasminogen activator induces opening of the blood-brain barrier via the ldl receptor-related protein. J Clin Invest. 2003;112:1533–40.
110. Montaner J, Molina CA, Monasterio J, Abilleira S, Arenillas JF, Ribo M, et al. Matrix metalloproteinase-9 pretreatment level predicts intracranial hemorrhagic complications after thrombolysis in human stroke. Circulation. 2003;107:598–603.
111. Jovin TG, Chamorro A, Cobo E, de Miquel MA, Molina CA, Rovira A, et al. Thrombectomy within 8 hours after symptom onset in ischemic stroke. N Engl J Med. 2015;372:2296–306.
112. Rodrigues FB, Neves JB, Caldeira D, Ferro JM, Ferreira JJ, Costa J. Endovascular treatment versus medical care alone for ischaemic stroke: systematic review and meta-analysis. BMJ. 2016;353:i1754.
113. Ciccone A, Valvassori L, SYNTHESIS Expansion Investigators. Endovascular treatment for acute ischemic stroke. N Engl J Med. 2013;368:2433–4.
114. Miao Z, Huo X, Gao F, Liao X, Wang C, Peng Y, et al. Endovascular therapy for Acute ischemic Stroke Trial (EAST): study protocol for a prospective, multicentre control trial in China. Stroke Vasc Neurol. 2016;1:e000022.
115. Pires PW, Rogers CT, McClain JL, Garver HS, Fink GD, Dorrance AM. Doxycycline, a matrix metalloprotease inhibitor, reduces vascular remodeling and damage after cerebral ischemia in stroke-prone spontaneously hypertensive rats. Am J Physiol Heart Circ Physiol. 2011;301:H87–97.
116. Fagan SC, Cronic LE, Hess DC. Minocycline development for acute ischemic stroke. Transl Stroke Res. 2011;2:202–8.
117. Chen W, Hartman R, Ayer R, Marcantonio S, Kamper J, Tang J, et al. Matrix metalloproteinases inhibition provides neuroprotection against hypoxia-ischemia in the developing brain. J Neurochem. 2009;111:726–36.
118. Lapchak PA, Chapman DF, Zivin JA. Metalloproteinase inhibition reduces thrombolytic (tissue plasminogen activator)-induced hemorrhage after thromboembolic stroke. Stroke. 2000;31:3034–40.
119. Lampl Y, Boaz M, Gilad R, Lorberboym M, Dabby R, Rapoport A, et al. Minocycline treatment in acute stroke: an open-label, evaluator-blinded study. Neurology. 2007;69:1404–10.
120. Kohler E, Prentice DA, Bates TR, Hankey GJ, Claxton A, van Heerden J, et al. Intravenous minocycline in acute stroke: a randomized, controlled pilot study and meta-analysis. Stroke. 2013;44:2493–9.

121. Gooyit M, Suckow MA, Schroeder VA, Wolter WR, Mobashery S, Chang M. Selective gelatinase inhibitor neuroprotective agents cross the blood-brain barrier. ACS Chem Neurosci. 2012;3:730–6.
122. Stroke Council American Heart Association, American Stroke Association. Statins after ischemic stroke and transient ischemic attack: an advisory statement from the Stroke Council, American Heart Association and American Stroke Association. Stroke. 2004;35:1023.
123. McFarland AJ, Anoopkumar-Dukie S, Arora DS, Grant GD, McDermott CM, Perkins AV, et al. Molecular mechanisms underlying the effects of statins in the central nervous system. Int J Mol Sci. 2014;15:20607–37.
124. Reuter B, Rodemer C, Grudzenski S, Meairs S, Bugert P, Hennerici MG, et al. Effect of simvastatin on mmps and timps in human brain endothelial cells and experimental stroke. Transl Stroke Res. 2015;6:156–9.
125. Shang J, Yamashita T, Kono S, Morihara R, Nakano Y, Fukui Y, et al. Effects of pretreatment with warfarin or rivaroxaban on neurovascular unit dissociation after tissue plasminogen activator thrombolysis in ischemic rat brain. J Stroke Cerebrovasc Dis. 2016;25:1997–2003.

Chapter 2
Characters of Ischemic Stroke and Recanalization Arteries

Qingqing Dai, Shujuan Li, and Junfa Li

Abstract Ischemic stroke (>80% stroke patients) is a major cause of mortality and long-term disability. To understand the pathophysiology and artery recanalization are important to find the potential therapeutic targets and improve the long-term outcomes of ischemic stroke in the clinical practice. In this chapter, the issues were reviewed as follows: the epidemiology, risk factors, imaging, diagnosis and pathophysiology of ischemic stroke, as well as intravenous thrombolytic and antiplatelet therapy, endovascular treatment, antioxidant and reperfusion injury after artery recanalization.

Keywords Ischemic stroke · Recanalization · Diagnosis · Therapy · Epidemiology · Pathophysiology

1 Introduction

Stroke is the second most common cause of death worldwide and a major global cause of disability. By 2050, more than 1.5 billion people in the world will be aged 65 years or older and the global burden of stroke will keep increasing with the aging population [1]. About 75% of stroke patients are over the age of 65 years old and about one-third of patients die of stroke within a year of onset [2, 3]. In China, stroke is already the major leading cause of death and disability [4, 5]. Ischemic stroke is one of the most important subtypes that cover more than 80% stroke patients [6]. Recanalization is strongly associated with improved functional

Q. Dai · J. Li (✉)
Department of Neurobiology and Center of Stroke, Beijing Institute for Brain Disorders, Capital Medical University, Beijing, China
e-mail: qqingdai@mail.ccmu.edu.cn; junfali@ccmu.edu.cn

S. Li
Department of Neurology, Beijing Chao-Yang Hospital, Capital Medical University, Beijing, China
e-mail: shujuanli@ccmu.edu.cn

© Springer International Publishing AG, part of Springer Nature 2018
W. Jiang et al. (eds.), *Cerebral Ischemic Reperfusion Injuries (CIRI)*, Springer Series in Translational Stroke Research, https://doi.org/10.1007/978-3-319-90194-7_2

15

outcomes and reduced mortality, which is also an significant parameter in early phase trials of thrombolytic treatment in acute ischemic stroke [7]. To understand the characters, especially the pathophysiology and artery recanalization are very important to find the potential therapeutic targets and improve the long-term outcomes of ischemic stroke in the clinical practice.

2   Characters of the Ischemic Stroke

2.1   *Epidemiology and Risk Factors of Ischemic Stroke*

To analyze the risk factors and the data of epidemiology about ischemic stroke could improve the understanding of the health consequences and prevent the disease effectively. Worldwide in 2010, the investigation shows that 11,569,538 ischemic stroke events took place (63% in low-income and middle-income countries). Moreover, totally 2,835,419 individuals died from ischemic stroke (57% in low-income and middle-income countries). The average age of patients in low-income and middle-income countries was 3–5 years younger than in high-income countries. In years of 1990–2010, the incidence of ischemic stroke ascended significantly in adults aged 20–64 year-old in low-income and middle-income countries [8]. In these countries, because of lacking the effective prevention strategies and better health care for chronic diseases such as hypertension, diabetes and heart disease, these factors have been the mainly leading cause of disease burden [9]. On the other hand, the distribution of the artery lesions is different between China and USA. Twenty-two general hospitals covering a wide geographic area in China participated a prospective and cohort study, 2864 patients were enrolled including ischemic stroke and transient ischemic attack (TIA). The study shows 37.5% patients had intracranial artery lesions, and the most common vascular lesion of complete occlusion, severe (70–99%) and moderate (50–69%) stenosis occurred in middle cerebral artery with the proportion of 14.18%, 6.04% and 9.39% respectively [10]. The DEFUSE 2 trial, a prospective and single-arm study in USA, including 100 patients with anterior circulation ischemic strokes illustrated that the proximal middle cerebral artery accounted for 58% in the occlusive arteries [11].

According to the epidemiological reports, the risk factors for ischemic stroke include non-modifiable factors, such as birth weight, gender, age and ethnic, and the modifiable factors including hypertension, diabetes, overweight, atrial fibrillation, smoking and alcohol consumption [12]. Based on the selected theoretical minimum risk exposure level (TMREL), the stroke burden associated with 17 risks were established as follows: ambient particulate matter pollution, household air pollution from solid fuels, lead exposure, diet high in sodium, diet high in sugar-sweetened beverages, diet low in fruits, diet low in vegetables, diet low in whole grains, alcohol consumption, low physical activity, smoking, second-hand smoking, high BMI, high fasting plasma glucose, high systolic blood pressure (SBP), high total cholesterol, and low glomerular filtration rate (GFR). Globally, during 1990–2013, these

Table 2.1 Clinical risk factors for stroke, transient ischemic attack, and systemic embolism in the CHA2DS$_2$-VASc score [16]

CHA2DS$_2$-VASc risk factor	Points
Congestive heart failure Signs/symptoms of heart failure or objective evidence of reduced left-ventricular ejection fraction	+1
Hypertension Resting blood pressure >140/90 mmHg on at least two occasions or current antihypertensive treatment	+1
Age 75 years or older	+2
Diabetes mellitus Fasting glucose >125 mg/dL (7 mmol/L) or treatment with oral hypoglycaemic agent and/or insulin	+1
Previous stroke, transient ischemic attack, or thromboembolism	+2
Vascular disease Previous myocardial infarction, peripheral artery disease, or aortic plaque	+1
Age 65–74 years	+1
Sex category (female)	+1

CHA2DS$_2$-VASc congestive heart failure, hypertension, age ≥75 (doubled), diabetes, stroke (doubled), vascular disease, age 65–74, and sex (female)

factors show the significant increase in the stroke-related disability-adjusted life-years (DALYs) except household air pollution and second-hand smoke [13]. The results of the study show that more than 90% of stroke burden is due to modifiable risk factors. To control the behavioral and metabolic risk factors could prevent about 75% of the global stroke burden [13].

In China, hypertension remains the most important risk factor in all types of strokes. A cohort study of 11 provinces in China, including 26,787 people with ages from 35 to 64 years, showed that there were 750 cases of hypertension with hypercholesterolemia and 250 cases of hypertension alone in 100,000 populations [14, 15]. Another important risk factor is atrial fibrillation (AF), the blood clots usually develop in the left atrial appendage. The fragments or the whole clot detach from the left atrial appendage and then embolize the arterial system, including cerebral arteries. Initially, the CHADS$_2$ score is used to evaluate the stroke risk as a rapid and easy-to-remember method. Due to the CHADS$_2$ score lacking many other influenced stroke risk factors, the more precise CHA2DS$_2$-VASc score (Table 2.1) is applied in assessing stroke risk in AF patients [16]. To control the risk factors and strength the primary and secondary stroke prevention could effectively low the burden of stroke.

2.2 The Imaging and Diagnosis of Ischemic Stroke

Several neuroimaging techniques were used to examine and research ischemic stroke, including computed tomography (CT), magnetic resonance imaging (MRI), conventional angiography, magnetic resonance angiography (MRA), CT

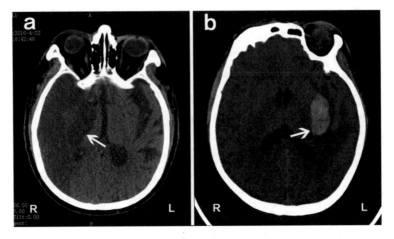

Fig. 2.1 Non-contrast CT showed the differences between ischemic stroke and haemorrhage. (**a**) Ischemic stroke with right temporal lobe and occipital lobe as the arrow indicates. (**b**) Haemorrhage with left basal ganglia as the arrow indicates

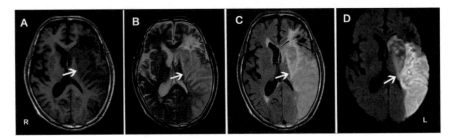

Fig. 2.2 Cardiogenic infarction on the left hemisphere as the arrows indicate. A 77-year-old male with atrial fibrillation suddenly developed right hemiparesis, aphasia. (**a**) T1-weighted MRI. (**b**) T2-weighted MRI. (**c**) Fluid attenuated inversion recovery (FlAIR) MRI. (**d**) Diffusion MRI

angiography (CTA), Doppler ultrasonography, diffusion-weighted MRI (DWI), gradient echo MRI (GRE) and perfusion-weighted MRI (PWI), high-resolution vessel wall MRI (HRMRI), and digital subtraction angiography (DSA). As shown in Fig. 2.1, CT is the convenient and common imaging tool in most hospital to exclude the presence of an acute cerebral hemorrhage. Especially, for the patient treated with intravenous (IV) alteplase (rTPA) in the 0–4.5 h time window, the high CT density (60–90 HU) is sensitive to exclude acute intracranial hemorrhage (ICH) [17]. With MRI shown in Fig. 2.2, diffusion-weighted imaging (DWI) can detect ischemia as early as minutes after stroke onset, which is much more sensitive than CT for identifying acute ischemia [18]. Perfusion-weighted imaging (PWI) is used to evaluate the extent of the tissue infarction before early reperfusion [19]. The apparent diffusion coefficient (ADC) indicates ischemic tissue by providing a quantitative measurement of water diffusion with low values. After

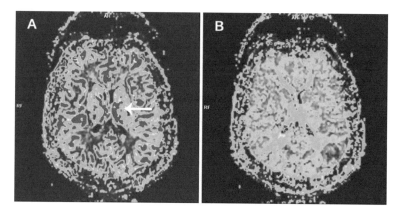

Fig. 2.3 Advanced imaging in ischemic stroke. A 37-year-old female with 10 years hypertension history. NIHSS 10 Score. It was diagnosed as acute ischemic stroke on left basal ganglia. (**a**) Left basal ganglia reduced CBF as arrow indicates. (**b**) Normal with CBV

ischemic stroke, the normalization of the ADC typically occurs about 3–5 days. During the period, because of the combination of true restriction and T2 signal for a variable time that relates to infarct size, DWI remains bright lesion. The lesion becomes larger from days to weeks [20].

As known, the ischemic infarct can be defined as "ischemic core" and "peri-infarct". The former is defined as the ischemic brain tissue that is irreversibly injured and will proceed to infarction even if following the reperfusion immediately. The "peri-infarct" is usually called as "penumbra" which represents functionally impaired ischemic brain tissue that has the potential to recover with early reperfusion, but the area will be the infarction if there isn't the early reperfusion [21, 22]. CT and MR definitions of "ischemic core" and "ischemic penumbra" are probabilistic. In clinical practice, ischemic core can be assessed using diffusion MRI or by threshold relative cerebral blood flow (CBF) on CT perfusion. The normal CBF is 60–80 mL/100 g/min, the CBF of penumbra is 12–20 mL/100 g/min and the infarct core is below 10–12 mL/100 g/min [17, 22]. The reduced CBF in the penumbra triggers energy-dependent autoregulatory mechanisms in order to keep the cerebral blood volume (CBV) normal, however, when CBV drop is a marker of infarction and correlates with the diffusion restriction only in hyperacute stroke [23] (Fig. 2.3). Perfusion imaging is much more sensitive in distal vessel occlusions and can more precisely measure the volume of dead and at risk brain tissue. About the diagnosis of ischemic stroke, the clinical symptoms are also the essential except imaging features. Typical symptoms of ischemic stroke include sudden unilateral weakness, numbness, or visual loss; diplopia; altered speech; ataxia; and non-orthostatic vertigo [24]. All in all, the symptoms are different because of the various brain lesions. The Face Arm and Speech Test (FAST) is a very simply, sensitive and rapid way to screen the stroke [25].

2.3   The Pathophysiology of Ischemic Stroke

2.3.1   Excitotoxicity

Excitotoxicity is attributed to the interaction between glutamate and neuronal gluta-mate receptors, especially N-methyl-D-aspartate (NMDA)-type glutamate recep-tors [26]. After stroke the neurotransmitter glutamate releasing is effected by the way mentioned above, further impairing the glutamate reuptake so that the gluta-mate uptake transporters are reversed [22] (Reverse glial glutamate uptake triggers neuronal cell death through extrasynaptic NMDA receptor activation). Glutamate excitotoxicity was identified as an ischemic injury mechanism in stroke long ago [21, 27]. Some gaseous signaling molecules, including nitric oxide and superoxide, were produced by NMDA receptor activation, which attributes to Ca^{2+} dependent processes. The study indicates that the communication between astrocytes and neu-rons depends on intracellular Ca^{2+} levels [28]. Ca^{2+} signals in astrocytes induce the release of chemical transmitters, which could regulate neuronal excitability and synaptic transmission [29]. Ca^{2+} over load could impair Ca^{2+} regulatory mechanisms and Ca^{2+} signals can also strengthen the effect of other relative death program, ulti-mately lead to the cells death [30]. There is the evidence about regulator mecha-nisms of neuronal NADPH oxidase (NOX) link with NMDA receptor activation to, and it's relevant to stroke pathogenesis [30]. Release of mitochondrial apoptogenic factors and cell death are mediated by casein kinase 2 (CK2) and NADPH oxidase. NOX2 isoform of NOX is considered as the primary source of excitotoxic neuronal superoxide production [31]. There are some evidences show the NOX2 activity is increased by ischemia–reperfusion, after knocking out NOX2 in the model mice, the infarct size in focal cerebral ischemia is reduced [32, 33]. Some researches indi-cated it's useful to apply NOX inhibitors in animal stroke models at the early stage of ischemic onset [32, 34–36]. Therefore, NOX inhibitors may have the neuropro-tective potential in treating acute ischemic stroke.

2.3.2   Inflammatory Response

The inflammatory response plays an important role in the pathophysiology process of ischemic stroke. The cytokines and harmful radicals are released by neutrophils, which would enforce the inflammatory response [37]. The proinflammatory cyto-kines such as IL-1β, IL-17 and TGF-α are released by the immune cells, which can up-regulate cell adhesion molecule expression and promote the leukocyte to accu-mulate on surface of the cerebral blood vessels of infarcted brain tissue, further increasing the inflammatory response [38]. As is known to all, the integrity of the blood brain barrier (BBB) is disrupted for up to 2 weeks after stroke [39]. After ischemia, astrocytes swell together to cause the increase of intracerebral pressure and the vascular hypoperfusion, and finally lead to aggravate the ischemia [40]. It can be said that astrocytes play some important roles in the central nervous system, which including the regulation of blood flow, maintaining the integrity of BBB, affecting the synaptic function by the release of growth factors, reducing the glutamate amount and alleviating excessive inflammatory response [41].

As a key membrane protein of BBB, aquaporins 4 (AQP4) is crucial in astrocyte swelling and cerebral edema process [42]. AQP4 is relevant with the expression of Agrin that accumulates in brain microvessels at the time of BBB tightening and is responsible for BBB integrity [43]. The AQP4 deletion could reduce astrocyte swelling and brain edema, as well as improve the neurological outcome of mice after focal ischemia [44]. Because of interaction between IL-17 and its receptor on endothelial cells promotes the progress of BBB breakdown by disrupting tight junctions [45]. As a critical effector of cerebral ischemia tissue, IL-17 secreting T cells could accumulate in the CNS and cross the BBB, further destroy BBB or directly damage neurons [46, 47]. In the middle cerebral artery occlusion (MCAO)-induced ischemic stroked mice, the brain tissue expression level of IL-17 elevated at 1 h after ischemia insult and peaked at day 6 [48]. However, the anti-inflammatory properties of IL-4 were reported that the ischemic damage were exacerbated in IL-4 knocked-out mice, which suggested its potential to attenuate ischemic injury and promote tissue repair [49]. Similarly, the regulatory T lymphocytes (Treg)-produced IL-10 is also regarded as the neuroprotective cytokine for inhibiting neurotoxic function of TNF-α and IFN-γ [50]. In addition, the essential immune roles of Toll-like receptors (TLR) including TLR2 and TLR4 were confirmed in the ischemic brain injury [51]. It was reported that TLR4 knock-out could protect the brain from high-mobility group protein box-1 (HMGB1) mediated ischemic damage [52], as well as both TLR2 and TLR4 knock-out alleviated ischemic brain injuries [53]. The proinflammatory cytokines-mediated inflammatory response is more complicated, hence the identification of their roles in neuroprotection may provide effective immune therapeutic strategies for ischemic stroke.

2.3.3 Apoptosis

There are two general pathways of apoptosis activated after ischemia: the Caspase-8-mediated extrinsic pathway and the Caspase-9 (mitochondria) or Caspase-12 (endoplasmic reticulum)-mediated intrinsic pathways. Intrinsic Caspase-9 pathway is initiated by the cytochrome C release of mitochondria, while the extrinsic pathway is due to the activation of cell surface death receptors, resulting in the stimulation of caspase-8 [54]. In the past decades, researchers focused on the apoptosis in the ischemic penumbra in order to find the potential target at the early stage of ischemic stroke onset. It suggests that mitochondria influence neuronal apoptosis primarily via the release of proapoptotic factors into the cytoplasm, including cytochrome C [55, 56]. Cytochrome C is involved to active Caspase-9, which is presumably an initiator of the cytochrome c-dependent Caspase cascade, then activates Caspase-3 [57]. Caspase-9 and Caspase-3 play the important roles in neuronal death after cerebral ischemia. The researches illustrated the cleaved Caspase-9 increased in brain tissue 12 h after ischemia and up-regulation of Caspase-3 mRNA in rat brain 1 h after the onset of focal ischemia [58, 59]. It has been also observed Caspase-3 was up-regulated in the human brain tissue after ischemia [60]. At the same time, some studies suggested that genetic deletion and pharmacological inhibition of Caspases have a strong neuroprotective effect in stroke animal models. Some evidences showed that cerebral ischemia triggers the extrinsic apoptotic signaling cascade. Tumor necrosis factor (TNF)-related apoptosis-inducing ligand, Fas and FasL expressions are up-regulated within 12 h after cerebral ischemia in the animal models and peaked between 24 and 48 h, which coincides with the time course of apoptotic death in neuronal cells [61, 62]. It was detected Caspase-3 and several other Caspase family members up-regulated in astrocytes and microglia after MCAO [63]. Although there are some studies discovering the mechanism of apoptosis after stroke, more detailed molecular process of apoptosis are needed to find the novel stroke therapies.

2.3.4 Autophagy

Programmed cell death includes type I (apoptosis), type II (autophagic cell death) and type III (necroptosis). The process of autophagy could degrade the damaged organelles and proteins through the autophagic vesicles [64]. More and more evidences indicate that autophagy may be involved in regulating neuronal death in ischemic stroke, including global and focal ischemia [47, 65–68]. Among all the autophagy related proteins, the microtubule-associated protein 1 light chain 3 I (LC3-I) is noteworthy. It could be hydrolyzed to LC3-II during the process of autophagy starts. Therefore, ratio of LC3-II/LC3-I is commonly used to evaluate the level of autophagy [69]. Turnover of LC3 I to II was increased in the ipsilateral hemisphere 24–72 h after cerebral hypoxia–ischemia [70]. In 1999, Beclin-1 is the first identified mammalian gene with a role in autophagy initiation [71]. Beclin-1 protein level was observed to increase as early as 4 h, appeared to peak at 24–72 h after ischemia injury [72]. 3-methyladenine (3-MA) is used to inhibit autophagy as the inhibitor, which has the time-dependent protective effect on neuronal death [73]. Applying it before ischemia could be better in protecting the global I/R injury. The previous study shows that conventional protein kinase C (cPKC)γ activation has neuroprotective effect against ischemic injuries [74]. Recent study finds that cPKCγ deficit could increase Beclin-1 expression level and the ratio of LC3-II/LC3-I, at the same time, BafA1 is regarded as the autophagy inhibitor, which could aggravate or alleviate OGD-induced ischemic injuries in cPKCγ wild-type and cPKCγ knock-out neurons, respectively [75]. All in all, as an essential role in ischemia, autophagy remains worth further study to develop newly effective therapeutic strategies for ischemic stroke.

2.3.5 Pyroptosis

Pyroptosis is an inflammasome-dependent programmed cell death. Inflammasomes activate Caspase-1 by the Caspase recruitment domain (CARD) or pyrin domain (PYD)-containing inflammasome, including nod-like receptor (NLRP1, 3, 6, 7, 12, NLRC4), absent in melanoma 2 (AIM2), or pyrin [76, 77]. And then Caspase-1 induces the activation and secretion of pro-IL-1b and pro-IL-18 [77]. It is the primary cause of anthrax-lethal- toxin-induced lung injury that pyroptosis induced by the NLRP1B inflammasome or a gain-of-function mutation in Nlrp1a [78]. The newly research indicated that interdomain cleavage of gasdermin D (GSDMD) by caspase-1/4/5/11 determines pyroptosis [79]. AIM2 contains a HIN200 domain and is specific for cytosolic DNA [80–83]. In 2014, the research firstly reported embryonic cortical neurons express a functional AIM2 inflammasome and identify pyroptosis as a cell death mechanism in neurons. It demonstrated pannexin1 as a cell death effector channel in pyroptosis and inhibition of pannexin1 by probenecid attenuates pyroptosis in neurons, which suggested that active Caspase-1 cleaves inflammatory cytokines, opens the pannexin1 pore, and induces cell death [84]. It is unclear of the pyroptosis mechanism in ischemic stroke. What is the role of pyroptosis and the relationship with apoptosis and autophagy? It is worthy to explore the role of pyroptosis deeply in the ischemic stroke models.

3 Artery Recanalization of Ischemic Stroke

3.1 Intravenous Thrombolytic and Antiplatelet Therapy Treatment

Thrombolytic treatment aims to remove the occlusion and to restore blood flow in the hypoperfused brain tissue. As we illustrated above, the penumbra region has the potential to recover, which is relevant with the clinical outcome. Intravenous recombinant tissue plasminogen activator (rt-PA) is the only proven effective treatment for acute ischemic stroke [85]. The preceding non-randomized studies illustrate the recanalization correlated with less infarct growth and better clinical outcomes [86]. In 1996, the US Food and Drug Administration (FDA) approved rt-PA for use in early acute ischemic stroke. Intravenous rt-PA, the recommend dose is 0.9 mg/kg within 4.5 h of ischemic stroke, with a bolus of 10% of the dose administered over 1 min, and the remainder infused over 60 min [85, 87]. The lower dose of rt-PA (0.6 mg/kg) could reduce the incidence of symptomatic intracerebral hemorrhage but it isn't better for the functional outcome at 90 days compared with standard-dose rt-PA [88]. The combination of the Barthel Index (BI), modified Rankin Scale (mRS), Glasgow Coma Scale (GCS) and National Institutes of Health Stroke Scale (NIHSS) are used to assess the functional outcome. According to the updated AHA/ASA guidelines for management of acute ischemic stroke, it was recommended that patients eligible for intravenous rt-PA should receive intravenous rt-PA even if endo-vascular treatments are being considered (Class I; Level of Evidence A) [89]. In recent years, some studies focus on trials to evaluate the ischemic stroke patients with small cores and substantial salvageable penumbra, as identified by CT or MR perfusion, whether could be benefit from rt-PA and endovascular thrombectomy beyond 6 h [90]. With regard to the antiplatelet therapy, the Clopidogrel in High-risk patients with Acute Non-disabling Cerebrovascular Events (CHANCE) trial indi-cates Clopidogrel-Aspirin therapy (loading dose of 300 mg of Clopidogrel on day 1, followed by 75 mg of Clopidogrel per day for 90 days, plus 75 mg of Aspirin per day for the first 21 days) or to the Aspirin-alone group (75 mg/day for 90 days) within 24 h after onset of minor stroke or high-risk TIA, which could reduce the risk of subsequent stroke persisted for the duration of 1-year of follow-up [91].

3.2 Endovascular Treatment for Acute Ischemic Stroke

Endovascular therapies include clot disruption or mechanical retrieval. The intra-arterial (IA) application of fibrinolytic agents was initially motivated by consider-ations of benefit ratio and extending the time window. The Middle Cerebral Artery Embolism Local Fibrinolytic Intervention Trial (MELT) implemented the patients with acute (0–6 h) M1 and M2 occlusions to either IA Urokinase or placebo. Although the primary endpoint did not show the statistical significance, the second-ary analyses suggested that IA treatment has the potential to increase the possibility of positive functional outcome. Later, IA treatment was commonly administered as an off-label therapy for stroke within 6 h of onset in the anterior circulation and up to 12–24 h after onset in the posterior circulation [92]. However the newest AHA/ASA guidelines suggest endovascular therapy with stent retrievers is recommended over intra-arterial fibrinolysis as the first-line therapy (Class I; Level of Evidence E). Intra-arterial fibrinolysis initiated within 6 h of stroke onset in carefully selected patients who have contraindications to the use of intravenous rt-PA might be consid-ered, but the consequences are unknown (Class IIb; Level of Evidence C).

So far rt-PA does not have FDA approval for intra-arterial use [89]. More and more prospective, randomized, open-label, blinded-end-point (PROBE) researches evaluate efficacy of recanalization after ischemic stroke. Intra-arterial Versus Systemic Thrombolysis for Acute Ischemic Stroke (SYNTHESIS Expansion) aimed to compare the intravenous rt-PA and endovascular therapy. And the score of mRS, symptomatic intracerebral hemorrhage (sICH), NIHSS were used to evaluate between the subgroups. While the result shows there are no significant differences in out-comes in subgroups [93]. The Interventional Management of Stroke Trial III (IMS III) allocated ischemic stroke patients (NIHSS score ≥10, within 3 h onset) to two groups, the one group receive the standard-dose intravenous rt-PA (0.9 mg/kg), the another group was treated with rt-PA (0.6 mg/kg). After therapy, if the arteries were still occlusion, a device and/or intra-arterial rt-PA were used follow-up. Recanalization

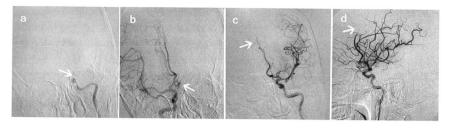

Fig. 2.4 Mechanical retrieval performed on the patient with cardiac source of emboli. A 50-year-old female with rheumatic heart disease, atrial flutter presented with right hemiparesis, aphasia and conscious disturbance. (**a**) Angiogram shows the left internal carotid artery occlusion as the arrow indicates. (**b**) There isn't any compensatory action from right internal carotid artery occlusion as the arrow indicates. (**c**) and (**d**) Recanalization of left internal carotid artery after receiving the endovascular therapy with a stent retriever. Anterior cerebral artery (A3), indicating by the arrow, is still occlusive

occurred 325 ± 52 min after stroke onset. Finally, there was no significant difference in mRS scores between the two groups [94]. The EPITHET study tests rt-PA given 3–6 h after stroke onset and PWI and DWI were done before and 3–5 days after therapy. It is the prospective, randomized, double blind, placebo-controlled and phase II trial research. The result shows that it could increase reperfusion beyond 3 h treatment, but it could not lower infarct growth. It is said reperfusion is associated with improved clinical outcomes [95]. The AHA/ASA guidelines suggest patients should receive endovascular therapy with a stent retriever if they meet all the following criteria (Class I; Level of Evidence A). (1) Prestroke mRS score 0–1. (2) Acute ischemic stroke receiving intravenous r-tPA within 4.5 h of onset according to guidelines from professional medical societies. (3) Causative occlusion of the ICA or proximal MCA (M1). (4) Age ≥18 years. (5) NIHSS score of ≥6. (6) ASPECTS of ≥6. (7) Treatment can be initiated (groin puncture) within 6 h of symptom onset [89] (Fig. 2.4).

3.3 Antioxidant and Other Treatment

The neuronal metabolism depends upon the lactate and activated glycolysis [96]. It has been shown that neurons synthesize energy mainly rely on the oxidative metabolism of glucose. Nevertheless, if the glucose and lactate are present at the same time, neurons can efficiently utilize lactate [97]. The superoxide anions, hydrogen peroxide, hydroxyl radicals, and peroxynitrite or nitrogen dioxide express in infiltrating phagocytes, vascular cells, and glial in the area of penumbra [98, 99]. Because of the high consumption of oxygen and its relatively low endogenous antioxidant capacity in the ischemic brain tissue, neutralisation of oxidative stresses is a potential and positive therapeutic way [100]. The research shows for the first time that mitochondrial metabolic oscillations happened in the vivo with the rapid time (10–15 s), while cellular reactive oxygen species (ROS) plays a role in promoting the NADH oscillations [101]. ROS is mainly produce by mitochondria, which could impair endothelium dependent vasodilator mechanisms, direct damage to biomolecules that result in necrosis, necroptosis and apoptosis. It could be said the mitochondria is very essential in pathogenesis of ischemic stroke [102, 103]. It seems a major cause of enhancing mitochondrial ROS production in stroke by impairing the function of respiratory chain complexes and ATP synthase [104].

In the penumbra region of ischemic stroke, ATP decreases obviously. Some of the ADP is further metabolised to AMP and ATP [105]. After hypoxic/ischemic injury, with the ROS is produced, antioxidant capacities are reduced. Because of the mitochondrial lipid peroxidation, Ca^{2+} overload so that mitochondrial membrane depolarization, which promote cytochrome c to be released and result in cell apoptosis [106–108]. As an antioxidant drug, Edaravone could scavenge hydroxyl, peroxyl and superoxide radicals. It was approved for use in patients with acute stroke in Japan in 2001 and the compound is widely used in China [109]. According to the research with cell culture and animal model, it has shown Edaravone inhibits microglia-induced neurotoxicity, chronic inflammation, lipo-oxygenase, oxidation of low-density lipoproteins, and altered expression of endothelial and neuronal proteins [110]. Some studies also focus on the non-pharmaceutical therapy for recanalization.

Brain is extremely sensitive to hypoxia or ischemia, while ischemic/hypoxic conditioning is a neuroprotective strategy for stroke. The study shows according the pathway of PI3 kinase (phosphatidylinositol 3-kinase)/Akt and the inhibitor-of-apoptosis proteins, hypoxic preconditioning (HPC) could inhibit apoptotic cell death in brain microvascular endothelial cells [111]. Another research indicates HPC could protect the BBB integrity in the wild-type mice, but the HPC-induced BBB protection would be damage in SphK (Sphingosine kinase)-knockout mice [112].

As the neuroprotective approach, hypothermia could decrease utilization of oxygen and glucose, keep the level of ATP and pH in ischemic tissues, reduce the downstream consequences of lactate overload and alleviate acidosis [113]. It is beneficial to mitochondrial function. Hypothermia could inhibit of cPKCδ translocation to the mitochondria, further hinder the process of reactive oxygen species and initiation of apoptosis [114]. On the other hand hypothermia could protect the BBB integrity according to decreasing MMP proteolytic activity and degrading the components of the extracellular matrix comprising the BBB [115]. But the efficacy, safety of hypothermia in clinical still requires further investigation.

4   Reperfusion Injury After Artery Recanalization

Through the medical or mechanical recanalization, the restoration of blood flow after ischemic stroke could be reached. In this process, ischemic tissues may be suffered the additional damage, which is called reperfusion injury. The concept was first described by Cerra et al. in 1975 [116]. In the early stage of ischemia, glycolysis provides ATP. After glycogen was burned out, lactate and other toxic metabolic products could activate the chemical mediators and enzymes [117]. One of the characters of reperfusion injury is that arachidonic metabolites promote the adhesion of leukocytes to vascular endothelium and increase the permeability of the postcapillary venules [118]. During reperfusion, with the accumulation of leukocytes, red blood cells and platelets, the capillaries may be blocked, which inhibits the restoration of blood flow. This phenomenon is called "no-reflow" which could cause the secondary cerebral ischemia [119–121]. Leukocytes release lysozymes, produce the reactive oxygen species and release leukocytes chemotactic agents, which involved in the process of reperfusion injury [122]. Inflammatory system is stimulated and nucleotides are released acting as signals to promote phagocytosis in the apoptotic tissue bed, further to aggravate reperfusion injury [123, 124]. As the important mediator of reperfusion injury, reactive radical oxide species could oxidize proteins, lipids, DNA and other biological molecules [125]. The activity of inducible nitric oxide synthase is incurred by cerebral ischemia and the production of nitric oxide, acting as the main reactive oxide species, is involved in the process of necrosis and apoptosis [126]. The study indicated that in middle cerebral artery occlusion animal models, the oxidative stress affects constantly on pericytes in the microvasculature, the injury lasts beyond the period of ischemia no matter whether reperfusion or not [127]. Because of the sustained effect of reactive oxygen on endothelium, it would influence the permeability of the vascular system and disrupt BBB [128]. Hemorrhagic occurs in the ischemic infarction is the most common reperfusion injury, which is impacted by sheer pressure and the cytotoxic nature of these blood products [129]. No matter mechanical and pharmacological recanalization therapies, the future of acute stroke therapy will focus on the advancement of revascularization, while not just the recanalization.

5   Conclusion

Ischemic stroke is a major cause of mortality and long-term disability, and contains complicated pathophysiological process. The understanding of molecular mechanisms underlying cerebral ischemic injury, artery recanalization-induced reperfusion injury and endogenous neuroprotective phenomenon, such as pre- or post-ischemia/hypoxia-induced neuroprotection, will provide new targets and strategies for the treatment of stroke in clinic. Although some ideas and efficiency on the animal models have been acquired, we need continue making efforts on the translational stroke research.Funding InformationThis work was supported by grants from the Beijing Natural Science Foundation (7132070 and 7141001) and National Natural Science Foundation of China (81301015, 31471142 and 31671205).

References

1. Feigin VL, Forouzanfar MH, Krishnamurthi R, Mensah GA, Connor M, Bennett DA, et al. Global and regional burden of stroke during 1990–2010: findings from the Global Burden of Disease Study 2010. Lancet. 2014;383(9913):245–54.
2. Anderson CS, Jamrozik KD, Broadhurst RJ, Stewart-Wynne EG. Predicting survival for 1 year among different subtypes of stroke. Results from the Perth Community Stroke Study. Stroke. 1994;25(10):1935–44.
3. Bonita R, Broad JB, Beaglehole R. Changes in stroke incidence and case-fatality in Auckland, New Zealand, 1981–91. Lancet. 1993;342(8885):1470–3.
4. Johnston SC, Mendis S, Mathers CD. Global variation in stroke burden and mortality: estimates from monitoring, surveillance, and modelling. Lancet Neurol. 2009;8(4):345–54.
5. Liu L, Wang D, Wong KS, Wang Y. Stroke and stroke care in China: huge burden, significant workload, and a national priority. Stroke. 2011;42(12):3651–4.
6. Harston GW, Rane N, Shaya G, Thandeswaran S, Cellerini M, Sheerin F, et al. Imaging biomarkers in acute ischemic stroke trials: a systematic review. AJNR Am J Neuroradiol. 2015;36(5):839–43.
7. Rha JH, Saver JL. The impact of recanalization on ischemic stroke outcome: a meta-analysis. Stroke. 2007;38(3):967–73.
8. Krishnamurthi RV, Feigin VL, Forouzanfar MH, Mensah GA, Connor M, Bennett DA, et al. Global and regional burden of first-ever ischaemic and haemorrhagic stroke during 1990–2010: findings from the Global Burden of Disease Study 2010. Lancet Glob Health. 2013;1(5):e259–81.
9. Yusuf S, Reddy S, Ounpuu S, Anand S. Global burden of cardiovascular diseases: Part I: General considerations, the epidemiologic transition, risk factors, and impact of urbanization. Circulation. 2001;104(22):2746–53.
10. Wang Y, Zhao X, Liu L, Soo YO, Pu Y, Pan Y, et al. Prevalence and outcomes of symptomatic intracranial large artery stenoses and occlusions in China: the Chinese Intracranial Atherosclerosis (CICAS) Study. Stroke. 2014;45(3):663–9.
11. Marks MP, Lansberg MG, Mlynash M, Kemp S, McTaggart RA, Zaharchuk G, et al. Angiographic outcome of endovascular stroke therapy correlated with MR findings, infarct growth, and clinical outcome in the DEFUSE 2 trial. Int J Stroke. 2014;9(7):860–5.
12. Gorelick PB, Sacco RL, Smith DB, Alberts M, Mustone-Alexander L, Rader D, et al. Prevention of a first stroke: a review of guidelines and a multidisciplinary consensus statement from the National Stroke Association. JAMA. 1999;281(12):1112–20.
13. Feigin VL, Roth GA, Naghavi M, Parmar P, Krishnamurthi R, Chugh S, et al. Global burden of stroke and risk factors in 188 countries, during 1990–2013: a systematic analysis for the Global Burden of Disease Study 2013. Lancet Neurol. 2016;15(9):913–24.
14. Wang YL, Wu D, Nguyen-Huynh MN, Zhou Y, Wang CX, Zhao XQ, et al. Antithrombotic management of ischaemic stroke and transient ischaemic attack in China: a consecutive cross-sectional survey. Clin Exp Pharmacol Physiol. 2010;37(8):775–81.
15. O'Donnell MJ, Xavier D, Liu L, Zhang H, Chin SL, Rao-Melacini P, et al. Risk factors for ischaemic and intracerebral haemorrhagic stroke in 22 countries (the INTERSTROKE study): a case-control study. Lancet. 2010;376(9735):112–23.
16. Kirchhof P, Benussi S, Kotecha D, Ahlsson A, Atar D, Casadei B, et al. 2016 ESC Guidelines for the management of atrial fibrillation developed in collaboration with EACTS: the task force for the management of atrial fibrillation of the European Society of Cardiology (ESC) developed with the special contribution of the European Heart Rhythm Association (EHRA) of the ESC Endorsed by the European Stroke Organisation (ESO). Europace. 2016;37:2893–962.
17. El-Koussy M, Schroth G, Brekenfeld C, Arnold M. Imaging of acute ischemic stroke. Eur Neurol. 2014;72(5–6):309–16.

18. Gonzalez RG, Schaefer PW, Buonanno FS, Schwamm LH, Budzik RF, Rordorf G, et al. Diffusion-weighted MR imaging: diagnostic accuracy in patients imaged within 6 hours of stroke symptom onset. Radiology. 1999;210(1):155–62.
19. Albers GW, Thijs VN, Wechsler L, Kemp S, Schlaug G, Skalabrin E, et al. Magnetic resonance imaging profiles predict clinical response to early reperfusion: the diffusion and perfusion imaging evaluation for understanding stroke evolution (DEFUSE) study. Ann Neurol. 2006;60(5):508–17.
20. Schaefer PW, Grant PE, Gonzalez RG. Diffusion-weighted MR imaging of the brain. Radiology. 2000;217(2):331–45.
21. Lassen NA. Pathophysiology of brain ischemia as it relates to the therapy of acute ischemic stroke. Clin Neuropharmacol. 1990;13(Suppl 3):S1–8.
22. Astrup J, Siesjo BK, Symon L. Thresholds in cerebral ischemia - the ischemic penumbra. Stroke. 1981;12(6):723–5.
23. Knash M, Tsang A, Hameed B, Saini M, Jeerakathil T, Beaulieu C, et al. Low cerebral blood volume is predictive of diffusion restriction only in hyperacute stroke. Stroke. 2010;41(12):2795–800.
24. Hankey GJ, Blacker DJ. Is it a stroke? BMJ. 2015;350:h56.
25. Whiteley WN, Wardlaw JM, Dennis MS, Sandercock PA. Clinical scores for the identification of stroke and transient ischaemic attack in the emergency department: a cross-sectional study. J Neurol Neurosurg Psychiatry. 2011;82(9):1006–10.
26. Lau A, Tymianski M. Glutamate receptors, neurotoxicity and neurodegeneration. Pflugers Archiv. 2010;460(2):525–42.
27. Gill R, Brazell C, Woodruff GN, Kemp JA. The neuroprotective action of dizocilpine (MK-801) in the rat middle cerebral artery occlusion model of focal ischaemia. Br J Pharmacol. 1991;103(4):2030–6.
28. Halassa MM, Fellin T, Haydon PG. The tripartite synapse: roles for gliotransmission in health and disease. Trends Mol Med. 2007;13(2):54–63.
29. Fellin T, Pascual O, Gobbo S, Pozzan T, Haydon PG, Carmignoto G. Neuronal synchrony mediated by astrocytic glutamate through activation of extrasynaptic NMDA receptors. Neuron. 2004;43(5):729–43.
30. Orrenius S, Zhivotovsky B, Nicotera P. Regulation of cell death: the calcium-apoptosis link. Nat Rev Mol Cell Biol. 2003;4(7):552–65.
31. Brennan-Minnella AM, Won SJ, Swanson RA. NADPH oxidase-2: linking glucose, acidosis, and excitotoxicity in stroke. Antioxid Redox Signal. 2015;22(2):161–74.
32. Jackman KA, Miller AA, De Silva TM, Crack PJ, Drummond GR, Sobey CG. Reduction of cerebral infarct volume by apocynin requires pretreatment and is absent in Nox2-deficient mice. Br J Pharmacol. 2009;156(4):680–8.
33. Kim GS, Jung JE, Narasimhan P, Sakata H, Yoshioka H, Song YS, et al. Release of mitochondrial apoptogenic factors and cell death are mediated by CK2 and NADPH oxidase. J Cereb Blood Flow Metab. 2012;32(4):720–30.
34. Kleinschnitz C, Grund H, Wingler K, Armitage ME, Jones E, Mittal M, et al. Post-stroke inhibition of induced NADPH oxidase type 4 prevents oxidative stress and neurodegeneration. PLoS Biol. 2010;8(9). https://doi.org/10.1371/journal.pbio.1000479.
35. Suh SW, Shin BS, Ma H, Van Hoecke M, Brennan AM, Yenari MA, et al. Glucose and NADPH oxidase drive neuronal superoxide formation in stroke. Ann Neurol. 2008;64(6):654–63.
36. Yoshioka H, Niizuma K, Katsu M, Okami N, Sakata H, Kim GS, et al. NADPH oxidase mediates striatal neuronal injury after transient global cerebral ischemia. J Cereb Blood Flow Metab. 2011;31(3):868–80.
37. George PM, Steinberg GK. Novel stroke therapeutics: unraveling stroke pathophysiology and its impact on clinical treatments. Neuron. 2015;87(2):297–309.
38. Moskowitz MA, Lo EH, Iadecola C. The science of stroke: mechanisms in search of treatments. Neuron. 2010;67(2):181–98.

39. Abulrob A, Brunette E, Slinn J, Baumann E, Stanimirovic D. Dynamic analysis of the blood-brain barrier disruption in experimental stroke using time domain in vivo fluorescence imaging. Mol Imaging. 2008;7(6):248–62.
40. Sykova E. Glial diffusion barriers during aging and pathological states. Prog Brain Res. 2001;132:339–63.
41. Sofroniew MV, Vinters HV. Astrocytes: biology and pathology. Acta Neuropathol. 2010;119(1):7–35.
42. Fukuda AM, Badaut J. Aquaporin 4: a player in cerebral edema and neuroinflammation. J Neuroinflammation. 2012;9:279.
43. Ueno M, Tomimoto H, Akiguchi I, Wakita H, Sakamoto H. Blood-brain barrier disruption in white matter lesions in a rat model of chronic cerebral hypoperfusion. J Cereb Blood Flow Metab. 2002;22(1):97–104.
44. Manley GT, Fujimura M, Ma T, Noshita N, Filiz F, Bollen AW, et al. Aquaporin-4 deletion in mice reduces brain edema after acute water intoxication and ischemic stroke. Nat Med. 2000;6(2):159–63.
45. Kebir H, Kreymborg K, Ifergan I, Dodelet-Devillers A, Cayrol R, Bernard M, et al. Human TH17 lymphocytes promote blood-brain barrier disruption and central nervous system inflammation. Nat Med. 2007;13(10):1173–5.
46. Shichita T, Sugiyama Y, Ooboshi H, Sugimori H, Nakagawa R, Takada I, et al. Pivotal role of cerebral interleukin-17-producing gammadeltaT cells in the delayed phase of ischemic brain injury. Nat Med. 2009;15(8):946–50.
47. Waisman A, Hauptmann J, Regen T. The role of IL-17 in CNS diseases. Acta Neuropathol. 2015;129(5):625–37.
48. Li GZ, Zhong D, Yang LM, Sun B, Zhong ZH, Yin YH, et al. Expression of interleukin-17 in ischemic brain tissue. Scand J Immunol. 2005;62(5):481–6.
49. Xiong X, Barreto GE, Xu L, Ouyang YB, Xie X, Giffard RG. Increased brain injury and worsened neurological outcome in interleukin-4 knockout mice after transient focal cerebral ischemia. Stroke. 2011;42(7):2026–32.
50. Liesz A, Suri-Payer E, Veltkamp C, Doerr H, Sommer C, Rivest S, et al. Regulatory T cells are key cerebroprotective immunomodulators in acute experimental stroke. Nat Med. 2009;15(2):192–9.
51. Shichita T, Ito M, Yoshimura A. Post-ischemic inflammation regulates neural damage and protection. Front Cell Neurosci. 2014;8:319.
52. Yang QW, Lu FL, Zhou Y, Wang L, Zhong Q, Lin S, et al. HMBG1 mediates ischemia-reperfusion injury by TRIF-adaptor independent Toll-like receptor 4 signaling. J Cereb Blood Flow Metab. 2011;31(2):593–605.
53. Wang Y, Ge P, Zhu Y. TLR2 and TLR4 in the brain injury caused by cerebral ischemia and reperfusion. Mediat Inflamm. 2013;2013:124614.
54. Broughton BR, Reutens DC, Sobey CG. Apoptotic mechanisms after cerebral ischemia. Stroke. 2009;40(5):e331–9.
55. Culmsee C, Zhu C, Landshamer S, Becattini B, Wagner E, Pellecchia M, et al. Apoptosis-inducing factor triggered by poly(ADP-ribose) polymerase and Bid mediates neuronal cell death after oxygen-glucose deprivation and focal cerebral ischemia. J Neurosci. 2005;25(44):10262–72.
56. Fujimura M, Morita-Fujimura Y, Kawase M, Copin JC, Calagui B, Epstein CJ, et al. Manganese superoxide dismutase mediates the early release of mitochondrial cytochrome C and subsequent DNA fragmentation after permanent focal cerebral ischemia in mice. J Neurosci. 1999;19(9):3414–22.
57. Sugawara T, Fujimura M, Noshita N, Kim GW, Saito A, Hayashi T, et al. Neuronal death/survival signaling pathways in cerebral ischemia. NeuroRx. 2004;1(1):17–25.
58. Asahi M, Hoshimaru M, Uemura Y, Tokime T, Kojima M, Ohtsuka T, et al. Expression of interleukin-1 beta converting enzyme gene family and bcl-2 gene family in the rat brain

following permanent occlusion of the middle cerebral artery. J Cereb Blood Flow Metab. 1997;17(1):11–8.

59. Sugawara T, Lewen A, Gasche Y, Yu F, Chan PH. Overexpression of SOD1 protects vulnerable motor neurons after spinal cord injury by attenuating mitochondrial cytochrome c release. FASEB J. 2002;16(14):1997–9.

60. Rami A, Sims J, Botez G, Winckler J. Spatial resolution of phospholipid scramblase 1 (PLSCR1), caspase-3 activation and DNA-fragmentation in the human hippocampus after cerebral ischemia. Neurochem Int. 2003;43(1):79–87.

61. Rosenbaum DM, Gupta G, D'Amore J, Singh M, Weidenheim K, Zhang H, et al. Fas (CD95/APO-1) plays a role in the pathophysiology of focal cerebral ischemia. J Neurosci Res. 2000;61(6):686–92.

62. Martin-Villalba A, Herr I, Jeremias I, Hahne M, Brandt R, Vogel J, et al. CD95 ligand (Fas-L/APO-1L) and tumor necrosis factor-related apoptosis-inducing ligand mediate ischemia-induced apoptosis in neurons. J Neurosci. 1999;19(10):3809–17.

63. Krupinski J, Lopez E, Marti E, Ferrer I. Expression of caspases and their substrates in the rat model of focal cerebral ischemia. Neurobiol Dis. 2000;7(4):332–42.

64. Shintani T, Klionsky DJ. Autophagy in health and disease: a double-edged sword. Science (New York, NY). 2004;306(5698):990–5.

65. Adhami F, Liao G, Morozov YM, Schloemer A, Schmithorst VJ, Lorenz JN, et al. Cerebral ischemia-hypoxia induces intravascular coagulation and autophagy. Am J Pathol. 2006;169(2):566–83.

66. Liu C, Gao Y, Barrett J, Hu B. Autophagy and protein aggregation after brain ischemia. J Neurochem. 2010;115(1):68–78.

67. Rami A, Langhagen A, Steiger S. Focal cerebral ischemia induces upregulation of Beclin 1 and autophagy-like cell death. Neurobiol Dis. 2008;29(1):132–41.

68. Koike M, Shibata M, Tadakoshi M, Gotoh K, Komatsu M, Waguri S, et al. Inhibition of autophagy prevents hippocampal pyramidal neuron death after hypoxic-ischemic injury. Am J Pathol. 2008;172(2):454–69.

69. Kabeya Y, Mizushima N, Ueno T, Yamamoto A, Kirisako T, Noda T, et al. LC3, a mammalian homologue of yeast Apg8p, is localized in autophagosome membranes after processing. EMBO J. 2000;19(21):5720–8.

70. Zhu C, Wang X, Xu F, Bahr BA, Shibata M, Uchiyama Y, et al. The influence of age on apoptotic and other mechanisms of cell death after cerebral hypoxia-ischemia. Cell Death Differ. 2005;12(2):162–76.

71. Liang XH, Jackson S, Seaman M, Brown K, Kempkes B, Hibshoosh H, et al. Induction of autophagy and inhibition of tumorigenesis by beclin 1. Nature. 1999;402(6762):672–6.

72. Carloni S, Buonocore G, Balduini W. Protective role of autophagy in neonatal hypoxia-ischemia induced brain injury. Neurobiol Dis. 2008;32(3):329–39.

73. Puyal J, Vaslin A, Mottier V, Clarke PG. Postischemic treatment of neonatal cerebral ischemia should target autophagy. Ann Neurol. 2009;66(3):378–89.

74. Zhang N, Yin Y, Han S, Jiang J, Yang W, Bu X, et al. Hypoxic preconditioning induced neuroprotection against cerebral ischemic injuries and its cPKCgamma-mediated molecular mechanism. Neurochem Int. 2011;58(6):684–92.

75. Wei H, Li Y, Han S, Liu S, Zhang N, Zhao L, et al. cPKCgamma-modulated autophagy in neurons alleviates ischemic injury in brain of mice with ischemic stroke through Akt-mTOR pathway. Transl Stroke Res. 2016;7(6):497–511.

76. Chae JJ, Cho YH, Lee GS, Cheng J, Liu PP, Feigenbaum L, et al. Gain-of-function Pyrin mutations induce NLRP3 protein-independent interleukin-1beta activation and severe autoinflammation in mice. Immunity. 2011;34(5):755–68.

77. von Moltke J, Ayres JS, Kofoed EM, Chavarria-Smith J, Vance RE. Recognition of bacteria by inflammasomes. Annu Rev Immunol. 2013;31:73–106.

78. Kovarova M, Hesker PR, Jania L, Nguyen M, Snouwaert JN, Xiang Z, et al. NLRP1-dependent pyroptosis leads to acute lung injury and morbidity in mice. J Immunol (Baltimore, Md: 1950). 2012;189(4):2006–16.
79. Shi J, Zhao Y, Wang K, Shi X, Wang Y, Huang H, et al. Cleavage of GSDMD by inflammatory caspases determines pyroptotic cell death. Nature. 2015;526(7575):660–5.
80. Roberts TL, Idris A, Dunn JA, Kelly GM, Burnton CM, Hodgson S, et al. HIN-200 proteins regulate caspase activation in response to foreign cytoplasmic DNA. Science (New York, NY). 2009;323(5917):1057–60.
81. Hornung V, Ablasser A, Charrel-Dennis M, Bauernfeind F, Horvath G, Caffrey DR, et al. AIM2 recognizes cytosolic dsDNA and forms a caspase-1-activating inflammasome with ASC. Nature. 2009;458(7237):514–8.
82. Fernandes-Alnemri T, Yu JW, Datta P, Wu J, Alnemri ES. AIM2 activates the inflammasome and cell death in response to cytoplasmic DNA. Nature. 2009;458(7237):509–13.
83. Burckstummer T, Baumann C, Bluml S, Dixit E, Durnberger G, Jahn H, et al. An orthogonal proteomic-genomic screen identifies AIM2 as a cytoplasmic DNA sensor for the inflammasome. Nat Immunol. 2009;10(3):266–72.
84. Adamczak SE, de Rivero Vaccari JP, Dale G, Brand FJ III, Nonner D, Bullock MR, et al. Pyroptotic neuronal cell death mediated by the AIM2 inflammasome. J Cereb Blood Flow Metab. 2014;34(4):621–9.
85. Marler J. Tissue plasminogen activator for acute ischemic stroke. The National Institute of Neurological Disorders and Stroke rt-PA Stroke Study Group. N Engl J Med. 1995;333(24):1581–7.
86. Schellinger PD, Jansen O, Fiebach JB, Heiland S, Steiner T, Schwab S, et al. Monitoring intravenous recombinant tissue plasminogen activator thrombolysis for acute ischemic stroke with diffusion and perfusion MRI. Stroke. 2000;31(6):1318–28.
87. Hacke W, Kaste M, Bluhmki E, Brozman M, Davalos A, Guidetti D, et al. Thrombolysis with alteplase 3 to 4.5 hours after acute ischemic stroke. N Engl J Med. 2008;359(13):1317–29.
88. Anderson CS, Robinson T, Lindley RI, Arima H, Lavados PM, Lee TH, et al. Low-dose versus standard-dose intravenous alteplase in acute ischemic stroke. N Engl J Med. 2016;374(24):2313–23.
89. Powers WJ, Derdeyn CP, Biller J, Coffey CS, Hoh BL, Jauch EC, et al. American Heart Association/American Stroke Association Focused Update of the 2013 guidelines for the early management of patients with acute ischemic stroke regarding endovascular treatment: a guideline for healthcare professionals from the American Heart Association/American Stroke Association. Stroke. 2015;46(10):3020–35.
90. Warach SJ, Luby M, Albers GW, Bammer R, Bivard A, Campbell BC, et al. Acute stroke imaging research roadmap III imaging selection and outcomes in acute stroke reperfusion clinical trials: consensus recommendations and further research priorities. Stroke. 2016;47(5):1389–98.
91. Wang Y, Pan Y, Zhao X, Li H, Wang D, Johnston SC, et al. Clopidogrel with aspirin in acute minor stroke or transient ischemic attack (CHANCE) trial: one-year outcomes. Circulation. 2015;132(1):40–6.
92. Ogawa A, Mori E, Minematsu K, Taki W, Takahashi A, Nemoto S, et al. Randomized trial of intraarterial infusion of urokinase within 6 hours of middle cerebral artery stroke: the middle cerebral artery embolism local fibrinolytic intervention trial (MELT) Japan. Stroke. 2007;38(10):2633–9.
93. Ciccone A, Valvassori L, Nichelatti M, Sgoifo A, Ponzio M, Sterzi R, et al. Endovascular treatment for acute ischemic stroke. N Engl J Med. 2013;368(10):904–13.
94. Broderick JP, Palesch YY, Demchuk AM, Yeatts SD, Khatri P, Hill MD, et al. Endovascular therapy after intravenous t-PA versus t-PA alone for stroke. N Engl J Med. 2013;368(10):893–903.

95. Davis SM, Donnan GA, Parsons MW, Levi C, Butcher KS, Peeters A, et al. Effects of alteplase beyond 3 h after stroke in the Echoplanar Imaging Thrombolytic Evaluation Trial (EPITHET): a placebo-controlled randomised trial. Lancet Neurol. 2008;7(4):299–309.
96. Falkowska A, Gutowska I, Goschorska M, Nowacki P, Chlubek D, Baranowska-Bosiacka I. Energy metabolism of the brain, including the cooperation between astrocytes and neurons, especially in the context of glycogen metabolism. Int J Mol Sci. 2015;16(11):25959–81.
97. Hertz L, Peng L, Dienel GA. Energy metabolism in astrocytes: high rate of oxidative metabolism and spatiotemporal dependence on glycolysis/glycogenolysis. J Cereb Blood Flow Metab. 2007;27(2):219–49.
98. Fabian RH, DeWitt DS, Kent TA. In vivo detection of superoxide anion production by the brain using a cytochrome c electrode. J Cereb Blood Flow Metab. 1995;15(2):242–7.
99. Fukuyama N, Takizawa S, Ishida H, Hoshiai K, Shinohara Y, Nakazawa H. Peroxynitrite formation in focal cerebral ischemia-reperfusion in rats occurs predominantly in the peri-infarct region. J Cereb Blood Flow Metab. 1998;18(2):123–9.
100. Chamorro A, Dirnagl U, Urra X, Planas AM. Neuroprotection in acute stroke: targeting excitotoxicity, oxidative and nitrosative stress, and inflammation. Lancet Neurol. 2016;15(8):869–81.
101. Porat-Shliom N, Chen Y, Tora M, Shitara A, Masedunskas A, Weigert R. In vivo tissue-wide synchronization of mitochondrial metabolic oscillations. Cell Rep. 2014;9(2):514–21.
102. Lee HL, Chen CL, Yeh ST, Zweier JL, Chen YR. Biphasic modulation of the mitochondrial electron transport chain in myocardial ischemia and reperfusion. Am J Physiol Heart Circ Physiol. 2012;302(7):H1410–22.
103. Kalogeris T, Bao Y, Korthuis RJ. Mitochondrial reactive oxygen species: a double edged sword in ischemia/reperfusion vs preconditioning. Redox Biol. 2014;2:702–14.
104. Niatsetskaya ZV, Sosunov SA, Matsiukevich D, Utkina-Sosunova IV, Ratner VI, Starkov AA, et al. The oxygen free radicals originating from mitochondrial complex I contribute to oxidative brain injury following hypoxia-ischemia in neonatal mice. J Neurosci. 2012;32(9):3235–44.
105. Onodera H, Iijima K, Kogure K. Mononucleotide metabolism in the rat brain after transient ischemia. J Neurochem. 1986;46(6):1704–10.
106. Murphy E, Steenbergen C. Mechanisms underlying acute protection from cardiac ischemia-reperfusion injury. Physiol Rev. 2008;88(2):581–609.
107. Wu CC, Bratton SB. Regulation of the intrinsic apoptosis pathway by reactive oxygen species. Antioxid Redox Signal. 2013;19(6):546–58.
108. Lemasters JJ, Theruvath TP, Zhong Z, Nieminen AL. Mitochondrial calcium and the permeability transition in cell death. Biochim Biophys Acta. 2009;1787(11):1395–401.
109. Feng S, Yang Q, Liu M, Li W, Yuan W, Zhang S et al. Edaravone for acute ischaemic stroke. Cochrane Database Syst Rev. 2011;(12):Cd007230.
110. Lapchak PA. A critical assessment of edaravone acute ischemic stroke efficacy trials: is edaravone an effective neuroprotective therapy? Expert Opin Pharmacother. 2010;11(10):1753–63.
111. Zhang Y, Park TS, Gidday JM. Hypoxic preconditioning protects human brain endothelium from ischemic apoptosis by Akt-dependent survivin activation. Am J Physiol Heart Circ Physiol. 2007;292(6):H2573–81.
112. Wacker BK, Freie AB, Perfater JL, Gidday JM. Junctional protein regulation by sphingosine kinase 2 contributes to blood-brain barrier protection in hypoxic preconditioning-induced cerebral ischemic tolerance. J Cereb Blood Flow Metab. 2012;32(6):1014–23.
113. Schaller B, Graf R. Hypothermia and stroke: the pathophysiological background. Pathophysiology. 2003;10(1):7–35.
114. Shimohata T, Zhao H, Sung JH, Sun G, Mochly-Rosen D, Steinberg GK. Suppression of deltaPKC activation after focal cerebral ischemia contributes to the protective effect of hypothermia. J Cereb Blood Flow Metab. 2007;27(8):1463–75.

115. Wagner S, Nagel S, Kluge B, Schwab S, Heiland S, Koziol J, et al. Topographically graded postischemic presence of metalloproteinases is inhibited by hypothermia. Brain Res. 2003;984(1–2):63–75.
116. Cerra FB, Lajos TZ, Montes M, Siegel JH. Hemorrhagic infarction: a reperfusion injury following prolonged myocardial ischemic anoxia. Surgery. 1975;78(1):95–104.
117. Khalil AA, Aziz FA, Hall JC. Reperfusion injury. Plast Reconstr Surg. 2006;117(3):1024–33.
118. Lehr HA, Guhlmann A, Nolte D, Keppler D, Messmer K. Leukotrienes as mediators in ischemia-reperfusion injury in a microcirculation model in the hamster. J Clin Invest. 1991;87(6):2036–41.
119. del Zoppo GJ, Schmid-Schonbein GW, Mori E, Copeland BR, Chang CM. Polymorphonuclear leukocytes occlude capillaries following middle cerebral artery occlusion and reperfusion in baboons. Stroke. 1991;22(10):1276–83.
120. del Zoppo GJ, Mabuchi T. Cerebral microvessel responses to focal ischemia. J Cereb Blood Flow Metab. 2003;23(8):879–94.
121. Hallenbeck JM, Dutka AJ, Tanishima T, Kochanek PM, Kumaroo KK, Thompson CB, et al. Polymorphonuclear leukocyte accumulation in brain regions with low blood flow during the early postischemic period. Stroke. 1986;17(2):246–53.
122. Breda MA, Drinkwater DC, Laks H, Bhuta S, Corno AF, Davtyan HG, et al. Prevention of reperfusion injury in the neonatal heart with leukocyte-depleted blood. J Thorac Cardiovasc Surg. 1989;97(5):654–65.
123. Chen GY, Nunez G. Sterile inflammation: sensing and reacting to damage. Nat Rev Immunol. 2010;10(12):826–37.
124. Elliott MR, Chekeni FB, Trampont PC, Lazarowski ER, Kadl A, Walk SF, et al. Nucleotides released by apoptotic cells act as a find-me signal to promote phagocytic clearance. Nature. 2009;461(7261):282–6.
125. Hess ML, Manson NH. Molecular oxygen: friend and foe. The role of the oxygen free radical system in the calcium paradox, the oxygen paradox and ischemia/reperfusion injury. J Mol Cell Cardiol. 1984;16(11):969–85.
126. Bolanos JP, Almeida A. Roles of nitric oxide in brain hypoxia-ischemia. Biochim Biophys Acta. 1999;1411(2–3):415–36.
127. Yemisci M, Gursoy-Ozdemir Y, Vural A, Can A, Topalkara K, Dalkara T. Pericyte contraction induced by oxidative-nitrative stress impairs capillary reflow despite successful opening of an occluded cerebral artery. Nat Med. 2009;15(9):1031–7.
128. Bektas H, Wu TC, Kasam M, Harun N, Sitton CW, Grotta JC, et al. Increased blood-brain barrier permeability on perfusion CT might predict malignant middle cerebral artery infarction. Stroke. 2010;41(11):2539–44.
129. Wang X, Lo EH. Triggers and mediators of hemorrhagic transformation in cerebral ischemia. Mol Neurobiol. 2003;28(3):229–44.

Chapter 3
Cerebral Ischemic Reperfusion Injury (CIRI) Cases

Weijian Jiang and Yi-Qun Zhang

Abstract Two recanalization stroke cases are presented, one led to harmful brain hemorrhage and the other beneficial effect and patient recovered smoothly. Discussed the potential mechanisms behind beneficial recanalization and reperfusion after recanalization.

Keywords Stroke · Reperfusion injury · Case studies

Cerebral blood flow (CBF) in the territory of a clotted artery may be restored by pharmaceutical or mechanical thrombolysis, such as intravenous r-tPA within 4.5 h of ischemic stroke onset or timely mechanical thrombectomy for occlusion of major intracranial artery. However, the disabling and death rates of 29–67% at 3 months were still high in ischemic stroke patients despite undergoing combined thrombectomy and medicine treatment [1]. The worse outcomes may be associated with cerebral ischemia-reperfusion injury (CIRI), and futile reperfusion as well. Therefore, it is necessary to seek out some effective strategies for prevention and treatment of the CIRI in the current era of recanalization for acute ischemic stroke. Here we will present two cases to start discussions on the CIRI and hope our rough ideas serve to invite brilliant contributions.

1 Case Description

Case 1. A 72-year-old right-handed female with history of atrial fibrillation was referred to our center due to right-side weakness, difficulty speaking, and then stupor for lasting of 3 h and 50 min. Brain CT ruled out intracranial hemorrhage (ICH).

W. Jiang (✉) · Y.-Q. Zhang
Vascular Neurosurgery Department, New Era Stroke Care and Research Institute, The PLA Rocket Force General Hospital, Beijing, China
e-mail: cjr.jiangweijian@vip.163.com

The patient never took any anticoagulation or antiplatelet medication before. On neurological examination, she presented with decreased consciousness level, left deviation of eyes with normal pupillary response, aphasia and right hemiparesis, with National Institutes of Health Stroke Scale (NIHSS) of 18. Blood pressure (BP) was 149/78 mmHg. So, a decision of combined intravenous thrombolysis and mechanical thrombectomy was made. Intravenous r-tPA (0.9 mg/kg) was initiated at 4 h after the onset.

Catheter angiography at 4.5 h discovered a complete occlusion of left terminal internal carotid artery (ICA). Mechanical thrombectomy with the use of a 4 × 20 mm Solitaire™ AB device (Covidien, Irvine, California) was then performed. After a total of five passes of the device, with the first at 5 h and 10 min and the last at 6 h, the occlusion was completely recanalized (Fig. 3.1).

Her NIHSS remained 18 immediately after procedure. BP was 137/59 mmHg. She received intravenous injection of nicardipine (a calcium channel blocker) 4.8 mg/h and esmolol (a beta receptor blocker) 60 mg/h in neurological intensive care unit, to maintain systolic BP between 110 and 120 mmHg. Edaravone (a free-radical scavenger) and rosuvastatin (a lipid-lowering agent) were also administered. However, the patient's status deteriorated at 9 h after the onset (i.e. 3 h after recanalization), with sudden vomiting and slowness of pupillary light reflex. Emergent CT showed an early cerebral edema with sulci effacement in the left hemisphere, followed by intravenous infusion of 20% mannitol 250 ml. At 13 h after the onset, her oxygen saturation dropped to 80%, which was restored to 98% after mechanical ventilation. The patient presented with deep coma at 16 h after the onset (i.e. 10 h after recanalization), with bilateral pupil dilation to 4 mm and light reflex disappearance. The brain edema was more severe on repeated CT, with midline shift (Fig. 3.1). Then, the patient got worse despite aggressive medical therapy, with central hyperpyrexia on day 2, central diabetes insipidus on day 3, circulatory collapse on day 4. The patient was deceased on day 5.

Case 2. A 60-year-old right-handed female with history of hypertension and cigarette smoking had episodes of left-sided weakness during the past 4 months despite aspirin treatment. Before 18 days on admission, she experienced a transient attack of difficulty finding words for about 10 min.

On admission, there was no abnormal finding on neurological examination. Blood clotting tests showed prothrombin time (PT) 9.9 s, activated partial thromboplastin time (APTT) 35 s, thromboplastin time (TT) 14.6 s, international normalized ratio (INR) 0.89, fibrinogen (Fib) 2.68 g/l, and D-Dimer 0.1 mg/l. A severe stenosis at bifurcation of right MCA was detected on magnetic resonance angiography. The stenosis rate was 80% on catheter angiography. The patient was medicated with aspirin 300 mg and clopidogrel 75 mg daily for antiplatelet, atorvastatin 20 mg daily for lipid-lowering, and valsartan/hydrochlorothiazide tablet (80 mg/12.5 mg) daily for antihypertension. The dual-antiplatelet therapy was effective, as thromboelastography (TEG) test after 12 days showed inhibition rate of arachidonic acid (AA) and adenosine diphosphate showed (ADP) of 90% and 54%, respectively.

On the operation day morning, her BP was 130/80 mmHg. The patient was treated by an experienced neurointerventionalist under general anesthesia. After

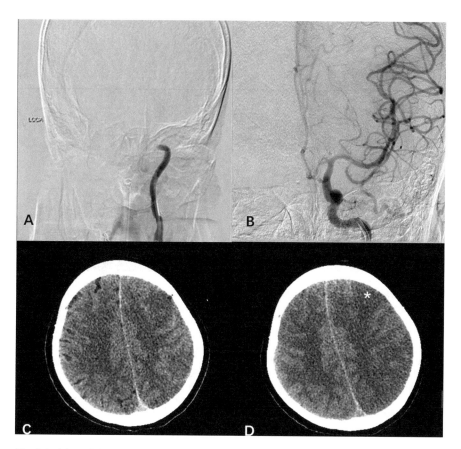

Fig. 3.1 A harmful emergent recanalization for acute ischemic stroke. This is a 72-year-old right-handed female with history of atrial fibrillation who had an acute occlusion of left terminal ICA verified by catheter angiography (**a**). The occlusion was timely revascularized by mechanical thrombectomy at 6 h of stroke onset (**b**). Despite aggressive management of cerebral ischemia-reperfusion injury, the patient's status deteriorated with sudden vomiting and slowness of pupillary light reflex at 3 h after recanalization, and the CT showed an early brain edema with sulci efface-ment in the left hemisphere (**c**). The edema developed with midline shift 10 h after recanalization (**d**), and finally resulted in the death. The hemisphere swelling was more likely to be vasogenic than cytotoxic, as the occlusive artery did not result in large infarct after early recanalization; and the edema peak time occurred within 24 h

intravenous bolus of 3000 U heparin, the stenosis was dilated by slowly inflation of a 2.0 mm × 15 mm Gateway balloon catheter (Boston Scientific, Fremont, CA), followed by implantation of a 4.5 mm × 22 mm Enterprise self-expanding stent (Codman Neurovascular, Raynham, Massachusetts). The stenosis rate was reduced to 10% from 80%, with TICI 3 blood flow, and no evidence of vessel injuries. CT immediately after stenting showed no ICH (Fig. 3.2).

Upon leaving operating room at 14:00, the patient was neurologically normal with NIHSS of 0 and BP of 123/56 mmHg. Her BP was ordered to be controlled at

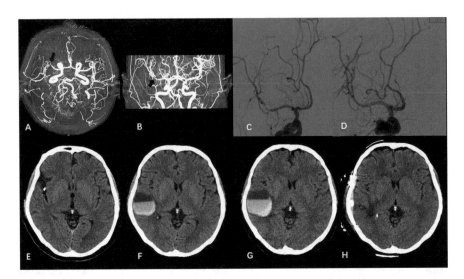

Fig. 3.2 A delayed ICH after elective stenting of intracranial stenosis. This is a 60-year-old right-handed female with episodes of left-sided weakness during the past 4 months despite aspirin therapy. A severe stenosis at bifurcation of right MCA was detected on magnetic resonance angiography (**a** and **b**). The stenosis rate was 80% on catheter angiography (**c**), and was reduced to 10% after stenting (**d**), with no ICH on CT immediately after stenting. However, the patient complained of headache and nausea at 15 h after stenting. A hematoma of 3.9 × 3.7 cm in the right temporal lobe was revealed by CT, with a "fluid level" of the superior hypo-density area (blood plasma) and the inferior hyper-density area (**f**). CT after 25.5 h of stenting showed enlargement (4.6 × 4.5 cm) of the hematoma (**g**), and then, the hematoma was completely evacuated surgically in the evening (**h**), with no difficulty in hemostasis. The hematoma was possibly related to procedure-related pseudoaneurysm, excessive antithrombotic therapy and cerebral hyperperfusion syndrome (CHS). The pseudoaneurysm was ruled out after catheter angiography. The excessive antithrombotic therapy was not supported by blood clotting tests with normal results, and no difficulty in hemostasis during surgery. So the CHS was most probably associated with the ICH

the level of 100–110/60–70 mmHg, about 80% of the baseline BP. Dalteparin sodium (a low-molecular-weight heparin, LMWH) 5000 IU per 12 h was subcutaneously injected. Dual-antiplatelet, lipid-lowering and antihypertension agents were continued. Edaravone (a free-radical scavenger) 30 mg was intravenously infused twice daily.

However, the patient complained of headache and nausea at 5:00 the next morning (i.e. 15 h post procedure), with no vomiting or neurological deficit. Her BP was 110–130/70–85 mmHg during the night. Emergent CT showed a hematoma of 3.9 × 3.7 cm in the right temporal lobe with a "fluid level" of the superior hypo-density area (blood plasma) and the inferior hyper-density area (blood cells). Pseudoaneurysm, hyperperfusion syndrome with hemorrhage, and overdose of antithrombotic agents were highly suspected. Antiplatelet and anticoagulation therapy was discontinued at once, and her BP was controlled at the level of about 100/60 mmHg. The hematoma was found to get enlarged to 4.6 × 4.5 cm on the follow-up CT at 15:30 (i.e. 25.5 h post procedure). After ruling out the pseudoaneurysm by emergent catheter angiography, the hematoma was surgically evacuated in the evening, with no difficulty in hemostasis (Fig. 3.2), followed by disappearance of the symptoms. Blood clotting tests after surgery showed normal results: PT 10.9 s, APTT 28.9 s, TT 14.1 s, INR 0.96, Fib 2.57 g/l, and D-Dimer 1.9 mg/l. The patient was given clopidogrel 75 mg/day again the next day to prevent in-stent thrombosis. She discharged with no sequela 2 weeks later.

2   Discussion

Case 1. This is a harmful emergent recanalization for occlusion of major intracranial artery within the time windows of intravenous r-tPA and mechanical thrombectomy.

In this case, the early and progressive cerebral edema with mass effect was extremely striking. The patient neurological function deteriorated at 3 h after recanalization (i.e., 9 h after stroke onset), which was related to the CT-confirmed early cerebral edema. The edema developed rapidly with cerebral midline shift at 16 h after the onset. The hemisphere swelling was more likely to be vasogenic than cytotoxic, as the occlusive artery did not result in large size of infarct after early recanalization; and the edema peak time occurred within 24 h. Vasogenic edema is supposed to be a consequence of increased hydrostatic pressure gradient in the territory of impaired autoregulation and disrupted blood-brain barrier (BBB), so it can occur soon after recanalization or hours later. Whereas, cytotoxic edema results from energetic and ionic failure with membrane depolarization, and peaks after 24 h of infarction, mostly within 72–96 h [2].

In clinical practice, embolic occlusion in the terminal ICA, like this case, is the most severe type among ischemic strokes and very challenging to manage. It may leave patients an extensive brain infarction for involvement of both middle and anterior cerebral artery if not recanalized timely. However, it leaves physicians a much narrower time window to treat for lack of ischemia preconditioning or good collateralization, which may exist with atherosclerotic occlusion. Meanwhile, its recanalization rate is very low after pharmaceutical or mechanical thrombolysis for large thrombus load. So, recanalization of such a lesion is a big technical challenge. Another challenging met by physicians is severe complications after recanalization, such as ICH, and hemisphere vasogenic edema. Of note, it was reported that both r-tPA thrombolysis and mechanical thrombectomy were associated with BBB disruption, which might contribute to vasogenic edema [3–8].

Cerebral vasogenic edema should be aggressively prevented and managed as early as possible. In this kind of edema, water is driven into extracellular space by high intra-vascular hydrostatic pressure, and high extra-vascular osmotic pressure as a result of protein extravasation via disrupted BBB. The driving forces of water can be reduced by BP control to relieve the vasogenic edema.

In this case, the preventive management BP control and usage of free radical scavenger, had been applied soon after timely recanalization. Unfortunately, the severe CIRI still developed and finally resulted in the death, suggesting that targeting to the CIRI be the main direction in the current era of recanalization for acute ischemic stroke.

Case 2. This is a delayed ICH 15 h after elective stenting for symptomatic MCA stenosis.

The symptomatic hematoma was possibly related to procedure-related pseudoaneurysm, excessive antithrombotic therapy and cerebral hyperperfusion syndrome (CHS) [9]. The pseudoaneurysm was ruled out after catheter angiography. The excessive antithrombotic therapy was not supported by blood clotting tests with normal results, and no difficulty in hemostasis during surgery. So the CHS was most probably associated with the ICH.

CHS with ICH after revascularization of chronic stenosis of extracranial or intracranial artery is not uncommon, occurring in 0.37% and 0.74% patients following carotid endarterectomy and stenting for ICA stenosis, respectively [9]. The pathogenesis may be associated with cerebral autoregulation impairment [10], that is, CBF varies passively with alteration in cerebral perfusion pressure (CPP) under the circumstance of the maximum dilation of small arteries and arterioles with the minimum cerebrovascular resistance (CVR). The CPP increases by eliminating pressure gradient across a severe stenosis after recanalization, which may cause CHS with or without ICH as a result of CBF upward breakthrough [11]. Furthermore, the cerebral autoregulation impairment may last for days and even weeks after recanalization. Thus, it is crucial to strictly control BP at the level of lower than 20% of the baseline BP immediately after revascularization, and for a couple of days at least.

In this case, the ICH could be explained by the sub-satisfactory BP control after stenting, which was 110–130/70–85 mmHg during the post-procedure night, higher than 80% of baseline level of 130/80 mmHg. Luckily, the hematoma was completely evacuated with no sequela.

References

1. Prabhakaran S, Ruff I, Bernstein RA. Acute stroke intervention: a systematic review. JAMA. 2015;313(14):1451–62.
2. Wijdicks EF, Sheth KN, Carter BS, et al. Recommendations for the management of cerebral and cerebellar infarction with swelling: a statement for healthcare professionals from the American Heart Association/American Stroke Association. Stroke. 2014;45(4):1222–38.
3. Simard JM, Kent TA, Chen M, et al. Brain oedema in focal ischaemia: molecular pathophysiology and theoretical implications. Lancet Neurol. 2007;6(3):258–68.
4. Hermann DM, Matter CM. Tissue plasminogen activator-induced reperfusion injury after stroke revisited. Circulation. 2007;116(4):363–5.
5. Warach S, Latour LL. Evidence of reperfusion injury, exacerbated by thrombolytic therapy, in human focal brain ischemia using a novel imaging marker of early blood-brain barrier disruption. Stroke. 2004;35(11 Suppl 1):2659–61.
6. Khatri R, McKinney AM, Swenson B, et al. Blood-brain barrier, reperfusion injury, and hemorrhagic transformation in acute ischemic stroke. Neurology. 2012;79(13 Suppl 1):S52–7.

7. Suzuki Y, Nagai N, Umemura K. A review of the mechanisms of blood-brain barrier permeability by tissue-type plasminogen activator treatment for cerebral ischemia. Front Cell Neurosci. 2016;10:2.
8. Renu A, Laredo C, Lopez-Rueda A, et al. Vessel wall enhancement and blood-cerebrospinal fluid barrier disruption after mechanical thrombectomy in acute ischemic stroke. Stroke. 2017;48(3):651–7.
9. Moulakakis KG, Mylonas SN, Sfyroeras GS, et al. Hyperperfusion syndrome after carotid revascularization. J Vasc Surg. 2009;49(4):1060–8.
10. Sekhon LH, Morgan MK, Spence I. Normal perfusion pressure breakthrough: the role of capillaries. J Neurosurg. 1997;86(3):519–24.
11. Hosoda K, Kawaguchi T, Shibata Y, et al. Cerebral vasoreactivity and internal carotid artery flow help to identify patients at risk for hyperperfusion after carotid endarterectomy. Stroke. 2001;32(7):1567–73.

Chapter 4
CIRI After Early Recanalization

Qingmeng Chen and Min Lou

Abstract Even with rapid reperfusion therapy for acute ischemic stroke, there is still a potential risk of clinical deterioration, such as cerebral hemorrhage, infarct growth or brain edema after early recanalization, which is so-called "reperfusion injury". Blood-brain barrier (BBB) damage, inflammatory responses and leukocyte recruitment are believed as main pathophysiological mechanisms. With the development of neuroimage, several novel neuroimaging markers had been reported. Hyperintense acute reperfusion marker (HARM) and quantitative methods including Patlak algorithm, relative recirculation (rR) on enhanced neuroimage can evaluate the damage of BBB, which may induce cerebral hemorrhage. Postischemic hyperperfusion and "no-reflow phenomenon", which correlate with infarct growth, are observed on perfusion image. Ischemic conditioning therapy, immunosuppressive agents and hypothermia treatment may have potential therapeutic values in future.

Keywords Stroke · Recanalization · Reperfusion injury

Without valid reperfusion therapy within limited time window, brain tissues would progress to infarction with subsequent neurological deficits. Early restoration of blood flow can reduce infarct growth and improve neurologic deficit after acute ischemic stroke, which has been proven by previous animals and humans studies. However, there is still a potential risk of developing cerebral hemorrhage, infarct growth, or brain edema, even with early recanalization [1, 2]. This phenomenon is so-called "early reperfusion injury" [3].

The mechanism of reperfusion injury has been identified in numerous ways, for example, activation of endothelium, excess production of oxygen free radicals, inflammatory responses and leukocyte recruitment, increment in cytokine production,

Q. Chen · M. Lou (✉)
Department of Neurology, School of Medicine, The 2nd Affiliated Hospital of Zhejiang University, Hangzhou, China

© Springer International Publishing AG, part of Springer Nature 2018
W. Jiang et al. (eds.), *Cerebral Ischemic Reperfusion Injuries (CIRI)*, Springer Series in Translational Stroke Research, https://doi.org/10.1007/978-3-319-90194-7_4

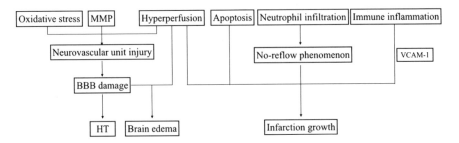

Fig. 4.1 The mechanisms and clinical outcomes of reperfusion injury. *MMP* matrix metalloproteinase, *BBB* blood-brain barrier, *VCAM-1* vascular cell adhesion molecule-1, *HT* hemorrhagic
transformation

as well as edema formation [4, 5]. However, among these mechanisms, the common
is brain-blood barrier (BBB) disruption [5] (Fig. 4.1).

In this chapter, we will review the current studies on basic mechanisms, explore
some neuroimaging markers for predicting reperfusion injury, and provide potential
treatments for reperfusion injury.

1 Hemorrhage Transformation (HT)

Hemorrhage transformation (HT) is supposed as one kind of reperfusion injury,
involving reperfusion-mediated injury with a biochemical cascade. As a result,
HT will lead to secondary deterioration of ischemic brain tissue and neurological
deficits [3].

The major pathophysiologic change of HT is the disturbance of vascular wall
integrity, which is also known as "BBB disruption". The permeability of BBB
in infarct region increases during the first hour after reperfusion [6], reaches the
two peaks after 3 and 48 h [7–9], then decreases gradually. The decrease process lasts for 4–5 weeks [10, 11]. There are several elements causing the
destruction of BBB, for instance, oxygen stress and the neurovascular unit
injury [12, 13].

1.1 Mechanisms

1.1.1 Oxidative Stress

The maintenance of the normal cerebral activation mainly depends on the abundant energy-supply of mitochondria in brain cells. While acute cerebral ischemia
takes place, numerous mitochondrial enzymes such as cytochrome oxidase and
manganese superoxide dismutase (MnSOD), will decrease their reactivities, then

inhibit oxidative phosphorylation of complex IV and the final electron chain [14]. On one hand, due to lack of cytochrome oxidase (one of the final electron acceptors), the remaining final electrons have to combine with proximal complexes, once cerebral blood perfusion is recovered. This process results in increasing reactive oxygen species (ROS). On the other hand, ischemic mitochondrial reperfusion easily makes complex I dysfunction, affects MnSOD expression, and reduces ROS clearance rate. A rapid increasing of the amount of ROS in the ischemic mitochondria causes changes in the cytoskeleton, and affects endothelial and epithelial cells—BBB permeability.

These processes have been found in several animal studies. During reperfusion after 2 h of ischemia, mitochondrial respiratory function recovers partially after the first hour. Whereas, after 2–4 h mitochondrial respiratory function will fall into secondary deterioration in the infarct tissues [14]. Folbergrova et al. also found these similar changes of energy metabolism around the infarct area, suggesting that impaired mitochondrial function is the key to this metabolic response [15].

1.1.2 Neurovascular Unit Injury

Matrix metalloproteinase (MMP) is recognized as an important protease. It affects the permeability of BBB [8], by acting on the basement membrane of the cerebral capillaries. At early stage of reperfusion, up-regulation of MMP-2 and MMP-9 expression degrades tight junction proteins, claudin-5 and occludin, then damages the BBB. Thus, early permeability peak of blood-brain barrier usually develops at 3 h after reperfusion [8, 16]. The underlying second expression peak of important factors (such as MMP, IL-1, TNF-a) may affect the later brain-blood barrier reopening [17].

Postischemic hyperperfusion, such as hyperemia or "luxury perfusion", has been defined as excess of blood flow or volume than what normal brain metabolic actually needs [18]. However, the underlying mechanisms of how it works and the function of postischemic hyperperfusion are still unclear. It is a double-edged sword, which largely depends on which time course it is [18]. The late postischemic reperfusion (12 h after onset), correlating with tissue necrosis, increases the hemorrhage transformation, infarct growth and edema [18]. The possible pathological mechanisms causing postischemic hyperperfusion may include the following two major aspects: (1) autoregulation dysfunction, that is, the accumulated ROS will cause delayed neuronal death and release vasoactive substances (such as lactic acid and adenosine), affecting vascular smooth muscles and leading to dilation of blood vessels [19, 20]; (2) BBB damage: If postischemic hyperperfusion starts immediately and persists for a long time, it often prompts malignant hyperperfusion [21]. Animal experiments have found that when malignant hyperperfusion occurred, there were severe edema and serious necrosis of astrocyte foot and endothelial cells, resulting in significant damage of BBB [22].

1.2 Predictive Imaging Markers

1.2.1 Brain-Blood Barrier Permeability

Hyperintense acute reperfusion marker (HARM), a novel marker of early BBB disruption, can be detected on delayed gadolinium enhancement of cerebrospinal fluid space on fluid-attenuated inversion recovery (FLAIR) [5]. Steven et al. found that among all of the acute ischemic stroke patients, 33% (47/144) of them had the HARM signal, which suggested the presence of early BBB damage. Early reperfusion is the strongest predictor of BBB damage (OR = 4.09, 95% CI = 1.28–13.1, p = 0.018). Comparing to those without intravenous thrombolysis, patients treated with intravenous thrombolysis showed significantly higher early BBB damage and rate of hemorrhagic transformation (55% vs. 25%, 31% vs. 14%). Besides, patients with hemorrhagic transformation (73% vs. 25%) was more prone to have early BBB destruction and achieved a poor prognosis (63% vs. 25%).

As for postischemic BBB permeability, there are several quantitative evaluation, including Patlak algorithm. Analysis of dynamic contrast-enhanced magnetic resonance imaging (DCE-MRI) using Patlak plots model can provide a quantitative approach to calculate the degree of BBB leakage damage [10, 23]. Parameters estimated in the Patlak method model include transfer constant (Ki) of Gd-DTPA and the distribution volume (Vp) of the mobile protons [24]. Abo-Ramadan et al. observed dynamic changes of parameters (Ki and Vp) from hyper-acute phase to chronic stage of ischemic stroke rat models and found that brain-blood barrier once opened at early stage and then closed, followed with secondary reopening [10]. With these parameters, it becomes possible to predictively depict the vulnerable regions in acute ischemic tissues where is destined for HT.

Lupo et al. pointed out that parameters like relative recirculation (rR) and recovery percentage (%R) can be used to estimate the degree of BBB tight connectivity on T2* images [25]. Thornhill et al. also analyzed the dynamic T2* imaging of 18 patients with acute cerebral ischemic stroke, among them 8 cases presented with HT. The average rR of those 8 patients with HT were apparently higher than patients without HT (0.22 ± 0.06 vs. 0.14 ± 0.06, p = 0.006), accompanied by a deceased trend between %R and HT (76 ± 6 versus $82 \pm 11\%$, p = 0.092), suggesting that both rR and %R were potential estimators for BBB leakage and HT identification [26].

1.2.2 Postischemic Hyperperfusion

Postischemic hyperperfusion is a common phenomenon after recanalization, which can be observed in perfusion images such as CT perfusion (CTP), magnetic resonance perfusion-weighted imaging (MRP), arterial spin labeling (ASL), and even earlier positron emission tomography (PET) [27, 28]. By means of cerebral blood flow (CBF) maps, which can be obtained from the above perfusion imagings, hyperperfusion is usually identified as visually perceivable regions with patchy increased CBF when compared with the homologous contralateral hemisphere [28] (Fig. 4.2).

Yu et al. studied 221 acute ischemic stroke patients due to middle cerebral artery occlusion, with a total of 361 ASL scans and found that postischemic hyperperfusion was more likely to appear in the patients who received reperfusion therapies, and was more prone to become HT [28]. Approximately 48% of patients who treated with reperfusion therapy had significant higher blood flow velocity (1.7 times on average) within or around the ischemic core areas than the contralateral side. During follow-up period, a correlation between HT and postischemic hyperperfusion was observed (OR = 3.5, 95% CI = 2.0–6.3, $p < 0.001$). About 47.6% of patients developed postischemic hyperperfusion and hemorrhagic transformation that occurred at the same time point. Late HT in hyperperfusion areas occurred in 35.7% of patients. The later time of hyperperfusion was related with the risk of higher grade of HT (Spearman's rank correlation of 0.44, $p = 0.003$).

2 Infarct Growth

It is known that progressive infarction or secondary ischemic damage in recirculated region is a common consequence of reperfusion injury. In one previous study, 19 Sprague-Dawley rats were divided into three groups, as transient middle cerebral artery occlusion of 30 min, 60 min, or 2.5 h. Based on a series of multi-model MRI, investigators found that DWI abnormalities were reversed transiently and recurred after 1 day, which could be attributed to secondary ischemic damage during early reperfusion period [29]. Moreover, this phenomenon is also found in several animal and human studies [27, 30]. The underlying reasons are probably apoptosis, vascular inflammation, postischemic hyperperfusion (mentioned in Sect. 1.1.2) and so on.

2.1 Mechanisms

2.1.1 Apoptosis

Apoptosis tends to occur in many vulnerable neurons, particularly in the penumbra area after reperfusion [31]. Explanation might be that the restored blood supply ensures energy-dependent caspase-activation cascades and thus develops full apoptosis [22]. Actually neurons in the infarct region present some features of early apoptosis, such as cytoplasmic and nuclear condensation, and specific caspases-cascades activation. When the energy impairment is severe, infarct neurons then activate calpains and shift to necrosis [22].

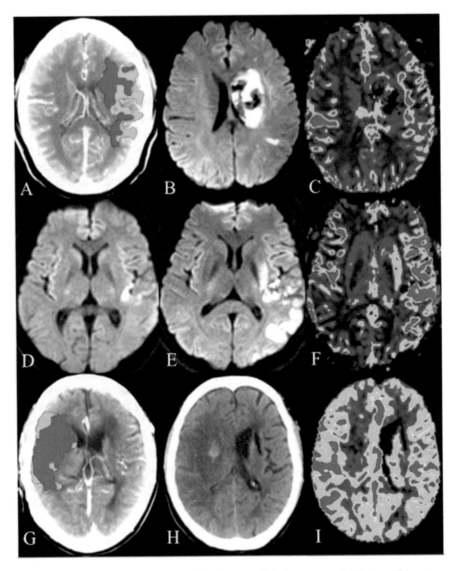

Fig. 4.2 Postischemic hyperperfusion with different clinical outcomes. A 38-year-old man presented with aphasia, right paralysis and was found to have a left MCA stroke with a baseline NIHSS of 20. IV tPA was given 6 h after onset and then clot retrieval was performed. Initial CTP (panel **a**) showed hypoperfusion in the left MCA region, with a follow-up DWI and CBF (panel **b**, **c**) showed left MCA hyperperfusion and HT. A 43-year-old man presented with slight coma, aphasia, right paralysis and was found to have a left MCA stroke with a baseline NIHSS of 25. IV tPA was given 3 h after onset. Initial DWI (panel **d**) showed right infarction in insular cortex, with a follow-up DWI and CBF (panel **e**, **f**) showed left MCA hyperperfusion and infarction growth. A 63-year-old man presented with slight coma, left paralysis and was found to have a right ICA stroke with a baseline NIHSS of 19. IV tPA was given 3.3 h after onset and then clot retrieval was performed. Initial CTP (panel **g**) showed hypoperfusion in the right ICA region, with a follow-up CTP (panel **h**, **i**) showed right MCA hyperperfusion and infarction and brain edema. *MCA* middle cerebral artery, *ICA* internal carotid artery, *NIHSS* National institutes of health stroke scale, *IV* intravenous, *tPA* tissue-type plasminogen activator, *CTP* computed tomography perfusion, *DWI* diffused weighted image, *CBF* cerebral blood flow

2.1.2 Neutrophil Infiltration and Immune Inflammation

Significant neutrophil accumulation is observed at 6 h after established recirculation in ischemic tissue [32]. A large number of active leukocytes flow into the brain tissue, interact with endothelial cells, resulting in a great accumulation of white blood cells, red blood cells and platelets in the microvascular bed. The obvious microvascular obstruction can generate the "no-reflow phenomenon" and give birth to secondary cerebral ischemia [32, 33]. Neutrophil infiltration exacerbates ischemic cells, until reaching the maximum of infarct expansion. Zhang et al. demonstrated that infarct growth and neutrophil infiltration were more dramatic in transient ischemia rats with 6 h and 24 h of reperfusion than those with 48 h permanent occlusion of middle cerebral artery [32]. In addition, other investigators obtained similar findings that during ischemia/reperfusion, neutrophil depletion had better recovery of regional blood flow and smaller infarct size in animal models induced by anti-neutrophil antibodies, compared to non-neutropenic groups [34].

T cells also play an important role in ischemia/reperfusion injury. Studies have shown that anti-$\alpha4$ integrin antibodies and vascular cell adhesion molecule 1 (VCAM 1) siRNA inhibited T-cell infiltration, so as to reduce infarct volume [35, 36]. Although early reports pointed out that in severe combined immunodeficient mice, lack of both T and B cells can decrease 40–70% of infarct size [3], there was no extra infarct volume in Rag1$^{-/-}$ mice (completely T- and B-cells deficient) with transplanted B cells. Infusion of wild type CD3+ T cells into Rag1$^{-/-}$ mice made it vulnerable to ischemia/reperfusion injury [37]. Hence, B cells did not enhance reperfusion injury alone. Moreover, complement system (e.g. C3, C1q) also takes part in this process.

2.2 Predictive Imaging Markers

2.2.1 No-Reflow Phenomenon

In addition to the mechanisms mentioned above, reperfusion therapy may lead to a specific pathway to aggravate the reperfusion injury. In real world, it would be perceivable as lack of appropriate capillary reperfusion despite of large vessel recanalization [38]. Diogo et al. retrospectively reviewed 60 acute ischemic stroke patients who achieved full reperfusion, defined as grade 3 or 2c of modified Thrombolysis In Cerebral Infarction (mTICI) and eventually found that 35% of patients had significant infarct growth, with an absolute infarct growth (30.6 ± 77.7 ml). It was suggested that initial embolus might crack into numerous particles after reperfusion therapy, and then obstruct the downstream small blood vessels and capillaries [39].

2.2.2   Cerebral Neutrophil Recruitment

C.J.S Price et al. [40] studied cerebral neutrophil recruitment of 15 acute ischemic stroke patients within 24 h of clinical onset. Indium-111 (^{111}In) troponolate-labeled neutrophils accumulation can be observed though single-photon emission computed tomography (SPECT), and the attenuating recruitment along with time is confirmed histologically by postmortem examination. In an exploratory analysis, neutrophil accumulation is found significantly related to infarct expansion, which needs further clinic practice.

2.2.3   Vascular Inflammation

According to the different ligand-conjugated microparticles of iron oxide (MPIO), different pathological mechanisms can be visualized and explained by molecular imaging. Thus, it makes molecular imaging a more accurate method to depict vascular inflammation [41], which is a vital characteristic mechanism of ischemia/ reperfusion injury in stroke. In experimental stroke rat models, ligand-targeted MPIO such as VCAM-MPIO for VCAM, Gd-DTPA-sLexA at both P- and E-selection [42] are commonly used to indicate vascular inflammation in vivo. Unfortunately, because of low sensitivity of contrast, molecular imaging has its own limitation when depicting the status of endothelial inflammation following ischemia/reperfusion injury [42]

2.2.4   Postischemic Hyperperfusion

Kidwell et al. reported that postischemic hyperperfusion areas developed into the final infarction during serial diffusion-perfusion MR studies [43]. Hyperperfusion was demonstrated in 5 of 12 patients within several hours after recanalization (mean volume, 18 ml) and in 6 of 11 patients at day 7 (mean volume, 28 ml), and 79% of voxels with hyperperfusion went into infarction at day 7 (Fig. 4.2).

3   Brain Edema

As the usual outcome of reperfusion injury, brain edema, as an abnormal accumulation of fluid within the brain parenchyma, sometimes may be life-threatening for patients.

3.1 Mechanisms and Edema Types

3.1.1 Brain-Blood Barrier Damage and Vasogenic Edema

Vasogenic edema results from the water shift from the intravascular to the extravascular environment. Therefore, every perturbation of microvascular integrity of the physiological barrier (brain-blood barrier) may cause leaking of intravascular fluid into the surrounding parenchyma. The brain-blood barrier consists of endothelial cells, cerebral microvessels, capillary basement membranes and glial cells [44, 45]. Hence, each mechanism that disturbs the stability of brain-blood barrier as above will lead to vasogenic edema [29].

3.1.2 Postischemic Hyperperfusion and Cytotoxic Edema

Cytotoxic edema, due to the translocation of interstitial water into the intracellular compartment, usually shows as an initial cellular swelling. It may occur in all cerebral cells like endothelial cells, neurons and glia. During ischemia/reperfusion, the movement of water molecules corresponds with intracellular electronic homeostasis. That is why postischemic hyperperfusion takes an essential impact to form cytotoxic edema [45] (Fig. 4.2).

4 Treatments

Since the limited knowledge of ischemia/reperfusion injury, there is a few therapeutic approaches used in clinic practice. On this occasion, we will provide some promising treatments for postischemic reperfusion injury.

4.1 Ischemic Conditioning Therapy

4.1.1 Ischemic Preconditioning Therapy

Ischemic preconditioning is an intrinsic protective mechanism. When a subinjurious ischemic exposure occurs, brain cells will initiate endogenous mechanisms to inhibit the activities of ROS and inflammation, alter gene expression, synthesis new proteins and activate Akt-pathway [3]. Thus, thanks to ischemic preconditioning therapy, if a subsequent ischemic stroke happens within one or several days, it would lessen neurological functional deficits and reduce infarct volume. However, ischemic preconditioning therapy has very limited clinical value because of its narrow effective period and the unpredictability of next ischemic attack. Previous study has shown BMS-191095 was an opener of the selective ATP-sensitive potassium (mito K_{ATP}) channel [46]. BMS-induced neuroprotection was attributed to its potential for ischemic preconditioning. Several important mechanisms were suggested such as mitochondrial depolarization without ROS, the activation of phosphoinositide 3-kinase and increased expression of catalase [47]. Mayanagi et al. suggested that BMS-191095 afforded remarkable neuroprotective effect on delayed ischemic preconditioning by selective opening of mitoK_{ATP} channels, without ROS generation [46]. Administration with BMS-191095 24 h before the initiation of 90 min transient ischemia in rats reduced total infarct volume by 32% and cortical infarct volume by 38%, while those treated at 30 or 60 min before the onset had no effect. However, given the unclear harmfulness, its prospect needs large trials to determine.

4.1.2 Ischemic Postconditioning Therapy

Different from ischemic preconditioning, ischemic postconditioning (PostC) is a neuroprotective strategy aiming to reduce ischemic damage as much as possible. According to its applied time course, there are rapid PostC (within a few seconds to minutes after reperfusion) and delayed PostC (within a few hours to days after reperfusion) [48]. In the early stage of reperfusion, rapidly alternated interruptions and restarts of cerebral blood flow are beneficial and operable for PostC [3]. Zhao et al. provided a PostC protocol with remarkable effect for reducing infarct volume [49]. Performing three cycles as 30 s of common carotid artery (CCA) reperfusion and 10 s of CCA occlusion on the special stroke rat models, final infarct volume reduced approximately 17–80%. Nowadays, general PostC comprises of limb remote ischemic PostC and chemical PostC with hypoxia, volatile anesthetic, CO_2, etc. [48].

4.2 Immunosuppressive Agents

As a key pathological pathway in ischemia/reperfusion injury, vascular inflammation should gain more insights on valid treatment. FTY720 (Fingolimod) is a kind of sphingosine-1-phosphate (S1P) receptor agonist. Liu et al. conducted a systematic analysis of previous animal studies, showing Fingolimod reduced the infarct volume (SMD = −1.31, 95% CI = −1.99 to −0.63), and improved the neurological functional outcome (SMD = −1.61, 95% CI = −2.17 to −1.05) [50]. Moreover, in a previous pilot study, the investigators found even combined with alteplase, fingolimod still showed its safety and neuroprotective effect [51]. Superior to patients who treated with alteplase alone, patients who received the combination of fingolimod with alteplase had smaller infarct lesion (10.1 vs. 34.3 ml, p = 0.04), better functional recovery at day 90 (73% vs. 23%, p < 0.01), and less hemorrhage volume (1.2

vs. 4.4 ml, p = 0.01). Since this study had a limited sample, the value of fingolimod to attenuate reperfusion injury needs to be tested in future.

4.3   Hypothermia Treatment

Cooling brain has a global neuroprotective effect. Hypothermia will reduce global cerebral metabolism so that it can suppresses BBB disruption. Besides its clinical usage on stroke common therapy, it is also one promising therapy in reducing reperfusion injury. Hong et al. performed a prospective cohort study to investigate the effects of therapeutic hypothermia (48-h hypothermia and 48-h rewarming) on recanalized stroke patients [52]. Compared to patients without hypothermia, patients who underwent a mild hypothermia therapy had less HT (14% vs. 39%, p = 0.016), less cerebral edema (17% vs. 54%, p = 0.001) and a better neurological recovery (45% vs. 23%, p = 0.017). Therapeutic hypothermia was also an independent predictor of good outcome (OR = 3.0, 95% CI = 1.0–8.9, p = 0.047). Moreover, Chen et al. recently reported that selective brain cooling by intra-arterial infusion of cold saline seemed to be feasible and safe in acute ischemic stroke [53]. Selective brain cooling could decrease 2 °C in cerebral tissue, whereas only 0.3 °C of systemic temperature.

5   Conclusion

Due to the rapid development of reperfusion therapies, reperfusion injury gradually plays an important role in neurological recovery. Numerous pathophysiological mechanisms induce reperfusion injury (HT, infarct growth and cerebral edema), such as oxidative stress, neurovascular injury, neutrophil infiltration and so on. HARM signal, postischemic hyperperfusion, "no-reflow phenomenon" are all practicably predictive image markers for reperfusion injury. Although there are few therapeutic approaches used in current clinical practice for prevention of reperfusion injury, some promising treatments, such as BMS-191095, fingolimod and hypothermia, may have potential therapeutic value in future.

References

1. Dalkara T, Arsava EM. Can restoring incomplete microcirculatory reperfusion improve stroke outcome after thrombolysis? J Cereb Blood Flow Metab. 2012;32:2091–9.
2. Molina CA, Saver JL. Extending reperfusion therapy for acute ischemic stroke: emerging pharmacological, mechanical, and imaging strategies. Stroke. 2005;36:2311–20.
3. Bai J, Lyden PD. Revisiting cerebral postischemic reperfusion injury: new insights in understanding reperfusion failure, hemorrhage, and edema. Int J Stroke. 2015;10(2):143–52.

4. Janardhan V, Qureshi AI. Mechanisms of ischemic brain injury. Curr Cardiol Rep. 2004;6:117–23.
5. Warach S, Latour LL. Evidence of reperfusion injury, exacerbated by thrombolytic therapy, in human focal brain ischemia using a novel imaging marker of early bloodbrain barrier disruption. Stroke. 2004;35:2659–61.
6. Kahles T, Luedike P, Endres M, et al. NADPH oxidase plays a central role in blood-brain barrier damage in experimental stroke. Stroke. 2007;38:3000–6.
7. Belayev L, Busto R, Zhao W, Ginsberg MD. Quantitative evaluation of blood-brain barrier permeability following middle cerebral artery occlusion in rats. Brain Res. 1996;739:88–96.
8. Rosenberg GA, Estrada EY, Dencoff JE. Matrix metalloproteinases and TIMPs are associated with blood-brain barrier opening after reperfusion in rat brain. Stroke. 1998;29:2189–95.
9. Kuroiwa T, Ting P, Martinez H, Klatzo I. The biphasic opening of the blood-brain barrier to proteins following temporary middle cerebral artery occlusion. Acta Neuropathol. 1985;68:122–9.
10. Abo-Ramadan U, Durukan A, Pitkonen M, et al. Post-ischemic leakiness of the blood-brain barrier: a quantitative and systematic assessment by Patlak plots. Exp Neurol. 2009;219:328–33.
11. Strbian D, Durukan A, Pitkonen M, et al. The blood-brain barrier is continuously open for several weeks following transient focal cerebral ischemia. Neuroscience. 2008;153:175–81.
12. Schaller B, Graf R. Cerebral ischemia and reperfusion: the pathophysiologic concept as a basis for clinical therapy. J Cereb Blood Flow Metab. 2004;24(4):351–71.
13. Shen Q, Du F. Spatiotemporal characteristics of postischemic hyperperfusion with respect to changes in T1, T2, diffusion, angiography, and blood-brain barrier permeability. J Cereb Blood Flow Metab. 2011;31:2076–85.
14. Lin LH, Cao S, Yu L, Cui J, Hamilton WJ, Liu PK. Upregulation of base excision repair activity for 8-hydroxy-2-deoxyguanosine in the mouse brain after forebrain ischemia-reperfusion. J Neurochem. 2000;74:1098–105.
15. Folbergrova J, Zhao Q, Katsura KI, Siesjo BK. N-tert-butylalpha phenylnitorne improves recovery of brain energy state in rats following transient focal ischemia. Proc Natl Acad Sci U S A. 1995;92:5057–61.
16. Yang Y, Estrada EY, Thompson JF, et al. Matrix metalloproteinase mediated disruption of tight junction proteins in cerebral vessels is reversed by synthetic matrix metalloproteinase inhibitor in focal ischemia in rat. J Cereb Blood Flow Metab. 2007;27:697–709.
17. Jordan J, Segura T, Brea D, Galindo MF, Castillo J. Inflammation as therapeutic objective in stroke. Curr Pharm Des. 2008;14:3549–64.
18. Lassen NA. The luxury-perfusion syndrome and its possible relation to acute metabolic acidosis localized within the brain. Lancet. 2:1113–5.
19. Berne RM, Rubio R. Regulation of coronary bloodlow. Adv Cardiol. 1974;12:303–17.
20. Cipolla MJ, Chan SL. Postischemic reperfusion causes smooth muscle calcium sensitization and vasoconstriction of parenchymal arterioles. Stroke. 2014;45(8):2425–30.
21. Graf R, Lottgen J. Dynamics of postischemic perfusion following transient MCA occlusion in cats determined by sequential PET. J Cereb Blood Flow Metab. 17:S323.
22. Chelluboina B, Klopfenstein JD. Temporal regulation of apoptotic and anti-apoptotic molecules after middle cerebral artery occlusion followed by reperfusion. Mol Neurobiol. 2014;49(1):50–65.
23. Patlak CS, Blasberg RG. Graphical evaluation of blood-to-brain transfer constants from multiple-time uptake data. Generalizations. J Cereb Blood Flow Metab. 1985;5:584–90.
24. Jiang Q, Ewing JR. Quantitative evaluation of BBB permeability after embolic stroke in rat using MRI. J Cereb Blood Flow Metab. 2005;25(5):583–92.
25. Lupo JM, Cha S, Chang SM, et al. Dynamic susceptibility-weighted perfusion imaging of high-grade gliomas: characterization of spatial heterogeneity. AJNR Am J Neuroradiol. 2005;26:1446–54.

26. Thornhill RE, Chen S. Contrast-enhanced MR imaging in acute ischemic stroke: T2* measures of blood-brain barrier permeability and their relationship to T1 estimates and hemorrhagic transformation. AJNR Am J Neuroradiol. 2010;31(6):1015–22.
27. Kidwell CS, Saver JL. Late secondary ischemic injury in patients receiving intraarterial thrombolysis. Ann Neurol. 2002;52(6):698–703.
28. Yu S, Liebeskind DS, UCLA Stroke Investigators. Postischemic hyperperfusion on arterial spin labeled perfusion MRI is linked to hemorrhagic transformation in stroke. J Cereb Blood Flow Metab. 2015;35(4):630–7.
29. Neumann-Haefelin T, Kastrup A. Serial MRI after transient focal cerebral ischemia in rats: dynamics of tissue injury, blood-brain barrier damage, and edema formation. Stroke. 2000;31(8):1965–72; discussion 1972–3.
30. Olah L, Wecker S. Secondary deterioration of apparent diffusion coefficient after 1-hour transient focal cerebral ischemia in rats. J Cereb Blood Flow Metab. 2000;20(10):1474–82.
31. Phan TG, Wright PM. Salvaging the ischaemic penumbra: more than just reperfusion? Clin Exp Pharmacol Physiol. 2002;29(1–2):1–10.
32. Zhang RL, Chopp M. Temporal profile of ischemic tissue damage, neutrophil response, and vascular plugging following permanent and transient (2H) middle cerebral artery occlusion in the rat. J Neurol Sci. 1994;125:3–10.
33. Fischer EG, Ames A III. Reassessment of cerebral capillary changes in acute global ischemia and their relationship to the "no-reflow phenomenon". Stroke. 1977;8(1):36–9.
34. Matsuo Y, Onodera H, Shiga Y, et al. Correlation between myeloperoxidase-quantified neutrophil accumulation and ischemic brain injury in the rat. Effects of neutrophil depletion. Stroke. 1994;25:1469–75.
35. Relton JK, Sloan KE, Frew EM, et al. Inhibition of alpha4 integrin protects against transient focal cerebral ischemia in normotensive and hypertensive rats. Stroke. 2001;32:199–205.
36. Liesz A, Zhou W, Mracsko E, et al. Inhibition of lymphocyte trafficking shields the brain against deleterious neuroinflammation after stroke. Brain. 2011;134(Pt 3):704–20.
37. Kleinschnitz C, Schwab N, Kraft P, et al. Early detrimental T-cell effects in experimental cerebral ischemia are neither related to adaptive immunity nor thrombus formation. Blood. 2010;115:3835–42.
38. Haussen DC, Nogueira RG. Infarct growth despite fullreperfusion in endovascular therapy for acute ischemic stroke. J Neurointerv Surg. 2016;8(2):117–21.
39. Cho TH, Nighoghossian N. Reperfusion within 6 hours outperforms recanalization in predicting penumbra salvage, lesion growth, final infarct, and clinical outcome. Stroke. 2015;46(6):1582–9.
40. Price CJ, Menon DK. Cerebral neutrophil recruitment, histology, and outcome in acute ischemic stroke: an imaging-based study. Stroke. 2004;35(7):1659–64.
41. McAteer MA, Akhtar AM. An approach to molecular imaging of atherosclerosis, thrombosis, and vascular inflammation using microparticles of iron oxide. Atherosclerosis. 2010;209(1):18–27.
42. Barber PA, Foniok T, Kirk D, et al. MR molecular imaging of early endothelial activation in focal ischemia. Ann Neurol. 2004;56:116–20.
43. Kidwell CS, Saver JL. Diffusion-perfusion MRI characterization of post-recanalization hyperperfusion in humans. Neurology. 2001;57(11):2015–21.
44. Khatri R, McKinney AM. Blood-brain barrier, reperfusion injury, and hemorrhagic transformation in acute ischemic stroke. Neurology. 2012;79(13 Suppl 1):S52–7.
45. Heo JH, Han SW. Free radicals as triggers of brain edema formation after stroke. Free Radic Biol Med. 2005;39(1):51–70.
46. Mayanagi K, Gáspár T. The mitochondrial K(ATP) channel opener BMS-191095 reduces neuronal damage after transient focal cerebral ischemia in rats. J Cereb Blood Flow Metab. 2007;27:348–55.
47. Gáspár T, Snipes JA. ROS-independent preconditioning in neurons via activation of mitoK(ATP) channels by BMS-191095. J Cereb Blood Flow Metab. 2008;28(6):1090–103.

48. Fan YY, Hu WW. Postconditioning-induced neuroprotection, mechanisms and applications in cerebral ischemia. Neurochem Int. 2017. https://doi.org/10.1016/j.neuint.2017.01.006.

49. Zhao H, Sapolsky RM. Interrupting reperfusion as a stroke therapy: ischemic postconditioning reduces infarct size after focal ischemia in rats. J Cereb Blood Flow Metab. 2006;26:1114–21.

50. Liu JF, Zhang CF. Systematic review and meta analysis of the efficacy of sphingosine-1-phosphate (S1P) receptor agonist FTY720 (fingolimod) in animal models of stroke. Int J Neurosci. 2013;123(3):163–9.

51. Zhu ZL, Fu Y. Combination of the immune modulator fingolimod with alteplase in acute ischemic stroke. Circulation. 2015;132:1104–12.

52. Hong JM, Lee JS. Therapeutic hypothermia after recanalization in patients with acute ischemic stroke. Stroke. 2014;45(1):134–40.

53. Chen J, Liu L. Endovascular hypothermia in acute ischemic stroke: pilot study of selective intra-arterial cold saline infusion. Stroke. 2016;47(7):1933–5.

Chapter 5
Programmed Cell Death in CIRI

Ruili Wei, Yang Xu, Jie Zhang, and Benyan Luo

Abstract Neurons in the ischemic penumbra or peri-infarct zone may undergo delayed cell death which called programmed cell death (PCD) and thus they are potentially recoverable for some time after the onset of stroke. There were three major morphologies of PCD in the cerebral ischemic injury, including apoptosis, autophagy and programmed necrosis (also known as necroptosis). In this review we will discuss the characteristics, molecular mechanism of each PCD mode and their role in cerebral ischemia and reperfusion injury (CIRI), furthermore crosstalk between various modes of PCD is also dicussed.

Keywords Programmed cell death · Apoptosis · Autophagy · Programmed necrosis · Necroptosis · Cerebral ischemia and reperfusion injury

1 Introduction

When blood flow to the brain is interrupted, cells undergo a series of molecular events which include excitotoxicity, mitochondrial dysfunction, acidotoxicity, ionic imbalance, oxidative stress and inflammation. These molecular events can lead to cell death and irreversible tissue injury [1, 2]. The fate of brain cells following cerebral ischemia depends upon the severity of the insult and vulnerability of the neurons. The severity of ischemia depends on the extent of cerebral blood flow (CBF) reduction that determines the degree and deprivation of oxygen and glucose from the cells, however a particular threshold do exist for various kinds of pathophysiologic tissue events. Moreover, the high sensitivity of the brain to blood flow changes and dependence on continuous blood flow are critical factors that make the brain particularly more vulnerable to ischemia. Under physiological conditions, the normal CBF is maintained around 50–60 mL/100 g/min but during

R. Wei · Y. Xu · J. Zhang · B. Luo (✉)
Department of Neurology, The First Affiliated Hospital, Zhejiang University School of Medicine, Hangzhou, Zhejiang, China
e-mail: luobenyan@zju.edu.cn

© Springer International Publishing AG, part of Springer Nature 2018
W. Jiang et al. (eds.), *Cerebral Ischemic Reperfusion Injuries (CIRI)*, Springer Series in Translational Stroke Research, https://doi.org/10.1007/978-3-319-90194-7_5

cerebral ischemia due to declining CBF, the ripples of damage spread from the center towards the periphery forming a gradient in such a way that maximum damage (infarction) is at the center (core). The CBF in this region falls to <7 mL/100 g/min and within minutes of a focal ischemic stroke occurring, the core of brain tissue exposed to the most dramatic blood flow reduction is fatally injured and subsequently undergoes necrotic cell death. The ischemic core is surrounded by region of moderate ischemic zone called ischemic penumbra (IP), with a CBF ranging from 7 to 17 mL/100 g/min [3], which remains metabolically active but electrically silent [4]. The ischemic penumbra region may comprise as much as half the total lesion volume during the initial stages of ischemia, and represents the region in which there is opportunity for salvage via poststroke therapy. Recent research has revealed that many neurons in the ischemic penumbra or peri-infarct zone may undergo delayed cell death, and thus they are potentially recoverable for some time after the onset of stroke.

This delayed cell death modal was usually called programmed cell death (PCD), which was different from the necrotic cell death that has been considered merely as an accidental uncontrolled form of cell death. PCD is defined as regulated cell death mediated by an intracellular program, which is a basic biological phenomenon that plays an important role during development, preservation of tissue homeostasis, and elimination of damaged cells. There were three major morphologies of programmed cell death in the ischemic injury, including type I, apoptosis; type II, autophagy; and type III, programmed necrosis (known as necroptosis) [5, 6].

Type I—apoptotic cell death—acts as part of a quality control and repair mechanism by elimination of unwanted, genetically damaged, or senescent cells, and as such is critically important for the development of organisms. Highly conserved in both plants and animals, it is also the cell death mechanism best characterised at both genetic and biochemical levels [7]. Type II—autophagic cell death—is a catabolic process conserved among all eukaryotes from yeast to mammals; it is a mechanism by which organelles are removed. Autophagic cell death is the primary degradation mechanism for long lived proteins, and thus maintains quality control for proteins and organelles to enhance survival under conditions of scarcity or starvation [8]. Type III—programmed necrosis [6, 9]— appears as a distinct entity, not by exclusive engagement of selected effectors, but rather, by combinatorial use of the effectors shared with other cell death outcomes.

PCD displays several cellular phenotypes affecting various intracellular organelles and membranes, and the cell nucleus. For example, the well characterised processes of cytoplasmic and chromatin condensation, nuclear fragmentation, membrane blebbing, and formation of membrane bound apoptotic bodies are part of apoptosis. Autophagy involves the formation of a double membrane vesicle which encapsulates cytoplasm and organelles, and fuses with lysosomes, thus resulting in the degradation of the vesicle contents. Programmed necrosis is characterised by the presence of swelling organelles followed by the appearance of "empty" spaces in the cytoplasm that merge and make connections with the extracellular space. The plasma membrane is fragmented, but the nucleus is relatively preserved (Fig. 5.1).

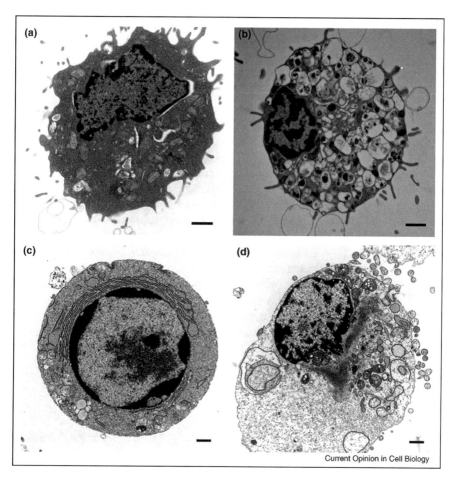

Fig. 5.1 Morphological (electron microscope) features of autophagic, apoptotic and necrotic cells. (**a**) Normal, (**b**) autophagic, (**c**) apoptotic (**d**) and necrotic cells. Whereas the morphologic features of apoptosis are well defined, the distinction between necrotic and autophagic death is less clear. The bioenergetic catastrophe that culminates in cellular necrosis also stimulates autophagy as the cell tries to correct the decline in ATP levels by catabolizing its constituent molecules. Thus, vacuolation of the cytoplasm is observed in both autophagic cells (**b**) and in cells stimulated to undergo programmed necrosis (**d**). By contrast, ATP levels are maintained in normal (**a**) and apoptotic cells (**c**) consistent with the limited number of autophagic vacuoles in their cytoplasm. The scale bar represents 1 mm (From "Death by design: apoptosis, necrosis and autophagy" by Aimee L Edinger and Craig B Thompson [10])

Up to now, the only available therapeutic strategy for ischemic stroke is to reopen an occluded artery by thrombolytic therapy to restore perfusion to the ischemic area during the first few hours, procedure which in itself can sometimes induce secondary damage, so this delayed cell death modalities "programmed cell death" after CIRI provide us a extended therapeutic time window, which is an important research direction in the ischemic stroke study.

2 Apoptotic Cell Death in CIRI

The morphology of apoptotic cells is characterized by vacuoles containing cytoplasm and intact organelles which are named apoptotic bodies. Before the loss of cell membrane integrity, the dying cell is gradually shrinking and absorbed by phagocytic uptake. To date, research indicates that there are two main apoptotic pathways: the intrinsic and extrinsic pathways [11, 12]. The former is also called the mitochondrial pathway because the disruption of mitochondria is pivotal in the process, which leads to the release of the cytochrome C and the downstream activation of caspases. The other pathway, referred as the extrinsic pathway, receptors can be activated by specific ligands that bind to cell surface death receptors.

There are lots of other factors influencing post-stroke apoptosis, including age and gender [13, 14]. It is said that immature brains are more sensitive to the induction of apoptosis because caspase-3 is activated much more in immature brains than in those of adults [13]. Besides, sex hormone exposure may lead to higher risk of cerebral ischemia for women. The pathways of cell death differ in sexual dimorphism, as caspase-dependent pathway is more involved in female whereas AIF translocation is more important in males [14].

It is a common physiological death mechanism in ischemic stroke; but it also causes further impairment under certain pathological conditions. Energetic stress is the consequence of cerebral ischemia, and then reperfusion is accompanied by abrupt ionic shifts and considerable oxidative stress. During above physiopathologic process, apoptosis plays a key role of the neurons.

2.1 Molecular Biology Mechanism of Apoptosis After Ischemic Stroke

2.1.1 Molecules Related to Apoptosis of Neurons

Caspase Family

There are totally 14 caspase proteins identified by researchers, and among them at least eight proteins participate in the cell apoptosis. Caspase related to apoptosis can be classified into two types, the trigger and the executor. Caspase-8, Caspase-9 and Caspase-10 belong to the triggers while Caspase-3 and Caspase-7 are the executors [15]. Caspase-3 has been identified as a key mediator of apoptosis in animal models of ischemic stroke [12]. Activation of Caspase 3 requires assembly of a large multimeric complex comprising Caspase 9, APAF1, and cytochrome c. Caspase-3 cleaves many substrate proteins, including poly (ADP-ribose) polymerase (PARP) [16]. PARP inactivation after cleavage by caspase-3 leads to DNA injury and subsequently to apoptotic cell death [17]. In brief, Caspase-3 and Caspase-7 are the main participate and executor when apoptosis is activated after ischemic brain injury.

Albeit the underlying mechanism is consistent, the severity of ischemia, tempo-ral and spatial heterogeneity may influence the specific condition of neuronal cell death. In the early stages of cerebral infarction, caspase-8 and caspase-1 are involved in the early apoptosis, contributing to the core. However, caspase-9 is related to the secondary expansion of the lesion in the penumbral area [18].

B-Cell Leukemia/Lymphoma 2 (Bcl-2) Family

The B-cell leukemia/lymphoma 2 (Bcl-2) family has the role of maintaining the integrity of the mitochondrial membrane. It has three subfamilies according to the molecular structure [19–21]. The first subtype is antiapoptotic protein, including Bcl-2, Bcl-xl (B-cell lymphoma-extra large) and Bcl-w. Proapoptotic protein is the second subtype, for instance, Bax (Bcl-2-associated X protein) and Bak (Bcl-2 homologous antagonist killer). The last is Bcl-2 homology domains 3 (BH3) domain protein including Bad (Bcl-2-associated death promoter), Bid (BH3 interacting-domain death agonist), Bim (Bcl-2-interacting mediator of cell death), Noxa and p53 [22]. Cerebral ischemia and reperfusion lead to intracellular stress originating from the mitochondria, the endoplasmic reticulum and the nucleus. The proteins from Bcl-2 family are sensitive to these stress factors after cerebrovascular events.

Tumor Necrosis Factor Receptor (TNFR) Superfamily

The Tumor necrosis factor receptor (TNFR) superfamily includes Fas and TNFR1 [23]. Fas is also called CD95 or Apo1. The Fas ligand (FasL) is a homotrimer, con-stituting microaggregate on the surface of cells. Caspase-8 is activated by death-inducing signaling complex (DISC), of which Fas is an important part.

Other Molecules

Besides, there are still other potential molecules that participate in the post-stroke apoptosis. Nuclear factor-Y transcription factor (NF-YC) [24], Secretory phospholipase A2 (sPLA2) [25], Bim [26], Numb [27] have been suggested to be correlated with apoptosis of neurons by experiments.

2.2 Pathways Related to Apoptosis of CIRI

2.2.1 Intrinsic Pathway

The stimulation by glutamate of N-methyl-D-aspartate (NMDA), amino-3-hydroxy-5-methyl-isoxazolpropionic acid (AMPA) receptors, or acid-sensing ion channels (ASICs) causes high-level intracellular calcium after cerebral ischemia [28, 29].

Then, the increased cytosolic calcium activates calpains and induces the cleavage of Bid. The truncated Bid (tBid) interacts with apoptotic proteins such as Bad and Bax at the mitochondrial membrane, which is called heterodimerization [30]. On the other hand, antiapoptotic Bcl-2 interacts with apoptotic proteins and neutralizes their effects. The above process involved Bax and Bcl-2 is the critical event in the mitochondrial-mediated pathway [31]. Mitochondrial transition pores (MTP) are opened after the heterodimerization. Cytochrome c (Cytc) is released from the pores into the cytosol. Then an apoptosome is constituted by Cytc, procaspase-9 and apoptotic protein-activating factor-1 (Apaf-1) [31]. The apoptosome plays the role of activating caspase family. Activated caspase-3 by caspase-9 exert the ultimate effect of nDNA damage and apoptosis through cleaving nDNA repair enzymes such as poly ADP-ribose polymerase (PARP). By contrast, apoptosis-inducing factor (AIF) mediates cell death by a caspase-independent method, which is also released from the pores and translocates rapidly to the nucleus. Phosphorylation and activation of p53 can also mediates the neuronal apoptosis by damaging DNA [32, 33]. Noticeably, secondary reperfusion injury carrying superoxide anions, can also cause DNA damage.

2.2.2 Extrinsic Pathway

There is considerable evidence from animal studies indicating that brain ischemia triggers the extrinsic apoptotic signaling cascade. Due to the initiating effect of death receptors on the plasma member, the extrinsic pathway is also named receptor-mediated pathway. The extracellular Fas ligand (FasL) binds to Fas death receptors (FasR), which triggers the recruitment of the Fas-associated death domain protein (FADD) [31]. FADD binds to procaspase-8 to create a death-inducing signaling complex (DISC), which activates caspase-8 [34]. Activated caspase-8 either mediates cleavage of Bid to truncated Bid (tBid), which integrates the different death pathways at the mitochondrial checkpoint of apoptosis, or directly activates caspase-3. At the mitochondrial membrane tBid interacts with Bax, which is usually neutralized by antiapoptotic B-cell leukemia/lymphoma 2 (Bcl-2) family proteins Bcl-2 or Bcl-xL. Dimerization of tBid and Bax leads to the opening of mitochondrial transition pores (MTP), thereby releasing cytochrome c (Cytc), which execute caspase 3-dependent cell death.

2.3 Significance of Apoptosis in CIRI

The molecular mechanisms of apoptosis after stroke enlighten the exploration of neuroprotective agents. Ischemic preconditioning in animals triggers activation of caspase-3 downstream and upstream of its target caspase-activated DNase (CAD) to prevent neuronal death [35]. Furthermore, enhanced formation of Apaf-1/caspase-9 complex is observed in the rat hippocampus 8–24 h after ischemia [36, 37]. Cao et al. have cloned a rat gene product, a specific Apaf-1 inhibitor of the Apaf-1/caspase-9 pathway that can be neuroprotective in CIRI [12, 35]. Therefore, Apaf-1 signaling pathway may be a legitimate therapeutic target for the treatment of ischemic brain injury [38]. Fas/FasL system acts as apoptosis inducer and triggers pro-inflammatory cytokine production, while the hematopoietic growth factor, erythropoietin (EPO) inhibits apoptosis and protects from ischemic neuronal damage [39]. These findings indicate that death receptors are critically engaged in the apoptosis induction after ischemia in the adult brain and that their suppression may improve the neuronal survival after ischemic injury [12, 40]. FTY720, another antiapoptotic agent, successfully decreased cleaved Caspase-3 expression by activation of sphingosine 1-phosphate-1 in rats after cerebral artery occlusion [41]. In global cerebral ischemia in the gerbils, treatment with a purified medicinal herb called baicalin remarkably promoted the expression of BDNF and inhibited the expression of caspase-3 at mRNA and protein levels [42]. Additionally, it is reported that different concentrations of normobaric oxygen can inhibit the apoptotic pathway by reducing caspase-3 and -9 expression, thereby promoting neurological functional recovery after CIRI [43]. These are various neuroprotective agents on the animal models and they are potential therapeutic targets in future clinical pharmacological research.

3 Necroptosis in CIRI

Necrosis was classified as non-programmed necrotic death previously which has been described as a response of extreme stress. However, in recent years, there is strong evidence to confirm that part of necrosis also contained program control, therefore proposed new concept as programmed necrosis or named necroptosis. Necroptosis are all classified as programmed cell death based on morphological and biochemical features [6, 44]. This phenomenon was observed in the ischemic stroke model.

3.1 Signal Pathway of Necroptosis

Caspase inhibition cannot blocked tumor necrosis factor (TNF) induced cell death completely, but rather switch to cell fate to necrotic death signal pathway like apoptosis [45, 46]. TNFα is the major trigger of necroptosis, which has capable of initiating caspase-8-dependent apoptosis and RIPK1 kinase-dependent necroptosis [47]. Caspase-8 plays a critical regulatory role in the switch. When FADD-caspase-8-FLIP complex functions inhibited, the cell death pathway switches from apoptosis to typical necroptosis features [48–51].

TNF-α induced necroptosis is the mostly intensively investigated. TNF receptor 1 (TNFR1) ligation leads to the recruitment of TRADD, TRAF2 and cIAP1/2, which is named as complex I [52]. The complex I activating death-inducing TNFR1 complex II via cylindromatosis (CYLD) [53]. In necrotic signal pathway, receptor-interacting kinase 1 (RIP1 or RIPK1) was the first molecule identified as the core components of the necroptotic machinery [54]. When RIPK1 and RIPK3 phosphor-ylated, then formed a necrosome through their homotypic interaction motif (RHIM) domains, and activates their kinase activities [55]. This RIPK1–RIPK3 interacts with mixed-lineage kinase domain-like (MLKL) phosphorylation [56]. Downstream of the necrosome are two splice variants of PGAM5, PGAM5S and PGAM5L. PGAM5L binds to the necrosome is not affected by the presence of the necrosis inhibitor necrosulfonamide (NSA). However, the binding of PGAM5S is blocked by NSA. Furthermore, mitochondrial fragmentation caused by the mito-chondrial phosphatase PGAM5S recruited the mitochondrial fission factor Drp1 may up-regulate ROS generation [57].

3.2 Necroptosis in Cerebral Ischemia Disease

Necroptosis delayed mouse ischemic brain injury in the absence of apoptotic signaling [58]. In hippocampal neurons oxygen-glucose deprivation (OGD) models RIP3 mRNA and protein levels upregulation nevertheless caspase-8 mRNA downregula-tion. Similar to RIP3, RIP1 protein level was correlated with the activation of neuronal death. Consistent with the classical procedural necroptosis cellular pathways, ischemic injury upregulated RIP1-RIP3 expression and decreased the caspase-8 expression, which may be available afterwards for activation of necroptotic signaling [59].

Global brain ischemia and reperfusion (I/R) injury is another form of brain cell injury, which the hippocampal CA1 layer is especially vulnerable [60]. As a marker of necroptosis, RIP3 upregulated and transferred into nucleus after cerebral ischemia and reperfusion injury. RIP1–RIP3 complex is necessary for TNF induced necropoptosis in cell cytosol. ATP depletion is one of the results of the mitochondrial permeability transition pore (mPTP) leads to mitochondrial swelling. CypD as a gatekeeper of mPTP, alleviated the levels of RIP1 and RIP3, which mediated mPTP opening may contribute to not only apoptosis but also necroptotic cell death in cerebral I/R injury [61]. RIP3 was activated after I/R injury, and then interacts with AIF in the cytoplasm. The nuclear translocation of AIF and RIP3 is critical to neuronal necropoptosis, and the nuclear translocation of AIF may be RIP3-dependent [51]. AIF is the mediating molecule that links caspase-independent PCD with the necroptotic pathway.

It was observed that nerve cell necrosis occurred following focal middle carotid artery occlusion/reperfusion (MCAO/R) ischemic stroke model. TNFR1 and RIP3 were positively expressed and significantly increased following the volume of cerebral infarction post-reperfusion. Pre-administration with Z-VAD-FMK (zVAD) significantly increased the protein level of RIP3 [62]. In addition to phosphorylation modification, RIP3 S-nitrosylation in ischemia and reperfusion paralleled with elevated phosphorylation. It means RIP3 could be regulated by its S-nitrosylation triggered by NMDAR-dependent nNOS activation [63].

3.3 The Regulation of Necroptosis in Cerebral Ischemic Model

The classic inhibitor is a small molecule compound NSA, which did not block necrosis-induced RIP1 and RIP3 interactions, it blocks necroptosis downstream of RIP3 activation. In human glioblastoma cells, NSA switch from necrosis to apoptosis in edelfosine-treated [64].

In the field of cerebral ischemia, Necrostatin-1 (NEC-1) is another inhibitor of necroptosis has been shown to ameliorate tissue damage in ischemic brain injury animal models [58]. NEC-1 has a selective primary cellular target responsible for the death domain receptor-associated adaptor kinase RIP1 activity [65, 66]. It not only inhibited the expression of RIP1, prevented upregulation and nuclear translocation of RIP3, but also decrease cathepsin-B releasing in globe cerebral ischemic model. CA074-me and 3-methyladenine (3-MA), as autophagy inhibitors [67], were used to determine whether beneficial for global cerebral ischemia in the process of necroptosis signal pathways. The mechanism of 3-MA is inhibiting the nuclear translocation and co-localization of RIP3 and AIF. As the nuclear translocation of RIP3-AIF complex is critical to ischemic neuronal DNA degradation and necroptosis [51] (Fig. 5.2). Beside this, CA074-me almost completely hampered the loss of mitochondrial membrane depolarization, phosphatidylserine (PS) translocation, and plasma membrane rupture [68].

4 Autophagic Cell Death in CIRI

Autophagy is the process by which a membrane engulfs organelles and cytosolic macromolecules to form an autophagosome, with the engulfed materials being delivered to the lysosome for degradation [69]. Briefly, autophagy proceeds through the capture of portions of cytoplasm containing target material inside expanding membranes, which finally enclose to form double-membrane vesicles called autophagosomes. Fully formed autophagosomes are shuttled along microtubules to lysosomes, whereupon fusion and degradation occur [70]. This removal and recycling serves as an emergency energy supply during starvation, but autophagy has also been linked to a diverse range of other protective roles [71, 72]. However, despite these pro-survival roles, autophagy has also been implicated as a mechanism of programmed cell death [73, 74]. Numerous studies have reported instances of dying cells displaying accumulated autophagosomes, which engulf large portions of the cell's cytoplasm and which have been presumed to lead to excessive destruction of vital components [75, 76]. "Autophagic cell death" is morphologically defined as a type of cell death (type II) that occurs in the absence of chromatin condensation but accompanied by massive autophagic vacuolization of the cytoplasm [77].

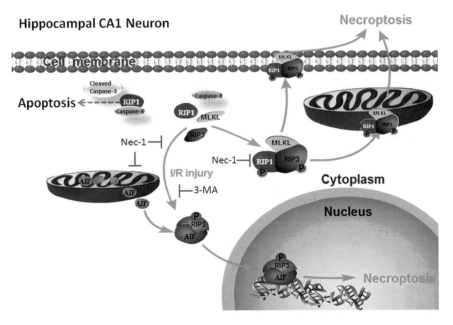

Fig. 5.2 RIP1 and RIP3 are activated (phosphorylated) and combine with each other after CIRI. AIF is released from mitochondria and combines with RIP3 (perhaps phosphorylated RIP3) to form RIP3-AIF complexes. The RIP3-AIF complexes translocate into the nucleus resulting in chromatin condensation and DNA degradation, and then the neurons are triggered to undergo programmed necrosis. All of these changes after I/R injury are inhibited by pre-treatment with Nec-1 and 3-MA, except for the release of AIF from mitochondria in the 3-MA pre-treatment group. In neurons, the findings that caspase-8 expression was undetectable and caspase-3 was not activated indicate that caspase-dependent apoptosis is not involved in this process. Another necroptosis pathway in the cytoplasm induced by RIP1-RIP3-MLKL complexes, described by others, may also participate in this process

4.1 Possible Autophagy Signaling Pathways in Cerebral Ischemia

The existence of autophagy in ischemic stroke has been found for many years; however, it is not sure whether autophagy plays a protective role in ischemic cerebral injury or not yet [78, 79]. Generally, in the neuronal system, moderate autophagy is thought to be neuroprotective because autophagy helps to clear aggregated-protein associated with neurodegeneration. Inadequate or defective autophagy may lead to neuronal cell death, while excess autophagy, often triggered by intensive stress, can also promote neuronal cell death.

Almost any signal can be a trigger for autophagy, some activating the pathway and some suppressing the pathway. By far, energy depletion and oxygen deficient environment are the most powerful triggers for stimulating autophagy, while the reverse environment factors, hormones, receptors with cytokine activities, receptors with tyrosine kinase activities and receptors that recognize pathogen ligands can

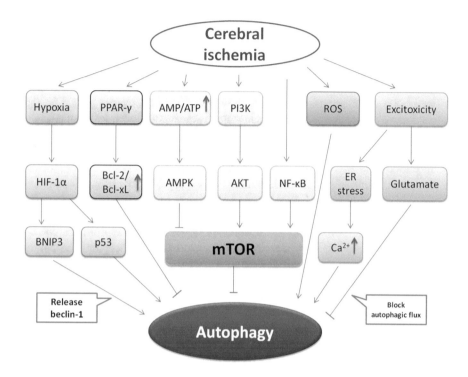

Fig. 5.3 Possible autophagy signaling pathways in CIRI

also activate autophagy. Cerebral ischemia can activate multiple signaling pathways that subsequently feed into the autophagy pathway (Fig. 5.3).

The figure shows the many different signaling pathways involved in the activation of autophagy during cerebral ischemia. When activated, Akt and NF-κB activate mTOR to inhibit autophagy in cerebral ischemia. However, the activation of AMPK could inhibit the activity of mTOR and induce autophagy. Hypoxia caused by cerebral ischemia activates HIF-1α and induces autophagy through BNIP3 and p53. Excitotoxicity could induce autophagy by ER stress and block autophagic flux by glutamate in cerebral ischemia. Autophagy could also be induced through ROS and inhibited through PPAR-γ. PPAR-γ: Peroxisome proliferator-activated receptor-γ; AMP: Adenosine 5′-monophosphate; PI3K: phosphatidylinositol 3-kinase; ROS: reactive oxygen species; HIF-1α: hypoxia inducible factor 1α; Bcl-2: B cell lymphoma/leukmia-2; Bcl-xL: B-cell lymphoma-extra large; AMPK: AMP-activated protein kinase; AMPK: AMP-activated protein kinase; Akt/PKB: protein kinase B; NF-κB: nuclear factor kappa B; ER: endoplasmic reticulum; BNIP3: Bcl-2 and adenovirus E1B 19 kDa interacting proteins 3; mTOR: mammalian target of rapamycin.

1. PI3K-Akt-mTORC1 mTOR is a 289 kDa serine/threonine protein kinase that regulates transcription, cytoskeleton organization, cell growth and cell survival. The mTOR is a high energy sensor, which on the other hand is a negative

regulator of autophagy. By binding to different co-factors, mTOR can form two distinct protein complexes, mTORC1 (mTOR complex 1) and mTORC2 (mTOR complex 2) [80]. mTORC1 is responsible for the inhibitory effect of rapamycin, more so than mTORC2. Recent studies suggest that the PI3K/Akt/mTOR pathway could regulate acute nervous system injury in cerebral hypoxia-ischemia [81]. PI3K consists of class I, class II and class III. Class I PI3K plays an important role in the PI3K-Akt-mTOR pathway. PI3K phosphorylates and activates Akt which in turn phosphorylates and inactivates tuberous sclerosis complex (TSC) 1/2. Inactivated TSC1/2 increases the activation of Rheb which is part of the Ras family GTP-binding protein, and mTOR is subsequently activated. Autophagy is inhibited by activating mTOR [82]. Beclin-1, a component of the class III PI3K, is essential for the initial steps of autophagy and could also induce autophagy via the interaction with other components of the class III PI3K pathway in cerebral ischemia [83]. Peroxisome prolif-erator-activated receptor-γ (PPAR-γ), a member of nuclear hormone receptor superfamily, is a ligand-activated transcription factor. PPAR-γ activation antagonizes beclin-1-mediated autophagy via upregulation of Bcl-2/Bcl-xl which interact with beclin-1 in cerebral ischemia/reperfusion [84].

2. AMPK-mTORC1 AMP-activated protein kinase (AMPK) is a serine/threonine protein kinase and consists of three subunits: a catalytic α-subunit and regulatory β and γ-subunits. Each subunit appears to have distinct functions. The most studied is the catalytic α-subunit which contains a threonine phosphorylation site that when phosphorylated, activates AMPK. The status of nutrient and energy depletion is sensed and modulated by kinase B1 (LKB1), Ca^{2+}/calmodulin-dependent kinase kinase beta (CaMKKβ) and transforming growth factor β activated kinase-1 (TAK1), resulting in phosphorylation of threonine residue at 172 position and activation of AMPK [85] and AMPK activation could subsequently inhibit the activity of mTOR to induce autophagy [86, 87].

3. Beclin 1-Bcl-2 complex.

Beclin 1 was identified as a Bcl-2-interacting protein through its BH3 domain [88]. The binding of Bcl-2 to Beclin 1 disrupts the association of Beclin 1 with PI3K, hVps34 and p150, therefore inhibiting autophagy [89]. Intriguingly, only ER-localized, but not mitochondria-localized, Bcl-2 inhibits autophagy [89]. Under stress conditions, Beclin 1 is released and induces autophagy [90, 91]. As previously demonstrated, the expression of Beclin 1 in neurons is dramatically increased in neonatal HI or focal cerebral ischemia [92, 93]. Ischemia stimulates autophagy through the AMPK–mTOR pathway, whereas ischemia/reperfusion stimulates autophagy through a Beclin 1-dependent but AMPK-independent pathway [94]. Although there are several different mechanisms to regulate the dissociation of Beclin 1 from Bcl-2 during autophagy in mammalian cells [95], the specific mechanism in cerebral ischemia is not yet established.

Hypoxia-inducible factor 1 (HIF-1) is a key transcriptional factor that is activated in response to hypoxia during cerebral ischemia [96]. HIF-1 is composed of a constitutively expressed HIF-1β subunit and an inducibly expressed HIF-1α subunit. Since ubiquitination is inhibited under hypoxic conditions, HIF-1α can accumulate and dimerize with HIF-1β. This dimer activates transcription of a number of downstream hypoxia-responsive genes, including vascular endothelial growth factor (VEGF), erythropoietin (EPO), glucose transporter 1, and glycolytic enzymes [97]. Bcl-2 and adenovirus E1B 19 kDa interacting proteins 3 (BNIP3) with a single Bcl-2 homology 3 (BH3) domain is a subfamily of Bcl-2 family proteins and also serves as an important target gene of HIF-1α [98]. BNIP3 can compete with beclin-1 for binding to Bcl-2 and beclin-1 is released to trigger autophagy [99]. BNIP3 also binds and inhibits Rheb, an upstream activator of mTOR, so it could activate autophagy by inhibiting mTOR activity. The induced p53 stabilization by up-regulation of HIF-1α also plays an important role in post-ischemic autophagy activation [97].

4. p53 The tumor suppressor and transcription factor p53 has been reported to be pivotal in neuronal apoptosis [100]. Crighton et al. demonstrated that p53 induced autophagy through the upregulation of damage-regulated autophagy modulator (DRAM), the p53 target gene encoding a lysosomal protein [101]. Other study also demonstrated that the NF-κB-regulated p53 pathway contributes to excitotoxic neuronal death by activating the autophagic process [102]. Overstimulation of N-methyl-D-aspartate receptors (NMDARs) induces the upregulation of p53, its target gene DRAM, and other autophagic proteins including LC3 and Beclin 1. Moreover, the NF-κB inhibitor SN50 inhibits the excitotoxin-induced upregulation of p53, its target gene DRAM, and other autophagic proteins.

    Nuclear factor kappa B (NF-κB) is a transcription factor that regulates expression of multiple genes [103]. Recent experiments have demonstrated that the knockout of p50 (NF-κB1) enhanced autophagy by repression of mTOR in cerebral ischemic mice [104]. NF-κB-dependent p53 signal transduction pathway is also associated with autophagy and apoptosis in the rat hippocampus after cerebral ischemia/reperfusion insult [105]. Mitogen-activated protein kinases (MAPKs) include extracellular signal-related kinase (ERK), Jun NH2 terminal kinase (JNK) and p38 [106]. MAPK is one upstream regulator of mTORC1 and autophagy could also be induced via MAPK-mTOR signaling pathway in cerebral ischemia/reperfusion [107].

5. Others.

    Autophagic cell death is activated in the nervous system in response to oxidative stress [108]. Oxidative stress can occur in cerebral ischemia and could increase reactive oxygen species such as superoxide, hydroxyl radical and hydrogen peroxide. Recent studies have reported that selenium provides neuroprotection through preserving mitochondrial function, decreasing reactive oxygen species production and reducing autophagy [109]. Autophagy can also be induced under conditions of excitotoxicity which can also occur in cerebral ischemia [110]. Although excitotoxic glutamate blocks autophagic flux, it could also induce autophagy in hippocampal neurons [111]. Sustained elevations of Ca^{2+} in the mitochondrial matrix are a major feature of the intracellular cascade of lethal events during cerebral ischemia. Recently, it was reported that endoplasmic reticulum stress is one of the effects of excitotoxicity [1]. When endoplasmic reticula were exposed to toxic levels of excitatory neurotransmitters, Ca^{2+} was released via the activation of both ryanodine receptors and IP3R, leading to mitochondrial Ca^{2+} overload and activation of apoptosis. During endoplasmic reticulum stress, Ca^{2+} increase seems to be required for activating autophagy.

4.2 The Dual Roles of Autophagy in Cerebral Ischemia

Numerous data have demonstrated that autophagy is activated by ischemic insult in various models, and the elevated autophagic activity could be regulated by a wide range of interventions, mainly including pharmacological and genetic methods. There is no question that disrupting the autophagic process in brain is deleterious, particularly for the lifespan of the animal, resulting in the accumulation of dysfunctional or aging macromolecules and organelles [112, 113]. However, upon the acute cerebral ischemia stress, whether autophagy plays a beneficial or harmful role in the survival of neuronal cells is not an easy question. Adhami et al. [114] showed for the first time that many damaged neurons displayed features of autophagic/lysosomal cell death, and very few cells completed the apoptosis process in cerebral ischemic stress. This result suggested that the damaged neuronal cells can exhibit multiple forms of cell death morphological features, and autophagy is only one kind of cell death during ischemic injury. Alternatively, autophagy may protect neurons by degrading damaged organelles to abrogate apoptosis or generating energy to delay the onset of ionic imbalance and necrosis after cerebral ischemia–hypoxia. However, these early reports did not determine the exact role of autophagy. Dozens of later investigations pointed out the complex effects of autophagy in cerebral ischemia. The autophagy and the controversial impacts of autophagy on cerebral ischemic injury as a double-edged sword have been uncovered.

4.2.1 Detrimental Role of Autophagy in Ischemic Cerebral Injury

Mice deficient in Atg7, the gene essential for autophagy induction, showed nearly complete protection from both hypoxia-ischemia-induced caspase-3 activation and neuronal death, indicating autophagy is essential in triggering neuronal death after hypoxia-ischemia injury [115]. Wen et al. [116] confirmed autophagy was activated in a permanent middle cerebral artery occlusion (MCAO) model. In their paper, the infarct volume, brain edema and motor deficits could be significantly reduced by administration of 3-MA (an autophagy inhibitor). The neuroprotective effects of 3-MA were associated with an inhibition of ischemia-induced upregulation of LC3-II, a marker of active autophagosomes and autophagolysosomes. Moreover, it was observed that the inhibition of autophagy, either by direct inhibitor 3-MA or by indirect inhibitor 2ME2 (an inhibitor of hypoxia inducible factor-1α; HIF-1α) might prevent pyramidal neuron death after ischemia [97].

4.2.2 Beneficial Role of Autophagy in Cerebral Ischemic Injury

Carloni et al. [117] suggested that in neonatal hypoxia-ischemia, autophagy may be part of an integrated pro-survival signaling complex that includes PI3K-Akt-mTOR. When either autophagy or PI3K-Akt-mTOR pathways were interrupted, cells underwent necrotic cell death. Wang et al. [118] reported that neuronal survival was promoted during cerebral ischemia when autophagy was induced by nicotinamide phosphoribosyltransferase (Nampt, also known as visfatin), which is the rate-limiting enzyme in mammalian NAD+ biosynthesis and regulates the TSC2-mTOR-S6K1 signaling pathway. These studies suggest that autophagy may be a potential target for post-ischemic neuronal protection.

4.3 The Factors Determining the Role of the Autophagy in Cerebral Ischemia

4.3.1 The Degree of Autophagy Determines the Fate of Cells in Cerebral Ischemia

Kang and Avery [119] proposed that levels of autophagy were critical for the survival or death of cells: physiological levels of autophagy promote survival, whereas insufficient or excessive levels of autophagy promote death. This hypothesis was confirmed in an oxygen and glucose deprivation model that observed dual roles of the autophagy inhibitor 3-MA in different stages of re-oxygenation [75]. Twenty-four hours prior to reperfusion, 3-MA triggered a high rate of neuronal death. However, during 48–72 h of reperfusion, 3-MA significantly protected neurons from death. It is possible that prolonged oxygen and glucose deprivation/reperfusion triggers excessive autophagy, switching its role from protection to deterioration.

4.3.2 The Time at Which Autophagy Is Induced Determines Its Role

Autophagy could play a protective role in ischemic preconditioning but have a different effect once ischemia/reperfusion has occurred [120]. Infarct volume, brain edema and motor deficits induced by permanent focal ischemia were significantly reduced after ischemic preconditioning treatment. 3-MA suppressed neuroprotection induced by ischemic preconditioning, while rapamycin reduced infarct volume, brain edema and motor deficits induced by permanent focal ischemia [121]. This hypothesis was supported by a study by Yan et al. [122] in which 3-MA administrated through intracere-broventricular injection before hyperbaric oxygen preconditioning, attenuated the neuroprotection of hyperbaric oxygen preconditioning against cerebral ischemia. Moreover, 3-MA treatment before middle cerebral artery occlusion aggravated subsequent cerebral ischemic injury. In contrast, Carloni et al.

[92, 117] found that when 3-MA and rapamycin were injected 20 min before hypoxia-ischemia, 3-MA inhibited autophagy, significantly reduced beclin-1 expression and caused neuronal death, while rapamycin increased autophagy and decreased brain injury. In addition, 3-MA administrated by intracerebroventricular injections strongly reduced the lesion volume (by 46%) even when given 4 h after the beginning of the ischemia [123]. Gao et al. [124] found that rapamycin applied at the onset of reperfusion might attenuate the neuroprotective effects of ischemic postconditioning. Conversely, 3-MA administered before reperfusion significantly reduced infarct size and abolished the increase of brain water content after ischemia. Targeting autophagy either pre- or post-treatment has different results and this may reflect the different effects of autophagy at early and late stages. The time of intervention could be related to the degree of autophagy at different stages of ischemia and further studies are necessary to confirm this.

4.3.3 Autophagy May Be Interrupted in Cerebral Ischemia

A common feature of many neurodegenerative diseases is the accumulation of an abnormally large number of autophagic vacuoles (autophagosomes and autolysosomes) or the frequent appearance of irregularly shaped autophagic vacuoles. Enhanced autophagosome formation seems to be reflected by increased density of autophagic vacuoles, but these increased autophagic vacuoles may also imply impaired autolysosomal degradation [125]. Rami et al. [93] also observed a dramatic up-regulation of Beclin-1 and LC3 in rats after cerebral ischemia. These results indicate that autophagy was activated in the brain following ischemia. Recently, however, it has been hypothesized that the increase in proteins may reflect a failure in lysosomal function leading to an accumulation of autophagosomes, or an improvement in the activity of autophagy [126]. Other studies found that accumulation of LC3-II was observed in sham-operated rats after treatment with lysosomal inhibitor-chloriquine, but the further change of LC3-II levels in post-ischemic brain tissues was not observed [127, 128]. The results indicated that accumulation of autophagy-associated protein following ischemia could be the result of failure of the autophagy pathway. Puyal and Clarke [129] found that lysosomal activity detected by LAMP-1 and cathepsin D was increased in neurons with punctate LC3 expression in neonatal focal cerebral ischemia model. The failure of autophagosome and lysosome fusion caused an increase of autophagosomes. The deficiency of acid phosphatase activity in the lysosome could lead to the increase of autophagosomes and autolysosomes. Further studies are required to verify whether the activity of autophagy is enhanced in cerebral ischemia.

5   Crosstalk Between Apoptosis, Autophagy and Necroptosis (Necrosis) After CIRI

PCD in vivo involves the complex interaction between apoptosis, autophagy, and necroptosis [130, 131] (Fig. 5.4). In some cases, a specific stimulus triggers only one type of programmed cell death, but in other situations, the same stimulus may initiate multiple cell death processes. Different types of mechanisms may co-exist and interact with each other within a cell, but ultimately, one mechanism dominates the others. The decision taken by a cell to undergo apoptosis, autophagy, or necroptosis is regulated by various factors, including the energy/ATP levels, the extent of damage or stress, and the presence of inhibitors of specific pathways (e.g., caspase inhibitors). ATP depletion activates autophagy. However, if autophagy fails to maintain the energy levels, necroptosis occurs [132]. Slight/moderate damage and low levels of death signaling typically induce apoptosis, whereas severe damage and high levels of the death signaling often result in necroptosis. Similarly, inhibition of caspase activity might change apoptosis to necrosis or autophagic cell death, whereas activation of calpain-mediated cleavage of autophagy-regulated protein, Atg-5, switches the mode of cell death from autophagy to apoptosis [133, 134]. Interestingly, although necroptosis, necrosis and secondary necrosis following apoptosis, represent different modes of cell death, all of them might eventually converge on similar cellular disintegration features, albeit with different kinetics [135]. Furthermore, apoptosis and autophagy differ from the necrosis by the feature of tissue inflammation [136]. Both apoptosis and autophagy do not exhibit tissue

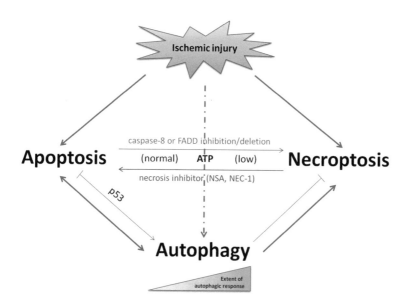

Fig. 5.4 Cross-talk between different modes of programmed cell death after CIRI. *FADD* Fas-associated death domain-containing protein, *NSA* necrosulfonamide, *NEC-1* necrostatin-1

inflammation, while the latter does. Thus, learning more about the molecular mechanisms regulating various cell death modalities and their cross-talk is very important, since they play a critical role in CIRI.

Although death-receptor mediated apoptosis represents a canonical apoptotic pathway, stimulation of death receptors under apoptotic deficient conditions is now known to activate necroptosis [58]. The activation of death receptors by their respective ligands, such as FasL (CD95L) and TNF-α, respectively, leads to the formation of DISC (death-inducing signaling complex) that includes the adaptor protein FADD (Fasassociated death domain), caspase-8 and death domain-containing kinase RIP1. In apoptotic proficient condition, the recruitment of caspase-8 leads to its activation which in turn activates downstream caspases, such as caspase-3, and mitochondrial damage by cleaving Bid [137]. In apoptotic deficient cells when caspases cannot be activated, however, stimulation of death receptors leads to the activation of RIP1 kinase and necroptosis [54, 65]. Activation of AKT also appears to act as a switch, in addition to facilitating the necroptotic response, it also acts to inhibit apoptosis [138, 139]. These results clearly illustrate that the molecular pathways regulating death ligand-induced apoptosis and necroptossis are intimately intertwined. They also firmly establish the paradigm that inhibition of caspase-dependent apoptosis primes cells towards necroptosis.

Crosstalk between apoptosis and autophagy in CIRI is also complex. It has been acknowledged that appropriately controlled autophagy can induce neuroprotection and can rescue neurons from apoptotic cell death in the cerebral ischemia. For instance, clearance of damaged mitochondria via autophagy prevented neurons from caspase-dependent apoptosis [140]. However, autophagy may also act as a pro-apoptotic mechanism [99] and is causally connected with the subsequent onset of apoptotic cell death [141–143]. Cathepsin B, a protease which is normally confined inside the lysosomal-endosomal compartment, leaked from the lysosomes into the cytoplasm, initiating and promoting the execution of apoptosis [141]. It has been hypothesized that when the autophagic flux impairs, autolysosomes would extensively accumulate and the autophagic stress would be induced [126]. This would lead to autolysosomes and lysosomes membrane destabilization, which results in leakage of hydrolases, and subsequently provoke apoptosis [144]. As a result, the initial autophagy, as a defensive reaction, when over-activated, is converted into a damage response [75]. In this case, the inhibition of autophagy attenuates apoptotic cascades in ischemic injury. These evidences demonstrated that elucidating the interrelationships between autophagy and apoptosis will present novel opportunities for discovering targets in the therapy for cerebral ischemic injury.

So far, with the identification of several key molecules (e.g., ATG, Bcl-2 family members, Beclin 1, and p53) [105, 145–150], the mechanisms underlying the autophagy-apoptosis conversation are beginning to be uncovered. However, current researches seemingly only reveal a tip of the iceberg among the intricate interactions between autophagic and apoptotic cascades during the cerebral ischemic injury. Thus, more studies about the crosstalk between autophagy and apoptosis are warranted in the future.

The autophagy and necrosis can be activated in parallel or sequentially, and have either common or opposite objectives. The molecular underpinnings of this relationship remain largely elusive and somewhat controversial; autophagy has been shown to either promote [123] or suppress necroptosis (necrosis) [92, 151]. However, the ability of autophagy to suppress various forms of necrotic cell death is considered to be one of the most important pro-survival functions of autophagy that is achieved either by blocking apoptosis or suppressing necrotic cell death.

Therefore, because of the complex crosstalk between cell death pathways, much effort should be put on the finding of biomarkers that may predict the risk of a hypoxic-ischemic condition during the CIRI to initiate the treatment in an early stage, allowing the possibility of using the preconditioning effect of putative drugs. These early treatments may be followed by endovascular recanalization therapy (thrombolytic therapy or arterial embolectomy), that potentially reduces both apoptosis and necrosis. Of course, a better understanding of the mechanisms responsible for the switch among the different cell death phenotypes and the development of new and more selective molecules that can act upstream of these putative checkpoints will help to find new pharmacological strategies that could be associated to endovascular recanalization therapy.

References

1. Ouyang YB, Giffard RG. Er-mitochondria crosstalk during cerebral ischemia: Molecular chaperones and er-mitochondrial calcium transfer. Int J Cell Biol. 2012;2012:493934.
2. Lipton P. Ischemic cell death in brain neurons. Physiol Rev. 1999;79:1431–568.
3. Baron JC. Mapping the ischaemic penumbra with pet: implications for acute stroke treatment. Cerebrovasc Dis. 1999;9:193–201.
4. Astrup J, Siesjo BK, Symon L. Thresholds in cerebral ischemia – the ischemic penumbra. Stroke. 1981;12:723–5.
5. Galluzzi L, Vitale I, Abrams JM, Alnemri ES, Baehrecke EH, Blagosklonny MV, et al. Molecular definitions of cell death subroutines: recommendations of the nomenclature committee on cell death 2012. Cell Death Differ. 2012;19:107–20.
6. Fuchs Y, Steller H. Programmed cell death in animal development and disease. Cell. 2011;147:742–58.
7. Hengartner MO. The biochemistry of apoptosis. Nature. 2000;407:770–6.
8. Klionsky DJ. The molecular machinery of autophagy: unanswered questions. J Cell Sci. 2005;118:7–18.
9. Conrad M, Angeli JP, Vandenabeele P, Stockwell BR. Regulated necrosis: disease relevance and therapeutic opportunities. Nat Rev Drug Discov. 2016;15:348–66.
10. Edinger AL, Thompson CB. Death by design: apoptosis, necrosis and autophagy. Curr Opin Cell Biol. 2004;16:663–9.
11. Elmore S. Apoptosis: a review of programmed cell death. Toxicol Pathol. 2007;35:495–516.
12. Broughton BR, Reutens DC, Sobey CG. Apoptotic mechanisms after cerebral ischemia. Stroke. 2009;40:e331–9.
13. Zhu C, Wang X, Xu F, Bahr BA, Shibata M, Uchiyama Y, et al. The influence of age on apoptotic and other mechanisms of cell death after cerebral hypoxia-ischemia. Cell Death Differ. 2005;12:162–76.

14. Zhu C, Xu F, Wang X, Shibata M, Uchiyama Y, Blomgren K, et al. Different apoptotic mechanisms are activated in male and female brains after neonatal hypoxia-ischaemia. J Neurochem. 2006;96:1016–27.

15. Rai NK, Tripathi K, Sharma D, Shukla VK. Apoptosis: a basic physiologic process in wound healing. Int J Low Extrem Wounds. 2005;4:138–44.

16. Namura S, Zhu J, Fink K, Endres M, Srinivasan A, Tomaselli KJ, et al. Activation and cleavage of caspase-3 in apoptosis induced by experimental cerebral ischemia. J Neurosci. 1998;18:3659–68.

17. Siegel C, McCullough LD. Nad+ depletion or par polymer formation: which plays the role of executioner in ischaemic cell death? Acta Physiol. 2011;203:225–34.

18. Benchoua A, Guegan C, Couriaud C, Hosseini H, Sampaio N, Morin D, et al. Specific caspase pathways are activated in the two stages of cerebral infarction. J Neurosci. 2001;21:7127–34.

19. Hardwick JM, Chen YB, Jonas EA. Multipolar functions of bcl-2 proteins link energetics to apoptosis. Trends Cell Biol. 2012;22:318–28.

20. Webster KA, Graham RM, Thompson JW, Spiga MG, Frazier DP, Wilson A, et al. Redox stress and the contributions of bh3-only proteins to infarction. Antioxid Redox Signal. 2006;8:1667–76.

21. Ashkenazi A, Fairbrother WJ, Leverson JD, Souers AJ. From basic apoptosis discoveries to advanced selective bcl-2 family inhibitors. Nat Rev Drug Discov. 2017;16:273–84.

22. Zhai D, Chin K, Wang M, Liu F. Disruption of the nuclear p53-gapdh complex protects against ischemia-induced neuronal damage. Mol Brain. 2014;7:20.

23. Loh KP, Huang SH, De Silva R, Tan BK, Zhu YZ. Oxidative stress: apoptosis in neuronal injury. Curr Alzheimer Res. 2006;3:327–37.

24. Wang Y, Wan C, Yu S, Yang L, Li B, Lu T, et al. Upregulated expression of nf-yc contributes to neuronal apoptosis via proapoptotic protein bim in rats' brain hippocampus following middle cerebral artery occlusion (mcao). J Mol Neurosci. 2014;52:552–65.

25. Armugam A, Cher CD, Lim K, Koh DC, Howells DW, Jeyaseelan K. A secretory phospholipase a2-mediated neuroprotection and anti-apoptosis. BMC Neurosci. 2009;10:120.

26. Li D, Li X, Wu J, Li J, Zhang L, Xiong T, et al. Involvement of the jnk/foxo3a/bim pathway in neuronal apoptosis after hypoxic-ischemic brain damage in neonatal rats. PLoS One. 2015;10:e0132998.

27. Ma M, Wang X, Ding X, Teng J, Shao F, Zhang J. Numb/notch signaling plays an important role in cerebral ischemia-induced apoptosis. Neurochem Res. 2013;38:254–61.

28. Mergenthaler P, Dirnagl U, Meisel A. Pathophysiology of stroke: lessons from animal models. Metab Brain Dis. 2004;19:151–67.

29. Simard JM, Tarasov KV, Gerzanich V. Non-selective cation channels, transient receptor potential channels and ischemic stroke. Biochim Biophys Acta. 2007;1772:947–57.

30. Culmsee C, Zhu C, Landshamer S, Becattini B, Wagner E, Pellecchia M, et al. Apoptosis-inducing factor triggered by poly(adp-ribose) polymerase and bid mediates neuronal cell death after oxygen-glucose deprivation and focal cerebral ischemia. J Neurosci. 2005;25:10262–72.

31. Sugawara T, Fujimura M, Noshita N, Kim GW, Saito A, Hayashi T, et al. Neuronal death/ survival signaling pathways in cerebral ischemia. NeuroRx. 2004;1:17–25.

32. Cho BB, Toledo-Pereyra LH. Caspase-independent programmed cell death following ischemic stroke. J Investig Surg. 2008;21:141–7.

33. Daugas E, Susin SA, Zamzami N, Ferri KF, Irinopoulou T, Larochette N, et al. Mitochondrio-nuclear translocation of aif in apoptosis and necrosis. FASEB J. 2000;14:729–39.

34. Love S. Apoptosis and brain ischaemia. Prog Neuro-Psychopharmacol Biol Psychiatry. 2003;27:267–82.

35. Tanaka H, Yokota H, Jover T, Cappuccio I, Calderone A, Simionescu M, et al. Ischemic preconditioning: neuronal survival in the face of caspase-3 activation. J Neurosci. 2004;24:2750–9.

36. Cao G, Xiao M, Sun F, Xiao X, Pei W, Li J, et al. Cloning of a novel apaf-1-interacting protein: a potent suppressor of apoptosis and ischemic neuronal cell death. J Neurosci. 2004;24:6189–201.

37. Fujimura M, Morita-Fujimura Y, Kawase M, Copin JC, Calagui B, Epstein CJ, et al. Manganese superoxide dismutase mediates the early release of mitochondrial cytochrome c and subsequent DNA fragmentation after permanent focal cerebral ischemia in mice. J Neurosci. 1999;19:3414–22.
38. Gao Y, Liang W, Hu X, Zhang W, Stetler RA, Vosler P, et al. Neuroprotection against hypoxic-ischemic brain injury by inhibiting the apoptotic protease activating factor-1 pathway. Stroke. 2010;41:166–72.
39. Siren AL, Fratelli M, Brines M, Goemans C, Casagrande S, Lewczuk P, et al. Erythropoietin prevents neuronal apoptosis after cerebral ischemia and metabolic stress. Proc Natl Acad Sci U S A. 2001;98:4044–9.
40. Martin-Villalba A, Herr I, Jeremias I, Hahne M, Brandt R, Vogel J, et al. Cd95 ligand (fas-l/apo-1l) and tumor necrosis factor-related apoptosis-inducing ligand mediate ischemia-induced apoptosis in neurons. J Neurosci. 1999;19:3809–17.
41. Hasegawa Y, Suzuki H, Sozen T, Rolland W, Zhang JH. Activation of sphingosine 1-phosphate receptor-1 by fty720 is neuroprotective after ischemic stroke in rats. Stroke. 2010;41:368–74.
42. Cao Y, Mao X, Sun C, Zheng P, Gao J, Wang X, et al. Baicalin attenuates global cerebral ischemia/reperfusion injury in gerbils via anti-oxidative and anti-apoptotic pathways. Brain Res Bull. 2011;85:396–402.
43. Chen S, Peng H, Rowat A, Gao F, Zhang Z, Wang P, et al. The effect of concentration and duration of normobaric oxygen in reducing caspase-3 and -9 expression in a rat-model of focal cerebral ischaemia. Brain Res. 2015;1618:205–11.
44. Fuchs Y, Steller H. Live to die another way: modes of programmed cell death and the signals emanating from dying cells. Nat Rev Mol Cell Biol. 2015;16:329–44.
45. Vercammen D, Beyaert R, Denecker G, Goossens V, Van Loo G, Declercq W, et al. Inhibition of caspases increases the sensitivity of l929 cells to necrosis mediated by tumor necrosis factor. J Exp Med. 1998;187:1477–85.
46. Kawahara A, Ohsawa Y, Matsumura H, Uchiyama Y, Nagata S. Caspase-independent cell killing by fas-associated protein with death domain. J Cell Biol. 1998;143:1353–60.
47. Vandenabeele P, Galluzzi L, Vanden Berghe T, Kroemer G. Molecular mechanisms of necroptosis: an ordered cellular explosion. Nat Rev Mol Cell Biol. 2010;11:700–14.
48. Gunther C, Martini E, Wittkopf N, Amann K, Weigmann B, Neumann H, et al. Caspase-8 regulates tnf-alpha-induced epithelial necroptosis and terminal ileitis. Nature. 2011;477:335–9.
49. Oberst A, Dillon CP, Weinlich R, McCormick LL, Fitzgerald P, Pop C, et al. Catalytic activity of the caspase-8-flip(l) complex inhibits ripk3-dependent necrosis. Nature. 2011;471:363–7.
50. Kaiser WJ, Upton JW, Long AB, Livingston-Rosanoff D, Daley-Bauer LP, Hakem R, et al. Rip3 mediates the embryonic lethality of caspase-8-deficient mice. Nature. 2011;471:368–72.
51. Xu Y, Wang J, Song X, Qu L, Wei R, He F, et al. Rip3 induces ischemic neuronal DNA degradation and programmed necrosis in rat via aif. Sci Rep. 2016;6:29362.
52. Zhang J, Yang Y, He W, Sun L. Necrosome core machinery: Mlkl. Cell Mol Life Sci. 2016;73:2153–63.
53. Hitomi J, Christofferson DE, Ng A, Yao J, Degterev A, Xavier RJ, et al. Identification of a molecular signaling network that regulates a cellular necrotic cell death pathway. Cell. 2008;135:1311–23.
54. Holler N, Zaru R, Micheau O, Thome M, Attinger A, Valitutti S, et al. Fas triggers an alternative, caspase-8-independent cell death pathway using the kinase rip as effector molecule. Nat Immunol. 2000;1:489–95.
55. Li J, McQuade T, Siemer AB, Napetschnig J, Moriwaki K, Hsiao YS, et al. The rip1/rip3 necrosome forms a functional amyloid signaling complex required for programmed necrosis. Cell. 2012;150:339–50.
56. Sun L, Wang H, Wang Z, He S, Chen S, Liao D, et al. Mixed lineage kinase domain-like protein mediates necrosis signaling downstream of rip3 kinase. Cell. 2012;148:213–27.

57. Wang Z, Jiang H, Chen S, Du F, Wang X. The mitochondrial phosphatase pgam5 functions at the convergence point of multiple necrotic death pathways. Cell. 2012;148:228–43.
58. Degterev A, Huang Z, Boyce M, Li Y, Jagtap P, Mizushima N, et al. Chemical inhibitor of nonapoptotic cell death with therapeutic potential for ischemic brain injury. Nat Chem Biol. 2005;1:112–9.
59. Vieira M, Fernandes J, Carreto L, Anuncibay-Soto B, Santos M, Han J, et al. Ischemic insults induce necroptotic cell death in hippocampal neurons through the up-regulation of endogenous rip3. Neurobiol Dis. 2014;68:26–36.
60. Yin B, Xu Y, Wei RL, He F, Luo BY, Wang JY. Inhibition of receptor-interacting protein 3 upregulation and nuclear translocation involved in necrostatin-1 protection against hippocampal neuronal programmed necrosis induced by ischemia/reperfusion injury. Brain Res. 2015;1609:63–71.
61. Fakharnia F, Khodagholi F, Dargahi L, Ahmadiani A. Prevention of cyclophilin d-mediated mptp opening using cyclosporine-a alleviates the elevation of necroptosis, autophagy and apoptosis-related markers following global cerebral ischemia-reperfusion. J Mol Neurosci. 2017;61:52–60.
62. Dong Y, Bao C, Yu J, Liu X. Receptor-interacting protein kinase 3-mediated programmed cell necrosis in rats subjected to focal cerebral ischemia-reperfusion injury. Mol Med Rep. 2016;14:728–36.
63. Miao W, Qu Z, Shi K, Zhang D, Zong Y, Zhang G, et al. Rip3 s-nitrosylation contributes to cerebral ischemic neuronal injury. Brain Res. 2015;1627:165–76.
64. Melo-Lima S, Celeste Lopes M, Mollinedo F. Necroptosis is associated with low procaspase-8 and active ripk1 and −3 in human glioma cells. Oncoscience. 2014;1:649–64.
65. Degterev A, Hitomi J, Germscheid M, Ch'en IL, Korkina O, Teng X, et al. Identification of rip1 kinase as a specific cellular target of necrostatins. Nat Chem Biol. 2008;4:313–21.
66. Vandenabeele P, Declercq W, Vanden Berghe T. Necrotic cell death and 'necrostatins': Now we can control cellular explosion. Trends Biochem Sci. 2008;33:352–5.
67. Li H, Gao A, Feng D, Wang Y, Zhang L, Cui Y, et al. Evaluation of the protective potential of brain microvascular endothelial cell autophagy on blood-brain barrier integrity during experimental cerebral ischemia-reperfusion injury. Transl Stroke Res. 2014;5:618–26.
68. Xu Y, Wang J, Song X, Wei R, He F, Peng G, et al. Protective mechanisms of ca074-me (other than cathepsin-b inhibition) against programmed necrosis induced by global cerebral ischemia/reperfusion injury in rats. Brain Res Bull. 2016;120:97–105.
69. Yoshimori T. Autophagy: paying Charon's toll. Cell. 2007;128:833–6.
70. Shibutani ST, Yoshimori T. A current perspective of autophagosome biogenesis. Cell Res. 2014;24:58–68.
71. Rabinowitz JD, White E. Autophagy and metabolism. Science. 2010;330:1344–8.
72. Russell RC, Yuan HX, Guan KL. Autophagy regulation by nutrient signaling. Cell Res. 2014;24:42–57.
73. Kourtis N, Tavernarakis N. Autophagy and cell death in model organisms. Cell Death Differ. 2009;16:21–30.
74. Clarke PG, Puyal J. Autophagic cell death exists. Autophagy. 2012;8:867–9.
75. Shi R, Weng J, Zhao L, Li XM, Gao TM, Kong J. Excessive autophagy contributes to neuron death in cerebral ischemia. CNS Neurosci Ther. 2012;18:250–60.
76. Shimizu S, Kanaseki T, Mizushima N, Mizuta T, Arakawa-Kobayashi S, Thompson CB, et al. Role of bcl-2 family proteins in a non-apoptotic programmed cell death dependent on autophagy genes. Nat Cell Biol. 2004;6:1221–8.
77. Yu L, Alva A, Su H, Dutt P, Freundt E, Welsh S, et al. Regulation of an atg7-beclin 1 program of autophagic cell death by caspase-8. Science. 2004;304:1500–2.
78. Smith CM, Chen Y, Sullivan ML, Kochanek PM, Clark RS. Autophagy in acute brain injury: feast, famine, or folly? Neurobiol Dis. 2011;43:52–9.
79. Yu L, Lenardo MJ, Baehrecke EH. Autophagy and caspases: a new cell death program. Cell Cycle. 2004;3:1124–6.

80. Jung CH, Ro SH, Cao J, Otto NM, Kim DH. Mtor regulation of autophagy. FEBS Lett. 2010;584:1287–95.
81. Chong ZZ, Shang YC, Wang S, Maiese K. A critical kinase cascade in neurological disorders: Pi 3-k, akt, and mtor. Future Neurol. 2012;7:733–48.
82. Glick D, Barth S, Macleod KF. Autophagy: cellular and molecular mechanisms. J Pathol. 2010;221:3–12.
83. Xingyong C, Xicui S, Huanxing S, Jingsong O, Yi H, Xu Z, et al. Upregulation of myeloid cell leukemia-1 potentially modulates beclin-1-dependent autophagy in ischemic stroke in rats. BMC Neurosci. 2013;14:56.
84. Xu F, Li J, Ni W, Shen YW, Zhang XP. Peroxisome proliferator-activated receptor-gamma agonist 15d-prostaglandin j2 mediates neuronal autophagy after cerebral ischemia-reperfusion injury. PLoS One. 2013;8:e55080.
85. Inoki K, Ouyang H, Zhu T, Lindvall C, Wang Y, Zhang X, et al. Tsc2 integrates wnt and energy signals via a coordinated phosphorylation by ampk and gsk3 to regulate cell growth. Cell. 2006;126:955–68.
86. Poels J, Spasic MR, Callaerts P, Norga KK. Expanding roles for amp-activated protein kinase in neuronal survival and autophagy. Bioessays. 2009;31:944–52.
87. Li J, McCullough LD. Effects of amp-activated protein kinase in cerebral ischemia. J Cereb Blood Flow Metab. 2010;30:480–92.
88. Oberstein A, Jeffrey PD, Shi Y. Crystal structure of the bcl-xl-beclin 1 peptide complex: Beclin 1 is a novel bh3-only protein. J Biol Chem. 2007;282:13123–32.
89. Pattingre S, Tassa A, Qu X, Garuti R, Liang XH, Mizushima N, et al. Bcl-2 antiapoptotic proteins inhibit beclin 1-dependent autophagy. Cell. 2005;122:927–39.
90. He C, Levine B. The beclin 1 interactome. Curr Opin Cell Biol. 2010;22:140–9.
91. Maiuri MC, Criollo A, Kroemer G. Crosstalk between apoptosis and autophagy within the beclin 1 interactome. EMBO J. 2010;29:515–6.
92. Carloni S, Buonocore G, Balduini W. Protective role of autophagy in neonatal hypoxia-ischemia induced brain injury. Neurobiol Dis. 2008;32:329–39.
93. Rami A, Langhagen A, Steiger S. Focal cerebral ischemia induces upregulation of beclin 1 and autophagy-like cell death. Neurobiol Dis. 2008;29:132–41.
94. Matsui Y, Takagi H, Qu X, Abdellatif M, Sakoda H, Asano T, et al. Distinct roles of autophagy in the heart during ischemia and reperfusion: roles of amp-activated protein kinase and beclin 1 in mediating autophagy. Circ Res. 2007;100:914–22.
95. Kang R, Zeh HJ, Lotze MT, Tang D. The beclin 1 network regulates autophagy and apoptosis. Cell Death Differ. 2011;18:571–80.
96. Althaus J, Bernaudin M, Petit E, Toutain J, Touzani O, Rami A. Expression of the gene encoding the pro-apoptotic bnip3 protein and stimulation of hypoxia-inducible factor-1alpha (hif-1alpha) protein following focal cerebral ischemia in rats. Neurochem Int. 2006;48:687–95.
97. Xin XY, Pan J, Wang XQ, Ma JF, Ding JQ, Yang GY, et al. 2-Methoxyestradiol attenuates autophagy activation after global ischemia. J Can Sci Neurol. 2011;38:631–8.
98. Cho B, Choi SY, Park OH, Sun W, Geum D. Differential expression of bnip family members of bh3-only proteins during the development and after axotomy in the rat. Mol Cells. 2012;33:605–10.
99. He S, Wang C, Dong H, Xia F, Zhou H, Jiang X, et al. Immune-related gtpase m (irgm1) regulates neuronal autophagy in a mouse model of stroke. Autophagy. 2012;8:1621–7.
100. Banasiak KJ, Haddad GG. Hypoxia-induced apoptosis: Effect of hypoxic severity and role of p53 in neuronal cell death. Brain Res. 1998;797:295–304.
101. Crighton D, Wilkinson S, O'Prey J, Syed N, Smith P, Harrison PR, et al. Dram, a p53-induced modulator of autophagy, is critical for apoptosis. Cell. 2006;126:121–34.
102. Wang Y, Dong XX, Cao Y, Liang ZQ, Han R, Wu JC, et al. P53 induction contributes to excitotoxic neuronal death in rat striatum through apoptotic and autophagic mechanisms. Eur J Neurosci. 2009;30:2258–70.

103. Chen AC, Arany PR, Huang YY, Tomkinson EM, Sharma SK, Kharkwal GB, et al. Low-level laser therapy activates nf-kb via generation of reactive oxygen species in mouse embryonic fibroblasts. PLoS One. 2011;6:e22453.
104. Li WL, Yu SP, Chen D, Yu SS, Jiang YJ, Genetta T, et al. The regulatory role of nf-kappab in autophagy-like cell death after focal cerebral ischemia in mice. Neuroscience. 2013;244:16–30.
105. Cui DR, Wang L, Jiang W, Qi AH, Zhou QH, Zhang XL. Propofol prevents cerebral ischemia-triggered autophagy activation and cell death in the rat hippocampus through the nf-kappab/p53 signaling pathway. Neuroscience. 2013;246:117–32.
106. Lien SC, Chang SF, Lee PL, Wei SY, Chang MD, Chang JY, et al. Mechanical regulation of cancer cell apoptosis and autophagy: roles of bone morphogenetic protein receptor, smad1/5, and p38 mapk. Biochim Biophys Acta. 2013;1833:3124–33.
107. Wang PR, Wang JS, Zhang C, Song XF, Tian N, Kong LY. Huang-lian-jie-du-decotion induced protective autophagy against the injury of cerebral ischemia/reperfusion via mapk-mtor signaling pathway. J Ethnopharmacol. 2013;149:270–80.
108. Kubota C, Torii S, Hou N, Saito N, Yoshimoto Y, Imai H, et al. Constitutive reactive oxygen species generation from autophagosome/lysosome in neuronal oxidative toxicity. J Biol Chem. 2010;285:667–74.
109. Mehta SL, Kumari S, Mendelev N, Li PA. Selenium preserves mitochondrial function, stimulates mitochondrial biogenesis, and reduces infarct volume after focal cerebral ischemia. BMC Neurosci. 2012;13:79.
110. Puyal J, Ginet V, Grishchuk Y, Truttmann AC, Clarke PG. Neuronal autophagy as a mediator of life and death: contrasting roles in chronic neurodegenerative and acute neural disorders. Neuroscientist. 2012;18:224–36.
111. Kulbe JR, Mulcahy Levy JM, Coultrap SJ, Thorburn A, Bayer KU. Excitotoxic glutamate insults block autophagic flux in hippocampal neurons. Brain Res. 2014;1542:12–9.
112. Komatsu M, Waguri S, Ueno T, Iwata J, Murata S, Tanida I, et al. Impairment of starvation-induced and constitutive autophagy in atg7-deficient mice. J Cell Biol. 2005;169:425–34.
113. Komatsu M, Waguri S, Chiba T, Murata S, Iwata J, Tanida I, et al. Loss of autophagy in the central nervous system causes neurodegeneration in mice. Nature. 2006;441:880–4.
114. Adhami F, Liao G, Morozov YM, Schloemer A, Schmithorst VJ, Lorenz JN, et al. Cerebral ischemia-hypoxia induces intravascular coagulation and autophagy. Am J Pathol. 2006;169:566–83.
115. Koike M, Shibata M, Tadakoshi M, Gotoh K, Komatsu M, Waguri S, et al. Inhibition of autophagy prevents hippocampal pyramidal neuron death after hypoxic-ischemic injury. Am J Pathol. 2008;172:454–69.
116. Wen YD, Sheng R, Zhang LS, Han R, Zhang X, Zhang XD, et al. Neuronal injury in rat model of permanent focal cerebral ischemia is associated with activation of autophagic and lysosomal pathways. Autophagy. 2008;4:762–9.
117. Carloni S, Girelli S, Scopa C, Buonocore G, Longini M, Balduini W. Activation of autophagy and akt/creb signaling play an equivalent role in the neuroprotective effect of rapamycin in neonatal hypoxia-ischemia. Autophagy. 2010;6:366–77.
118. Wang P, Guan YF, Du H, Zhai QW, Su DF, Miao CY. Induction of autophagy contributes to the neuroprotection of nicotinamide phosphoribosyltransferase in cerebral ischemia. Autophagy. 2012;8:77–87.
119. Kang C, Avery L. To be or not to be, the level of autophagy is the question: dual roles of autophagy in the survival response to starvation. Autophagy. 2008;4:82–4.
120. Ravikumar B, Sarkar S, Davies JE, Futter M, Garcia-Arencibia M, Green-Thompson ZW, et al. Regulation of mammalian autophagy in physiology and pathophysiology. Physiol Rev. 2010;90:1383–435.
121. Sheng R, Zhang LS, Han R, Liu XQ, Gao B, Qin ZH. Autophagy activation is associated with neuroprotection in a rat model of focal cerebral ischemic preconditioning. Autophagy. 2010;6:482–94.

122. Yan W, Zhang H, Bai X, Lu Y, Dong H, Xiong L. Autophagy activation is involved in neuro-protection induced by hyperbaric oxygen preconditioning against focal cerebral ischemia in rats. Brain Res. 2011;1402:109–21.
123. Puyal J, Vaslin A, Mottier V, Clarke PG. Postischemic treatment of neonatal cerebral ischemia should target autophagy. Ann Neurol. 2009;66:378–89.
124. Gao L, Jiang T, Guo J, Liu Y, Cui G, Gu L, et al. Inhibition of autophagy contributes to ischemic postconditioning-induced neuroprotection against focal cerebral ischemia in rats. PLoS One. 2012;7:e46092.
125. Komatsu M, Ueno T, Waguri S, Uchiyama Y, Kominami E, Tanaka K. Constitutive autophagy: vital role in clearance of unfavorable proteins in neurons. Cell Death Differ. 2007;14:887–94.
126. Xu F, Gu JH, Qin ZH. Neuronal autophagy in cerebral ischemia. Neurosci Bull. 2012;28:658–66.
127. Luo T, Park Y, Sun X, Liu C, Hu B. Protein misfolding, aggregation, and autophagy after brain ischemia. Transl Stroke Res. 2013;4:581–8.
128. Liu C, Gao Y, Barrett J, Hu B. Autophagy and protein aggregation after brain ischemia. J Neurochem. 2010;115:68–78.
129. Puyal J, Clarke PG. Targeting autophagy to prevent neonatal stroke damage. Autophagy. 2009;5:1060–1.
130. Long JS, Ryan KM. New frontiers in promoting tumour cell death: targeting apoptosis, necroptosis and autophagy. Oncogene. 2012;31:5045–60.
131. Zhivotovsky B, Orrenius S. Cell death mechanisms: cross-talk and role in disease. Exp Cell Res. 2010;316:1374–83.
132. Amaravadi RK, Thompson CB. The roles of therapy-induced autophagy and necrosis in cancer treatment. Clin Cancer Res. 2007;13:7271–9.
133. Festjens N, Vanden Berghe T, Vandenabeele P. Necrosis, a well-orchestrated form of cell demise: signalling cascades, important mediators and concomitant immune response. Biochim Biophys Acta. 2006;1757:1371–87.
134. Wang N, Pan W, Zhu M, Zhang M, Hao X, Liang G, et al. Fangchinoline induces autophagic cell death via p53/sestrin2/ampk signalling in human hepatocellular carcinoma cells. Br J Pharmacol. 2011;164:731–42.
135. Vanden Berghe T, Vanlangenakker N, Parthoens E, Deckers W, Devos M, Festjens N, et al. Necroptosis, necrosis and secondary necrosis converge on similar cellular disintegration features. Cell Death Differ. 2010;17:922–30.
136. Wallach D, Kang TB, Kovalenko A. Concepts of tissue injury and cell death in inflammation: a historical perspective. Nat Rev Immunol. 2014;14:51–9.
137. Li J, Yuan J. Caspases in apoptosis and beyond. Oncogene. 2008;27:6194–206.
138. Jeong SJ, Dasgupta A, Jung KJ, Um JH, Burke A, Park HU, et al. Pi3k/akt inhibition induces caspase-dependent apoptosis in htlv-1-transformed cells. Virology. 2008;370:264–72.
139. Fresno Vara JA, Casado E, de Castro J, Cejas P, Belda-Iniesta C, Gonzalez-Baron M. Pi3k/akt signalling pathway and cancer. Cancer Treat Rev. 2004;30:193–204.
140. Zhang X, Yan H, Yuan Y, Gao J, Shen Z, Cheng Y, et al. Cerebral ischemia-reperfusion-induced autophagy protects against neuronal injury by mitochondrial clearance. Autophagy. 2013;9:1321–33.
141. Zhang ZB, Li ZG. Cathepsin b and phospo-jnk in relation to ongoing apoptosis after transient focal cerebral ischemia in the rat. Neurochem Res. 2012;37:948–57.
142. Canu N, Tufi R, Serafino AL, Amadoro G, Ciotti MT, Calissano P. Role of the autophagic-lysosomal system on low potassium-induced apoptosis in cultured cerebellar granule cells. J Neurochem. 2005;92:1228–42.
143. Grishchuk Y, Ginet V, Truttmann AC, Clarke PG, Puyal J. Beclin 1-independent autophagy contributes to apoptosis in cortical neurons. Autophagy. 2011;7:1115–31.
144. Heitz S, Grant NJ, Leschiera R, Haeberle AM, Demais V, Bombarde G, et al. Autophagy and cell death of purkinje cells overexpressing doppel in ngsk prnp-deficient mice. Brain Pathol. 2010;20:119–32.

145. Radoshevich L, Murrow L, Chen N, Fernandez E, Roy S, Fung C, et al. Atg12 conjugation to atg3 regulates mitochondrial homeostasis and cell death. Cell. 2010;142:590–600.
146. He G, Xu W, Tong L, Li S, Su S, Tan X, et al. Gadd45b prevents autophagy and apoptosis against rat cerebral neuron oxygen-glucose deprivation/reperfusion injury. Apoptosis. 2016;21:390–403.
147. Qi Z, Dong W, Shi W, Wang R, Zhang C, Zhao Y, et al. Bcl-2 phosphorylation triggers autophagy switch and reduces mitochondrial damage in limb remote ischemic conditioned rats after ischemic stroke. Transl Stroke Res. 2015;6:198–206.
148. Delgado M, Tesfaigzi Y. Bh3-only proteins, bmf and bim, in autophagy. Cell Cycle. 2013;12:3453–4.
149. Luo S, Rubinsztein DC. Apoptosis blocks beclin 1-dependent autophagosome synthesis: an effect rescued by bcl-xl. Cell Death Differ. 2010;17:268–77.
150. Luo S, Rubinsztein DC. Bcl2l11/bim: a novel molecular link between autophagy and apoptosis. Autophagy. 2013;9:104–5.
151. Balduini W, Carloni S, Buonocore G. Autophagy in hypoxia-ischemia induced brain injury: evidence and speculations. Autophagy. 2009;5:221–3.

Chapter 6
Reactive Astrocytes in Cerebral Ischemic Reperfusion Injury

Abhishek Mishra, Rachana Nayak, and Dandan Sun

Abstract Currently, limited stroke treatments are available due to a short time window for effective treatment. Previous research on stroke therapies has focused on neurons as therapeutic targets, with little emphasis on manipulation of other brain cells. Today, research is increasingly finding evidence of the potentials of astrocytes for stroke therapies. Here, we present a review of the roles of astrocytes in the healthy brain as well as the altered functions of astrocytes in the ischemic and post-ischemic brain that modulate neuronal recovery. Astrocytic regulation of neuronal function occurs in both healthy and diseased brains as a result of their close association in the tripartite synapse. We will place an emphasis on the astrocytic properties that promote neural protection and restoration, and will also discuss the hurdles reactive astrocytes and glial scarring pose to functional neuronal recovery. To overcome these challenges, therapeutic advances have been made by exploring drug treatments that target astrocyte function to modulate reactive astrogliosis. Continued research into reactive astrocyte biology will allow us to better understand their potential as targets for stroke therapies.

Keywords Cerebral ischemia · Astrocyte · Reperfusion

1 Introduction

Astrocytes are mature glial cells. In humans, the glia/neuron ratio does not increase uniformly with brain size during the course of human development. While glia are frequently reported to outnumber neurons 10:1, this statement has drawn increasing criticism in recent years as there is no clear evidence that glial cells are the most abundant brain cells [1]. Regardless of glia to neuron ratio, it is evident that proper astrocyte function is obligatory for normal brain function.

Supported by NIH R01 NS048216, NS038118 (D. Sun).

A. Mishra · R. Nayak · D. Sun (✉)
Department of Neurology, University of Pittsburgh, Pittsburgh, PA, USA
e-mail: sund@upmc.edu

There are two main types of astrocytes: protoplasmic and fibrous. The former astrocytes are largely found in all gray matter and encompass synapses in a globular distribution. Fibrous type astrocytes are found in all white matter and contact the node of Ranvier in a fiber-like distribution [2]. Astrocytes play a crucial role in regulating a healthy central nervous system (CNS) and participate in synaptic and vascular processes [2]. Astrocyte endfeet can tightly envelope both pre- and post-synaptic neuronal terminals to form the "tripartite synapse", which allows astrocytes to exert modulatory functions over synaptic transmission and neuronal potentiation [3]. This is accomplished, in part, by astrocytic secretion of neuromodulatory substance (gliotransmitters) as well as the uptake of ions and neurotransmitters from active synaptic terminals [4]. In addition, tightly grouped astrocytes can form boundaries to guide migrating axons and neuroblasts during development or form a barrier between neurons and the blood supply [5]. Astrocytes have drawn more attentions in stroke research because they are abundant and interact with, influence and regulate CNS functions [6].

A stroke occurs when blood flow is stopped to a part of the brain, thereby depriving brain cells in that area of oxygen and initiating cell death pathways. Stroke is a leading cause of death worldwide, and in the United States alone, its prevalence is around 3% of the adult population. Ischemic stroke is caused by a blockage of a blood vessel via a blood clot while hemorrhagic stroke is caused by a rupture of an artery in the brain. It is estimated that 87% of strokes are ischemic strokes [7]. Furthermore, stroke alters astrocyte function. Astrocytes in the ischemic penumbra region show dysfunction and delayed death while astrocytes directly in the core region of the cardiovascular bed are more likely to die sooner, as they cannot get blood supply from nearby arteries [8]. In vitro studies have shown that astrocytes typically die after 4–6 h after oxygen-glucose deprivation (OGD), whereas neurons typically die after 5–20 min [9, 10]. After a period of ischemia, astrocytes become detached from basal cell lamina. The attachment of the basal lamina to astrocytes is critical because evidence suggests that the connection helps maintains the permeability of the blood brain barrier and maintains impermeability of cerebral microvessels [11, 12].

After a period of ischemia has occurred, astrocytes increase their release of lactate to surrounding neurons to fuel aerobic energy metabolism [13]. Furthermore, after ischemia astrocytes protect damaged tissue from becoming further damaged, rebuild the blood brain barrier, produce neurotrophic and growth factors, and take up excess glutamate from the extracellular space to prevent excitotoxicity [6, 14]. Thus, astrocytes play a crucial long-term role in neuroprotection, neurorestoration, neuroregeneration and brain plasticity after stroke, and their dysfunction may inhibit neuronal metabolism, survival and regeneration, resulting in neurological impairment [6, 9]. This review will discuss changes of astrocytes after ischemia/reperfusion injury and its significance to ischemic brain damage.

2   Astrocyte Function in Normal Brains

2.1   Astrocytes "Save" Dying Neurons by Providing Them with Antioxidant Defense

Oxidative stress arises when the body is not able to detoxify the harmful effects of reactive oxygen species (ROS) such as hydrogen peroxide, which can ultimately lead to brain dysfunction. ROS neutralization is achieved through antioxidants, for example, glutathione. The brain is especially vulnerable to oxidative stress because it consumes a lot of oxygen, has unsaturated lipids and at the same time the brain has fewer ROS protective mechanisms than most other organs and muscles. Neurons are particularly sensitive to ROS and rely on astrocytes to provide free radical scavengers, thereby mitigating ROS damage under both healthy and ischemic conditions. This process occurs via gap junction communication among astrocytes and bidirectionality with neurons, where astrocytes release gliotransmitters and propagate calcium Ca^{2+} waves [15]. Astrocytes detoxify hydrogen peroxide and oxidative stress via catalase and the superoxide dismutase (SOD). Additional detoxification comes from the astrocytic glutathione, or GSH system in which gluthionine reductase (GR) reduces glutathione disulfide (GSSG) into the antioxidant glutathione (GSH) [16]. The cAMP signaling pathway has been linked to glutathione system gene upregulation and transportation of antioxidant vitamins in astrocytes. By overexpressing an antioxidant enzyme, the cAMP pathway restricts astrocyte activation and increases the survival rate of some neurons such as the CA1 neurons after a period of ischemia [17].

2.2   The Function of Glucose Transporters in Astrocytes

After a period of ischemia, energy demand for the brain increases [2]. These neighboring neurons perform lactate oxidation to produce energy and exhibit a strong preference for lactate over glucose as an energy substrate [18]. This energy cycle begins when neurons take up glucose via GLUT1 transporters. Glucose uptake in astrocyte endfeet occurs via GLUT1 and GLUT 5 and possibly GLUT4 [19]. Research has shown that in the rat brain, astrocytes are responsible for up to 50% of the glucose absorbed by the brain from the blood stream, especially during astrocyte activation. If glycolytic pathways are overactived in neurons, cell apoptosis and oxidative stress can occur [18]. This is contrary to the glycolytic metabolism of astrocytes, in that astrocytes always show a high level of glycolytic activity. After a period of neuronal activity, astrocytic glutamate transporters import excess extracellular glutamate, which is co-transported with Na^+ ions. Increased $[Na^+]i$ concentrations stimulate Na-K-ATPase activity, which is fueled by an increase in astrocytic glycolysis. The lactate produced by glycolysis can then be shuttled to neurons via MCT1 and MCT4 as a neuronal energy source, a process known as the astrocyte-neuron lactate shuttle [20]. Once lactate is produced, it is transported across the blood brain barrier (BBB) by MCT1, an electroneutral monocarboxylate transporter that cotransports lactate, among other monocarboxylates with H^+ [18, 21]. Then, neuronal uptake of lactate takes place via MCT-2 at a 60% saturated and slower rate. Due to their diverse functions, astrocytes show to be vital for neural cell proliferation and differentiation [22].

2.3   Roles of Astrocytes in Synaptogenesis and Synaptic Pruning

Research in the last two decades has focused on the role astrocytes in synapse formation, modification and connectivity. The close proximity of perisynaptic astrocytic endfeet to synapses allows astrocytes to both monitor and respond to neuronal activity [3]. Astrocytes stimulate synapse formation by sending positive signals, and they also provide negative cues to inhibit synaptogenesis in adult brains. On the other hand, contact with astrocytes has shown to be critical for the neuron's ability to form synapses. Astrocytes are also important players in the timing of structural synapse formation, because many times synapses cannot be formed unless an astrocyte touches them [23]. SPARC (secreted protein acidic and rich in cysteine) and TSP (thrombospondins) are astrocyte factors and have broad functions including regulation of synaptic formation, presynaptic plasticity, and postsynaptic receptor levels [24]. SPARC decreases the buildup of excess AMPARs (AMPA glutamate receptors) and TSP increases synaptic glycine receptors and likewise decreases the buildup of AMPARs, therefore these two molecules act as inhibitors to synapse formation [25].

2.4   Astrocytes Affect Synaptic Transmission by Releasing Gliotransmitters

Astrocytes are in constant bidirectional communication with neurons through the tripartite synapse. Astrocytes affect synaptic transmission and participate in information processing by releasing gliotransmitters, such as D-serine, ATP, GABA, glutamate, prostaglandins and neuropeptides. Such chemicals are released in response to changes in synaptic activity, and their release can either stimulate or dampen neuronal excitability. Many potential mechanisms for the process of gliotransmission have been proposed and are not mutually exclusive. The most widely accepted theory is that gliotransmission is initiated by intracellular calcium signaling. This calcium may be released from internal stores or enter the cell through transmembrane ion channels [26, 27].

In conclusion, astrocytes have a wide range of functions including synapse-astrocyte communication that includes modifying synapse plasticity and affecting synaptic transmission, neuron-astrocyte communication that includes providing neurons with antioxidant defense and upregulated glucose transport. Thus, maintaining normal functions of astrocytes in an ischemic brain is crucial for ischemic brain recovery and function [2].

3 Reactive Astrocytes in Ischemic Reperfusion

3.1 Formation of Reactive Astrocytes

Proliferating reactive astrocytes are consistently found along borders between healthy tissues and pockets of damaged tissue and inflammatory cells. This is usually found after a rapid, locally triggered inflammatory response to acute traumatic injury in the spinal cord and brain.

Animal model studies of cerebral ischemic injuries show that although other various glial cell types like microglia and macrophages are activated in the glial scar, astrocytes account for the majority of the scar tissue, up to 1/5th of all cells [28]. Astrocytes in the penumbra become hypertrophic and elongate their processes into the infarct core. These astrocytes strongly upregulate GFAP, a hallmark of astrogliosis. GFAP is upregulated as early as 1 day after injury and the number of reactive astrocytes is significantly increased in the peri-lesion area at 3–5 days after injury, however, astrocytes remain absent from the lesion core [29]. Ablating reactive astrocytes in a mouse model showed a significant increase in lesion size and tissue damage, indicating the beneficial role of reactive gliosis [2, 30]. No increase in the number of reactive cells in the adjacent normal tissue was found in the human ischemic brain [31]. GFAP-positive reactive astrocytes were significantly increased in the cortical peri-infarct regions of human ischemic brain, compared to adjacent normal tissues and control subjects [32]. The glial scars form the borders around the lesion sites and act as protective barriers to infectious agents and inflammatory cells. However, scar formation results in less advantageous outcomes in the long-term recovery after insults. This is partially because reactive gliosis inhibits axonal outgrowth and cellular migration because of astrocyte secreted chondroitin sulphate proteoglycans (CSPG) and other neurogenesis inhibiting molecules. Based on the findings in rodents, reactive gliosis may play a dual role in brain repair after injuries, which will be discussed in subsequent sections.

3.2 Altered Functions in Reactive Astrocyte

Under different conditions of stimulation, astrocytes can produce intercellular effector molecules that alter the expression of molecules regulating cell structure, energy metabolism, intracellular signaling, and membrane transporters and pumps [33, 34].

These signaling cascades induce changes in regional blood flow, availability of energy substrates, and release of neuromodulatory substances. The full complexity and interaction of these pathways is still being explored, however, astrocytes have the capacity to both enhance and undermine the repair process after ischemic injury.

Connexin43 (Cx43) is one of the most abundant gap junction proteins in the CNS. Cx43 has been detected in regions with astrogliosis induced by various brain pathologies including brain ischemia and epilepsy [35]. As gap junctions form channels that allow passage of small molecules such as ATP and glutamate between adjacent cells [35], they are especially suited to play a pivotal role in intercellular communication in a diseased state [35]. In addition, gap junction proteins can also form hemichannels that connect the cytoplasm directly to the extracellular space. Gap junctions can expand injury in the process of bystander death, the process in which undamaged cells exhibit adverse effects because of signals received from nearby damaged cells [36]. Cellular injury is linked to an upregulation of Cx43 expression in a variety of cells and tissues [37]. Cx43 is upregulated in cortical astrocytes after ischemia [38]. In addition, aquaporin 4 (AQP4), an astrocyte water channel, plays a crucial role in cytotoxic edema and determining outcomes after stroke [30]. This takes place predominantly in astrocyte endoot processes at the borders between the brain parenchyma and major fluid compartments. This suggests associations between AQP4-regulated water flux and neuroinflammation [39]. Ischemic stroke can trigger both cytotoxic and vasogenic edemas. Cytotoxic edema evolves over minutes to hours and may be reversible, while the vasogenic phase occurs over hours to days, and is considered an irreversibly damaging process [40]. In cytotoxic edema, AQP4 deletion slows the rate of water entry into brain, whereas in vasogenic edema, AQP4 deletion reduces the rate of water outflow from brain parenchyma [41]. A central role of AQP4 in neuroinflammation is in the established astrocyte proliferative response in ischemia [42]. AQP4 was also detected on reactive microglia following lipopolysaccharide (LPS) injection in rats, but the significance of this expression is poorly understood [43].

Pro-inflammatory mediators, including IL-1 and TNF-α, released by both activated microglia and neutrophils have been shown to stimulate the release of matrix metalloproteinases (MMPs) from astrocytes in cultures [44, 45]. MMPs are primary components of the neuroinflammatory response being partially responsible for BBB disruption [10]. The neuroinflammation produced by ischemia leads to upregulation of MMPs which cause AQP4-orthogonal arrays of particles (OAP) disruption, and possibly exacerbate BBB disturbance and worsen the edema [10]. MMP-9 degrades agrin while MMP-3 degrades dystroglycan [46], both of which play a pivotal role in OAP formation and AQP4 assembly in astroglial endfoot membranes [47, 48]. Development of vasogenic edema also amplifies BBB disruption due to increased hydrostatic pressure. AQP4 may also have a role in neuroinflammation via edema resolution. Up-regulation of perivascular AQP4 causes enhanced resorption of extracellular edema fluid which eases hydrostatic pressure and BBB disruption. Thus, there is less neutrophil infiltration, less pro-inflammatory cytokine production and less MMP activation [39]. Stretch-activated Cl$^-$ channels expressed on microglia are activated to a lesser extent because of pressure differences due to resorption of edema fluid. This leads to less microglial activation and a decrease in pro-inflammatory cytokine release [39].

There is a body of evidence [49] pointing to negative consequences of reactive gliosis when it does not get resolved within the post-acute and the early chronic stage after injury. Reactive gliosis and glial scarring have inhibitory effects on CNS regeneration as shown in several experimental models with a wide range of molecules implicated in this process [50–54]. Currently, these regeneration-inhibiting effects represent the cons that are tied in with other beneficial effects of reactive astrocytes. Notably, the effective handling of the acute stage of an injury through reactive astrocytosis reduces cellular and tissue stress and provides effective neuroprotection together with the beneficial isolation of the lesion area from the rest of the CNS [55–57].

3.3   Impact of Reactive Astrocytes on Neural Protection and Repair

In the glial scar, reactive astrocytes express a broad range of inhibitory molecules against axonal regeneration, such as chondroitin sulfate proteoglycans. However, the glial scar may also seclude the injury site from healthy tissue, preventing a cascading wave of uncontrolled tissue damage. In addition, reactive astrocytes take up excess glutamate and produce neurotrophic factors, to protect the neurons from ischemic lesion. Thus, the reactivity of astrocytes after stroke may potentially play both detrimental and beneficial roles under certain spatio-temporal conditions.

In response to brain damage of stroke or trauma, reactive astrocytes up-regulate GFAP and vimentin and re-express nestin. In mice lacking both GFAP and vimentin ($GFAP^{-/-}Vim^{-/-}$), reactive gliosis and the glial scar are increased after neurotrauma [58]. $GFAP^{-/-}Vim^{-/-}$ astrocytes exposed to oxygen-glucose deprivation and reperfusion exhibit increased cell death and confer lower degree of protection to co-cultured neurons than WT astrocytes [59], suggesting that reactive astrocytes are protective during brain ischemia.

Proliferating reactive astrocytes are critical to scar formation and function to reduce the spread and persistence of inflammatory cells, to enhance the repair of the BBB, to decrease tissue damage and lesion size, and to decrease neuronal loss and demyelination [31, 57]. Reactive astrocytes defend against oxidative stress through glutathione production and have the responsibility of protecting CNS cells from NH_4^+ toxicity [30]. They protect CNS cells and tissue through various methods, such as uptake of potentially excitotoxic glutamate, adenosine release, and degradation of amyloid β peptides [30, 60, 61]. During reperfusion, the return of oxygenated blood to the ischemic area challenges the BBB with oxidative stress. In experimental studies, BBB opening is biphasic; the initial breakdown is most likely caused by oxidative stress and is followed by a partial BBB recovery before the second increase in BBB permeability leads to neutrophil infiltration through tight junction redistribution, the exact causes for this are still unknown [62, 63]. Whether this order of events is also relevant in stroke patients still needs to be confirmed.

Reactive astrogliosis can lead to the appearance of newly proliferated astrocytes and scar formation in response to severe tissue damage or inflammation. Molecular triggers that lead to this scar formation include epidermal growth factor (EGF), fibroblast growth factor (FGF), endothelin 1 and adenosine triphosphate (ATP). Mature astrocytes can re-enter the cell cycle and proliferate during scar formation. Some proliferating reactive astrocytes are derived from NG2 progenitor cells in the local parenchyma after injury or stroke [34]. There are also multipotent progenitors in subependymal tissue that express glial fibrillary acidic protein (GFAP) and generate progeny cells that migrate towards sites of injury after trauma or stroke [34].

However, scar formation shows less advantageous outcome in the long-term recovery after insults. Axon regeneration does not occur in areas with an increase in GFAP and vimentin [64, 65]. Reactive gliosis inhibits axonal outgrowth and cellular migration by secreting chondroitin sulphate proteoglycans (CSPGs) and other molecules inhibitory to neurogenesis [66].

3.4   Reactive Astrocytes in Regulation of Inflammation

Reactive astrocytes are involved in the complex regulation of CNS inflammation that is likely to be context-dependent and regulated by multimodal extra- and intracellular signaling events. They have the capacity to make different types of molecules with either pro- or anti-inflammatory potential in response to different types of stimulation. Astrocytes interact extensively with microglia and play a key role in CNS inflammation. Reactive astrocytes can then affect healthy astrocytes and affect their regulation and response to inflammation [34, 67].

Reactive scar-forming astrocytes help reduce the spread of inflammatory cells during locally-initiated inflammatory responses to traumatic injury or during peripherally-initiated adaptive immune responses, thus serving an anti-inflammatory function. However, certain molecules in astrocytes are associated with an increase in inflammation after traumatic injury [30, 60]. For example, GFAP, vimentin and nestin, as well as altered expression of many other genes [68].

Several decades of investigation has shown that astrocytes can produce numerous pro-inflammatory molecules, including diverse cytokines, chemokines, growth factors and small molecules such as prostaglandin E (PGE) and nitric oxide (NO). Recent technologies that allow cell type-specific transcriptome analysis have begun to define specific contexts in which astrocytes produce a broad repertoire of pro-inflammatory molecules in vivo and to identify combinations of molecular triggers that regulate their production in vitro [33, 69, 70]. For example, analysis of astrocyte transcriptome profiles indicates that astrocyte exposure either in vivo or in vitro to PAMPs such as lipopolysaccharide (LPS) and associated cytokines markedly skews astrocyte transcriptome changes towards pro-inflammatory and potentially cytotoxic profiles [33, 69, 70]. By contrast, the ischemia caused by experimental stroke in vivo shifts the astrocyte transcriptome towards neuroprotective mechanisms. Transgenic approaches such as Cre–loxP-mediated cell type-specific loss-of-function models are being used to dissect intracellular signalling cascades and to identify intercellular effector molecules that mediate astrocyte pro-inflammatory functions. For example, nuclear factor-κB (NF-κB) and suppressor of cytokine signalling 3 (SOCS3) are pro-inflammatory transcriptional regulators in astrocytes during CNS traumatic injury and CNS autoimmune inflammation [71–73]. As intercellular effectors, CC-chemokine ligand 2 (CCL2) and CXC-chemokine ligand 10 (CXCL10) released specifically by astrocytes are important recruiters of perivascular leukocytes in CNS autoimmune inflammation [43–45]. CCL2 and CCL7 production by astrocytes is heterogeneous [70], which may contribute to the selective direction of leukocyte migration in CNS parenchyma. Multimolecular signaling cascades are also being defined. For example, in response to stimulation by the pro-inflammatory cytokine interleukin-1β (IL-1β), astrocytes generate and release vascular endothelial growth factor (VEGF), which increases BBB permeability and promotes leukocyte extravasation [74, 75]. ACT1 (also known as TRAF3IP2) signaling in astrocytes is critical for mediating IL-17 inflammatory gene induction in CNS autoimmune inflammation [76].

Essential anti-inflammatory roles of astrocytes have now been demonstrated by numerous transgenic loss-of-function experiments in diverse models of CNS injury and disease, and specific molecular mechanisms are gradually being identified. An early observation using transgenic models indicated that BBB repair after traumatic injury is critically dependent on the presence of newly proliferated scar-forming astrocytes [28, 30]. Astrocyte-produced molecules that exert or regulate anti-inflammatory functions activate diverse anti-inflammatory signaling mechanisms [33, 69, 70, 77–79]. Intracellular signaling factors STAT3, A20, GAL9 and CRYAB Suppress pro-inflammatory signaling mechanisms [72, 80–85]. As discussed above, astrocyte borders and scars form functional barriers that restrict leukocyte migration after diverse CNS insults, including trauma, ischemia, autoimmune attack and neurodegeneration [55, 58, 60, 86–88]. Transgenic loss-of-function approaches are beginning to define intracellular signaling cascades that mediate astrocyte anti-inflammatory functions. Studies from multiple laboratories indicate that the GP130 (also known as IL-6Rβ)–Janus kinase 2 (JAK2)–signal transducer and activator of transcription 3 (STAT3) signaling pathway is a crucial regulator of astrocyte anti-inflammatory functions after various CNS insults, by mediating scar formation and barrier functions that restrict the spread of microbial pathogens after traumatic injury [72, 80, 81], infection [89] and autoimmune attack [90]. In this regard, it is interesting that GFAP and vimentin [91] are required for astrocyte scar formation [65], and their absence markedly exacerbates the inflammation and tissue pathologies associated with autoimmune attack [92], stroke [58, 88] and neurodegeneration owing to lipid storage defects [93] or amyloid-β accumulation [60]. Intercellular effector molecules that mediate astrocyte anti-inflammatory functions are also being identified. For example, in response to IL-10, astrocytes release molecules such as transforming growth factor-β (TGFβ) that promote resolution of inflammation [94–96]. Certain molecules released by astrocytes have anti-inflammatory effects on microglia and monocytes [97, 98]. Retinoic acid released from astrocytes is implicated in protecting BBB function and attenuating inflammation [66]. An indirect form of anti-inflammatory regulation also seems to be operational in the form of intracellular signaling mechanisms that suppress proinflammatory activators. For example, tumour necrosis factor-α (TNFα) signaling not only induces production of the pro-inflammatory transcriptional regulator NF-κB but also promotes production of the ubiquitin-modifying protein A20, which acts to suppress NF-κB signalling [67].

Although initially protective, over time, dense glial scars form potent cell migration barriers that impair regeneration [60]. This paradox is hypothesized to be the result of an evolutionary adaptation that led to unintended consequence [60]. CNS injury responses have favored mechanisms that keep small injuries uninfected, thus inhibition of the migration of inflammatory cells and infectious agents would have promoted overall survival. However, the persistence of the scar leads to the accidental byproduct of axon regeneration inhibition [30, 57, 60].

4   Modulating Astrogliosis to Remove Detriments and Retain Benefits

4.1   Potential Targets for Improvement of Astrocyte Functions

Counter to the formation of a glial scar, reactive astrocytes can also positively affect the later post-ischemic stages by secreting vascular endothelial growth factor (VEGF) and thrombospondin to promote the formation of new blood vessels and synapses [99–101]. Thrombospondin 4 has recently been shown to control protective astrogenesis in the adult subventricular zone in response to neurotrauma [102], and its expression was proposed to depend on astrocyte reactivity and reactive astrogliosis [103]. VEGF secretion by reactive astrocytes can serve as a good example of a context-dependent response: the positive, stimulatory effect of VEGF on the formation of vessels and synapses in ischemic stroke [99, 104] contrasts with its induction of blood-brain barrier breakdown and lymphocyte infiltration in autoimmune CNS inflammation [74, 75]. Treatment with glial cell line-derived neurotrophic factor (GDNF) enhances neuronal survival in in vitro mouse ischemic brain tissue [105–107]. Although the expression of GDNF is low in an unchallenged brain, it is upregulated in reactive astrocytes in the peri-infarct region [108–110]. Thus, therapy targeting the enhancement of GDNF expression by astrocytes is an attractive approach to treating brain ischemia.

Treatment of stroke with bone marrow stromal cells (BMSCs) via intra-arterial injection enhances expression of bone morphogenetic proteins (BMPs) and Cx43 gap junction protein [111] Additionally, administration of BMSCs intra-arterially increased sub ventricular zone cell proliferation and induced differentiation of proliferating cells into astrocytes in the ischemic boundary zone (IBZ) [111]. This was shown to promote synaptogenesis and improve neurological functional recovery after stroke [111]. Administration of BMSCs intra-arterially at 1 day after MCAO significantly improved neurological functional recovery in rats at 14 days, and extended functional improvement to 21 and 28 days after MCAO, but did not significantly decrease lesion volume. Interestingly, treatment was initiated at 24 h after stroke, a time at which the ischemic lesion is relatively mature and neuroprotective agents are not effective in reducing volume of infarction. Although the detailed mechanism of this protection remains to be determined, BMSCs are an additional promising therapeutic strategy [111].

Functional modulation of neuronal synapses by astrocytes may be a new therapeutic goal in stroke patients. For example, a key negative regulator of neuronal activity is the inhibitory neurotransmitter γ-aminobutyric acid (GABA) that exerts its fast effects through synaptic receptors. Through extrasynaptic receptors, GABA that spills over at active synapses negatively regulates the background activity of neurons, a phenomenon is known as tonic inhibition. GABA is normally removed from the extracellular space by neuronal and astrocyte transporters. The expression of GABA transporters on astrocytes is reduced in the ischemic penumbra, and tonic inhibition by excessive amounts of GABA limits both neural plasticity and functional recovery [112]. In mouse models, appropriately timed administration of GABA inhibitors results in a faster recovery of function [112]. However, when administered too early, such treatment can increase the amount of brain tissue lost in the ischemic stroke. Similarly, VEGF is beneficial in the late post-stroke period, but leads to increased capillary permeability and larger infarctions when administered in the early post-ischemic phase [113]. Thus, the time of intervention is crucial. Though the molecular mechanisms are not yet fully understood, astrocytes seem to control several aspects of synaptic plasticity. It is predominantly synaptic plasticity in the weeks and months after stroke which determines the degree of functional recovery and which can be enhanced through specific neurorehabilitation programs, noninvasive brain stimulation, and pharmacological modulation [114–117]. For example, ischemic stroke triggers ephrin-5A expression in reactive astrocytes, and this leads to the inhibition of axonal sprouting and motor recovery [66]. Pharmacological blockage of ephrin-A5 combined with forced use of the affected limb was shown to promote new and widespread axonal projections within the entire cortical hemisphere at the side of experimentally induced ischemic stroke [66].

Other potential targets for improving outcome by manipulating functions and effects related to reactive astrogliosis also include manipulation of AQP4 channels during different forms of cerebral edema [61], attenuation of NF-κB [71, 73], or augmentation of STAT3 [81] signaling mechanisms to reduce inflammation. Utilization of advanced screening technologies will also hasten the identification of novel therapeutic targets. For example, using a novel quantitative proteomics approach, Hauck et al. [118] showed that Müller cells in the retina produce a pool of neuroprotective factors such as osteoponin and connective tissue growth factor. Application of such screening strategies to astrocytes exposed to a variety of stimuli will enable better characterization of these cells and their responses under different conditions to improve our understanding of the heterogeneity of astroglia.

5 Conclusion

During the early time after ischemic injury, the main function of reactive astrocytes is to preserve the integrity of the nervous tissue. This defensive reaction of astrocytes is conceivably aimed at handling the acute stress, limiting tissue damage, and restoring homeostasis. However, with time, the process becomes increasingly unregulated and maladaptive. The lingering glial scar further accentuates inflammation, generating a highly toxic microenvironment that inhibits migrating axons and interferes with long-term motor functional recovery. Understanding the multifaceted roles of astrocytes in the healthy and diseased CNS will undoubtedly contribute to the development of treatment strategies that will, in a context-dependent manner and at appropriate time points, modulate reactive astrogliosis to promote brain repair and reduce the neurological impairment. Given the dualistic nature of reactive astrogliosis, a comprehensive understanding of reactive astrogliosis is extremely vital to the designing strategies that can aid in modulating reactive astrogliosis for healing tissue. The primary goal of research will be to enhance and retain the therapeutically beneficial functions of reactive astrocytes while eliminating undesired outcomes that are also related to reactive astrogliosis. Modulating the signaling pathways in astrocytes to establish a neurogenic niche in the infarct area is an attractive strategy for stroke therapy.

Acknowledgment The authors wish to thank Dr. Karen Carney for her critical review of the manuscript, R01 NIH NS038118 grant.

References

1. Herculano-Houzel S. The glia/neuron ratio: how it varies uniformly across brain structures and species and what that means for brain physiology and evolution. Glia. 2014;62(9):1377–91.
2. Sofroniew MV, Vinters HV. Astrocytes: biology and pathology. Acta Neuropathol. 2010;119(1):7–35.
3. Araque A, Parpura V, Sanzgiri RP, Haydon PG. Tripartite synapses: glia, the unacknowledged partner. Trends Neurosci. 1999;22(5):208–15.
4. Ota Y, Zanetti AT, Hallock RM. The role of astrocytes in the regulation of synaptic plasticity and memory formation. Neural Plast. 2013;2013:185463.
5. Powell EM, Geller HM. Dissection of astrocyte-mediated cues in neuronal guidance and process extension. Glia. 1999;26(1):73–83.
6. Li Y, Liu Z, Xin H, Chopp M. The role of astrocytes in mediating exogenous cell-based restorative therapy for stroke. Glia. 2014;62(1):1–16.
7. Roger VL, Go AS, Lloyd-Jones DM, Adams RJ, Berry JD, Brown TM, et al. Heart disease and stroke statistics—2011 update: a report from the American Heart Association. Circulation. 2011;123(4):e18–e209.
8. Dirnagl U, Iadecola C, Moskowitz MA. Pathobiology of ischaemic stroke: an integrated view. Trends Neurosci. 1999;22(9):391–7.
9. Danilov CA, Fiskum G. Hyperoxia promotes astrocyte cell death after oxygen and glucose deprivation. Glia. 2008;56(7):801–8.

10. Rosell A, Ortega-Aznar A, Alvarez-Sabin J, Fernandez-Cadenas I, Ribo M, Molina CA, et al. Increased brain expression of matrix metalloproteinase-9 after ischemic and hemorrhagic human stroke. Stroke. 2006;37(6):1399–406.
11. Milner R, Hung S, Wang X, Spatz M, del Zoppo GJ. The rapid decrease in astrocyte-associated dystroglycan expression by focal cerebral ischemia is protease-dependent. J Cereb Blood Flow Metab. 2008;28(4):812–23.
12. del Zoppo GJ, Mabuchi T. Cerebral microvessel responses to focal ischemia. J Cereb Blood Flow Metab. 2003;23(8):879–94.
13. Carpenter KL, Jalloh I, Hutchinson PJ. Glycolysis and the significance of lactate in traumatic brain injury. Front Neurosci. 2015;9:112.
14. Malarkey EB, Parpura V. Mechanisms of glutamate release from astrocytes. Neurochem Int. 2008;52(1–2):142–54.
15. Wade JJ, McDaid LJ, Harkin J, Crunelli V, Kelso JA. Bidirectional coupling between astrocytes and neurons mediates learning and dynamic coordination in the brain: a multiple modeling approach. PLoS One. 2011;6(12):e29445.
16. Liddell JR, Robinson SR, Dringen R, Bishop GM. Astrocytes retain their antioxidant capacity into advanced old age. Glia. 2010;58(12):1500–9.
17. Paco S, Hummel M, Pla V, Sumoy L, Aguado F. Cyclic AMP signaling restricts activation and promotes maturation and antioxidant defenses in astrocytes. BMC Genomics. 2016;17:304.
18. Falkowska A, Gutowska I, Goschorska M, Nowacki P, Chlubek D, Baranowska-Bosiacka I. Energy metabolism of the brain, including the cooperation between astrocytes and neurons, especially in the context of glycogen metabolism. Int J Mol Sci. 2015;16(11):25959–81.
19. Shah K, Desilva S, Abbruscato T. The role of glucose transporters in brain disease: diabetes and Alzheimer's disease. Int J Mol Sci. 2012;13(10):12629–55.
20. Magistretti PJ. Neuron-glia metabolic coupling and plasticity. J Exp Biol. 2006;209(Pt 12):2304–11.
21. Sims NR, Nilsson M, Muyderman H. Mitochondrial glutathione: a modulator of brain cell death. J Bioenerg Biomembr. 2004;36(4):329–33.
22. Li M, Sun L, Luo Y, Xie C, Pang Y, Li Y. High-mobility group box 1 released from astrocytes promotes the proliferation of cultured neural stem/progenitor cells. Int J Mol Med. 2014;34(3):705–14.
23. Chung WS, Allen NJ, Eroglu C. Astrocytes control synapse formation, function, and elimination. Cold Spring Harb Perspect Biol. 2015;7(9):a020370.
24. Kucukdereli H, Allen NJ, Lee AT, Feng A, Ozlu MI, Conatser LM, et al. Control of excitatory CNS synaptogenesis by astrocyte-secreted proteins Hevin and SPARC. Proc Natl Acad Sci U S A. 2011;108(32):E440–9.
25. Jones EV, Bernardinelli Y, Tse YC, Chierzi S, Wong TP, Murai KK. Astrocytes control glutamate receptor levels at developing synapses through SPARC-beta-integrin interactions. J Neurosci. 2011;31(11):4154–65.
26. Araque A, Carmignoto G, Haydon PG, Oliet SH, Robitaille R, Volterra A. Gliotransmitters travel in time and space. Neuron. 2014;81(4):728–39.
27. Haydon PG, Nedergaard M. How do astrocytes participate in neural plasticity? Cold Spring Harb Perspect Biol. 2014;7(3):a020438.
28. Barreto GE, Sun X, Xu L, Giffard RG. Astrocyte proliferation following stroke in the mouse depends on distance from the infarct. PLoS One. 2011;6(11):e27881.
29. Kawano H, Kimura-Kuroda J, Komuta Y, Yoshioka N, Li HP, Kawamura K, et al. Role of the lesion scar in the response to damage and repair of the central nervous system. Cell Tissue Res. 2012;349(1):169–80.
30. Sofroniew MV. Molecular dissection of reactive astrogliosis and glial scar formation. Trends Neurosci. 2009;32(12):638–47.
31. Nowicka D, Rogozinska K, Aleksy M, Witte OW, Skangiel-Kramska J. Spatiotemporal dynamics of astroglial and microglial responses after photothrombotic stroke in the rat brain. Acta Neurobiol Exp (Wars). 2008;68(2):155–68.

32. Huang L, Wu ZB, Zhuge Q, Zheng W, Shao B, Wang B, et al. Glial scar formation occurs in the human brain after ischemic stroke. Int J Med Sci. 2014;11(4):344–8.
33. John GR, Lee SC, Song X, Rivieccio M, Brosnan CF. IL-1-regulated responses in astrocytes: relevance to injury and recovery. Glia. 2005;49(2):161–76.
34. Eddleston M, Mucke L. Molecular profile of reactive astrocytes—implications for their role in neurologic disease. Neuroscience. 1993;54(1):15–36.
35. Theodoric N, Bechberger JF, Naus CC, Sin WC. Role of gap junction protein connexin43 in astrogliosis induced by brain injury. PLoS One. 2012;7(10):e47311.
36. Sun CL, Kim E, Crowder CM. Delayed innocent bystander cell death following hypoxia in Caenorhabditis elegans. Cell Death Differ. 2014;21(4):557–67.
37. Daleau P, Boudriau S, Michaud M, Jolicoeur C, Kingma JG Jr. Preconditioning in the absence or presence of sustained ischemia modulates myocardial Cx43 protein levels and gap junction distribution. Can J Physiol Pharmacol. 2001;79(5):371–8.
38. Cotrina ML, Kang J, Lin JH, Bueno E, Hansen TW, He L, et al. Astrocytic gap junctions remain open during ischemic conditions. J Neurosci. 1998;18(7):2520–37.
39. Fukuda AM, Badaut J. Aquaporin 4: a player in cerebral edema and neuroinflammation. J Neuroinflammation. 2012;9:279.
40. Schaefer PW, Buonanno FS, Gonzalez RG, Schwamm LH. Diffusion-weighted imaging discriminates between cytotoxic and vasogenic edema in a patient with eclampsia. Stroke. 1997;28(5):1082–5.
41. Papadopoulos MC, Verkman AS. Aquaporin-4 and brain edema. Pediatr Nephrol. 2007;22(6):778–84.
42. Kuppers E, Gleiser C, Brito V, Wachter B, Pauly T, Hirt B, et al. AQP4 expression in striatal primary cultures is regulated by dopamine—implications for proliferation of astrocytes. Eur J Neurosci. 2008;28(11):2173–82.
43. Tomas-Camardiel M, Venero JL, de Pablos RM, Rite I, Machado A, Cano J. In vivo expression of aquaporin-4 by reactive microglia. J Neurochem. 2004;91(4):891–9.
44. Candelario-Jalil E, Yang Y, Rosenberg GA. Diverse roles of matrix metalloproteinases and tissue inhibitors of metalloproteinases in neuroinflammation and cerebral ischemia. Neuroscience. 2009;158(3):983–94.
45. Xia W, Han J, Huang G, Ying W. Inflammation in ischaemic brain injury: current advances and future perspectives. Clin Exp Pharmacol Physiol. 2010;37(2):253–8.
46. Wolburg-Buchholz K, Mack AF, Steiner E, Pfeiffer F, Engelhardt B, Wolburg H. Loss of astrocyte polarity marks blood-brain barrier impairment during experimental autoimmune encephalomyelitis. Acta Neuropathol. 2009;118(2):219–33.
47. Fallier-Becker P, Sperveslage J, Wolburg H, Noell S. The impact of agrin on the formation of orthogonal arrays of particles in cultured astrocytes from wild-type and agrin-null mice. Brain Res. 2011;1367:2–12.
48. Noel G, Tham DK, Moukhles H. Interdependence of laminin-mediated clustering of lipid rafts and the dystrophin complex in astrocytes. J Biol Chem. 2009;284(29):19694–704.
49. Burda JE, Sofroniew MV. Reactive gliosis and the multicellular response to CNS damage and disease. Neuron. 2014;81(2):229–48.
50. Alilain WJ, Horn KP, Hu H, Dick TE, Silver J. Functional regeneration of respiratory pathways after spinal cord injury. Nature. 2011;475(7355):196–200.
51. Busch SA, Silver J. The role of extracellular matrix in CNS regeneration. Curr Opin Neurobiol. 2007;17(1):120–7.
52. Davies SJ, Goucher DR, Doller C, Silver J. Robust regeneration of adult sensory axons in degenerating white matter of the adult rat spinal cord. J Neurosci. 1999;19(14):5810–22.
53. Fitch MT, Silver J. CNS injury, glial scars, and inflammation: Inhibitory extracellular matrices and regeneration failure. Exp Neurol. 2008;209(2):294–301.
54. Goldshmit Y, Galea MP, Wise G, Bartlett PF, Turnley AM. Axonal regeneration and lack of astrocytic gliosis in EphA4-deficient mice. J Neurosci. 2004;24(45):10064–73.
55. Faulkner JR, Herrmann JE, Woo MJ, Tansey KE, Doan NB, Sofroniew MV. Reactive astrocytes protect tissue and preserve function after spinal cord injury. J Neurosci. 2004;24(9):2143–55.

56. Myer DJ, Gurkoff GG, Lee SM, Hovda DA, Sofroniew MV. Essential protective roles of reactive astrocytes in traumatic brain injury. Brain. 2006;129(Pt 10):2761–72.
57. Sofroniew MV. Reactive astrocytes in neural repair and protection. Neuroscientist. 2005;11(5):400–7.
58. Li L, Lundkvist A, Andersson D, Wilhelmsson U, Nagai N, Pardo AC, et al. Protective role of reactive astrocytes in brain ischemia. J Cereb Blood Flow Metab. 2008;28(3):468–81.
59. de Pablo Y, Nilsson M, Pekna M, Pekny M. Intermediate filaments are important for astrocyte response to oxidative stress induced by oxygen-glucose deprivation and reperfusion. Histochem Cell Biol. 2013;140(1):81–91.
60. Bush TG, Puvanachandra N, Horner CH, Polito A, Ostenfeld T, Svendsen CN, et al. Leukocyte infiltration, neuronal degeneration, and neurite outgrowth after ablation of scar-forming, reactive astrocytes in adult transgenic mice. Neuron. 1999;23(2):297–308.
61. Zador Z, Stiver S, Wang V, Manley GT. Role of aquaporin-4 in cerebral edema and stroke. Handb Exp Pharmacol. 2009;190:159–70.
62. Pillai DR, Dittmar MS, Baldaranov D, Heidemann RM, Henning EC, Schuierer G, et al. Cerebral ischemia-reperfusion injury in rats—a 3 T MRI study on biphasic blood-brain barrier opening and the dynamics of edema formation. J Cereb Blood Flow Metab. 2009;29(11):1846–55.
63. Suzuki R, Yamaguchi T, Kirino T, Orzi F, Klatzo I. The effects of 5-minute ischemia in Mongolian gerbils: I. Blood-brain barrier, cerebral blood flow, and local cerebral glucose utilization changes. Acta Neuropathol. 1983;60(3–4):207–16.
64. Fawcett JW, Asher RA. The glial scar and central nervous system repair. Brain Res Bull. 1999;49(6):377–91.
65. Pekny M, Nilsson M. Astrocyte activation and reactive gliosis. Glia. 2005;50(4):427–34.
66. Overman JJ, Clarkson AN, Wanner IB, Overman WT, Eckstein I, Maguire JL, et al. A role for ephrin-A5 in axonal sprouting, recovery, and activity-dependent plasticity after stroke. Proc Natl Acad Sci U S A. 2012;109(33):E2230–9.
67. Farina C, Aloisi F, Meinl E. Astrocytes are active players in cerebral innate immunity. Trends Immunol. 2007;28(3):138–45.
68. Ridet JL, Malhotra SK, Privat A, Gage FH. Reactive astrocytes: cellular and molecular cues to biological function. Trends Neurosci. 1997;20(12):570–7.
69. Zamanian JL, Xu L, Foo LC, Nouri N, Zhou L, Giffard RG, et al. Genomic analysis of reactive astrogliosis. J Neurosci. 2012;32(18):6391–410.
70. Hamby ME, Coppola G, Ao Y, Geschwind DH, Khakh BS, Sofroniew MV. Inflammatory mediators alter the astrocyte transcriptome and calcium signaling elicited by multiple G-protein-coupled receptors. J Neurosci. 2012;32(42):14489–510.
71. Brambilla R, Bracchi-Ricard V, Hu WH, Frydel B, Bramwell A, Karmally S, et al. Inhibition of astroglial nuclear factor kappaB reduces inflammation and improves functional recovery after spinal cord injury. J Exp Med. 2005;202(1):145–56.
72. Okada S, Nakamura M, Katoh H, Miyao T, Shimazaki T, Ishii K, et al. Conditional ablation of Stat3 or Socs3 discloses a dual role for reactive astrocytes after spinal cord injury. Nat Med. 2006;12(7):829–34.
73. Brambilla R, Persaud T, Hu X, Karmally S, Shestopalov VI, Dvoriantchikova G, et al. Transgenic inhibition of astroglial NF-kappa B improves functional outcome in experimental autoimmune encephalomyelitis by suppressing chronic central nervous system inflammation. J Immunol. 2009;182(5):2628–40.
74. Argaw AT, Asp L, Zhang J, Navrazhina K, Pham T, Mariani JN, et al. Astrocyte-derived VEGF-A drives blood-brain barrier disruption in CNS inflammatory disease. J Clin Invest. 2012;122(7):2454–68.
75. Argaw AT, Gurfein BT, Zhang Y, Zameer A, John GR. VEGF-mediated disruption of endothelial CLN-5 promotes blood-brain barrier breakdown. Proc Natl Acad Sci U S A. 2009;106(6):1977–82.

76. Kang Z, Altuntas CZ, Gulen MF, Liu C, Giltiay N, Qin H, et al. Astrocyte-restricted ablation of interleukin-17-induced Act1-mediated signaling ameliorates autoimmune encephalomyelitis. Immunity. 2010;32(3):414–25.

77. Meeuwsen S, Persoon-Deen C, Bsibsi M, Ravid R, van Noort JM. Cytokine, chemokine and growth factor gene profiling of cultured human astrocytes after exposure to proinflammatory stimuli. Glia. 2003;43(3):243–53.

78. Jensen CJ, Massie A, De Keyser J. Immune players in the CNS: the astrocyte. J Neuroimmune Pharmacol. 2013;8(4):824–39.

79. Cooley ID, Chauhan VS, Donneyz MA, Marriott I. Astrocytes produce IL-19 in response to bacterial challenge and are sensitive to the immunosuppressive effects of this IL-10 family member. Glia. 2014;62(5):818–28.

80. Wanner IB, Anderson MA, Song B, Levine J, Fernandez A, Gray-Thompson Z, et al. Glial scar borders are formed by newly proliferated, elongated astrocytes that interact to corral inflammatory and fibrotic cells via STAT3-dependent mechanisms after spinal cord injury. J Neurosci. 2013;33(31):12870–86.

81. Herrmann JE, Imura T, Song B, Qi J, Ao Y, Nguyen TK, et al. STAT3 is a critical regulator of astrogliosis and scar formation after spinal cord injury. J Neurosci. 2008;28(28):7231–43.

82. Wang X, Deckert M, Xuan NT, Nishanth G, Just S, Waisman A, et al. Astrocytic A20 ameliorates experimental autoimmune encephalomyelitis by inhibiting NF-kappaB- and STAT1-dependent chemokine production in astrocytes. Acta Neuropathol. 2013;126(5):711–24.

83. Shao W, Zhang SZ, Tang M, Zhang XH, Zhou Z, Yin YQ, et al. Suppression of neuroinflammation by astrocytic dopamine D2 receptors via alphaB-crystallin. Nature. 2013;494(7435):90–4.

84. Steelman AJ, Smith R III, Welsh CJ, Li J. Galectin-9 protein is up-regulated in astrocytes by tumor necrosis factor and promotes encephalitogenic T-cell apoptosis. J Biol Chem. 2013;288(33):23776–87.

85. Sarafian TA, Montes C, Imura T, Qi J, Coppola G, Geschwind DH, et al. Disruption of astrocyte STAT3 signaling decreases mitochondrial function and increases oxidative stress in vitro. PLoS One. 2010;5(3):e9532.

86. Voskuhl RR, Peterson RS, Song B, Ao Y, Morales LB, Tiwari-Woodruff S, et al. Reactive astrocytes form scar-like perivascular barriers to leukocytes during adaptive immune inflammation of the CNS. J Neurosci. 2009;29(37):11511–22.

87. Toft-Hansen H, Fuchtbauer L, Owens T. Inhibition of reactive astrocytosis in established experimental autoimmune encephalomyelitis favors infiltration by myeloid cells over T cells and enhances severity of disease. Glia. 2011;59(1):166–76.

88. Liu Z, Li Y, Cui Y, Roberts C, Lu M, Wilhelmsson U, et al. Beneficial effects of gfap/vimentin reactive astrocytes for axonal remodeling and motor behavioral recovery in mice after stroke. Glia. 2014;62(12):2022–33.

89. Drogemuller K, Helmuth U, Brunn A, Sakowicz-Burkiewicz M, Gutmann DH, Mueller W, et al. Astrocyte gp130 expression is critical for the control of Toxoplasma encephalitis. J Immunol. 2008;181(4):2683–93.

90. Haroon F, Drogemuller K, Handel U, Brunn A, Reinhold D, Nishanth G, et al. Gp130-dependent astrocytic survival is critical for the control of autoimmune central nervous system inflammation. J Immunol. 2011;186(11):6521–31.

91. Middeldorp J, Hol EM. GFAP in health and disease. Prog Neurobiol. 2011;93(3):421–43.

92. Liedtke W, Edelmann W, Chiu FC, Kucherlapati R, Raine CS. Experimental autoimmune encephalomyelitis in mice lacking glial fibrillary acidic protein is characterized by a more severe clinical course and an infiltrative central nervous system lesion. Am J Pathol. 1998;152(1):251–9.

93. Macauley SL, Pekny M, Sands MS. The role of attenuated astrocyte activation in infantile neuronal ceroid lipofuscinosis. J Neurosci. 2011;31(43):15575–85.

94. Norden DM, Fenn AM, Dugan A, Godbout JP. TGFbeta produced by IL-10 redirected astrocytes attenuates microglial activation. Glia. 2014;62(6):881–95.

95. Cekanaviciute E, Dietrich HK, Axtell RC, Williams AM, Egusquiza R, Wai KM, et al. Astrocytic TGF-beta signaling limits inflammation and reduces neuronal damage during central nervous system toxoplasma infection. J Immunol. 2014;193(1):139–49.

96. Cekanaviciute E, Fathali N, Doyle KP, Williams AM, Han J, Buckwalter MS. Astrocytic transforming growth factor-beta signaling reduces subacute neuroinflammation after stroke in mice. Glia. 2014;62(8):1227–40.

97. Min KJ, Yang MS, Kim SU, Jou I, Joe EH. Astrocytes induce hemeoxygenase-1 expression in microglia: a feasible mechanism for preventing excessive brain inflammation. J Neurosci. 2006;26(6):1880–7.

98. Kostianovsky AM, Maier LM, Anderson RC, Bruce JN, Anderson DE. Astrocytic regulation of human monocytic/microglial activation. J Immunol. 2008;181(8):5425–32.

99. Beck H, Plate KH. Angiogenesis after cerebral ischemia. Acta Neuropathol. 2009;117(5):481–96.

100. Christopherson KS, Ullian EM, Stokes CC, Mullowney CE, Hell JW, Agah A, et al. Thrombospondins are astrocyte-secreted proteins that promote CNS synaptogenesis. Cell. 2005;120(3):421–33.

101. Zhang ZG, Chopp M. Neurorestorative therapies for stroke: underlying mechanisms and translation to the clinic. Lancet Neurol. 2009;8(5):491–500.

102. Benner EJ, Luciano D, Jo R, Abdi K, Paez-Gonzalez P, Sheng H, et al. Protective astrogenesis from the SVZ niche after injury is controlled by Notch modulator Thbs4. Nature. 2013;497(7449):369–73.

103. Andersson D, Wilhelmsson U, Nilsson M, Kubista M, Stahlberg A, Pekna M, et al. Plasticity response in the contralesional hemisphere after subtle neurotrauma: gene expression profiling after partial deafferentation of the hippocampus. PLoS One. 2013;8(7):e70699.

104. Winter CG, Saotome Y, Levison SW, Hirsh D. A role for ciliary neurotrophic factor as an inducer of reactive gliosis, the glial response to central nervous system injury. Proc Natl Acad Sci U S A. 1995;92(13):5865–9.

105. Horita Y, Honmou O, Harada K, Houkin K, Hamada H, Kocsis JD. Intravenous administration of glial cell line-derived neurotrophic factor gene-modified human mesenchymal stem cells protects against injury in a cerebral ischemia model in the adult rat. J Neurosci Res. 2006;84(7):1495–504.

106. Kobayashi T, Ahlenius H, Thored P, Kobayashi R, Kokaia Z, Lindvall O. Intracerebral infusion of glial cell line-derived neurotrophic factor promotes striatal neurogenesis after stroke in adult rats. Stroke. 2006;37(9):2361–7.

107. Ikeda T, Xia XY, Xia YX, Ikenoue T, Han B, Choi BH. Glial cell line-derived neurotrophic factor protects against ischemia/hypoxia-induced brain injury in neonatal rat. Acta Neuropathol. 2000;100(2):161–7.

108. Kitagawa H, Sasaki C, Zhang WR, Sakai K, Shiro Y, Warita H, et al. Induction of glial cell line-derived neurotrophic factor receptor proteins in cerebral cortex and striatum after permanent middle cerebral artery occlusion in rats. Brain Res. 1999;834(1–2):190–5.

109. Kokaia Z, Airaksinen MS, Nanobashvili A, Larsson E, Kujamaki E, Lindvall O, et al. GDNF family ligands and receptors are differentially regulated after brain insults in the rat. Eur J Neurosci. 1999;11(4):1202–16.

110. Wei G, Wu G, Cao X. Dynamic expression of glial cell line-derived neurotrophic factor after cerebral ischemia. Neuroreport. 2000;11(6):1177–83.

111. Zhang C, Li Y, Chen J, Gao Q, Zacharek A, Kapke A, et al. Bone marrow stromal cells upregulate expression of bone morphogenetic proteins 2 and 4, gap junction protein connexin-43 and synaptophysin after stroke in rats. Neuroscience. 2006;141(2):687–95.

112. Clarkson AN, Huang BS, Macisaac SE, Mody I, Carmichael ST. Reducing excessive GABA-mediated tonic inhibition promotes functional recovery after stroke. Nature. 2010;468(7321):305–9.

113. van Bruggen N, Thibodeaux H, Palmer JT, Lee WP, Fu L, Cairns B, et al. VEGF antagonism reduces edema formation and tissue damage after ischemia/reperfusion injury in the mouse brain. J Clin Invest. 1999;104(11):1613–20.

114. Carmichael ST. Plasticity of cortical projections after stroke. Neuroscientist. 2003;9(1):64–75.
115. Krakauer JW, Carmichael ST, Corbett D, Wittenberg GF. Getting neurorehabilitation right: what can be learned from animal models? Neurorehabil Neural Repair. 2012;26(8):923–31.
116. Nilsson M, Pekny M. Enriched environment and astrocytes in central nervous system regeneration. J Rehabil Med. 2007;39(5):345–52.
117. Pekna M, Pekny M, Nilsson M. Modulation of neural plasticity as a basis for stroke rehabilitation. Stroke. 2012;43(10):2819–28.
118. Hauck SM, von Toerne C, Ueffing M. The neuroprotective potential of retinal Muller glial cells. Adv Exp Med Biol. 2014;801:381–7.

Chapter 7
Oxidative Stress and Nitric Oxide in Cerebral Ischemic Reperfusion Injury

Junning Ma, Zhong Liu, and Zhongsong Shi

Abstract Cerebral ischemic reperfusion injury is a heterogeneous phenomenon with a multi-factorial etiology, and characterized as a cascade of neurochemical processes evolving in time and space after restriction or sudden interruption of cerebral blood flow. It has been suggested that oxidative stress and nitrosative stress are important mechanisms in cerebral ischemic reperfusion. The concept of oxidative and nitrosative stress stem from the generation of the reactive oxygen species (ROS) involving the nicotinamide adenine dinucleotide phosphate (NADPH) oxidases (NOX) family and the reactive nitrogen species (RNS) including nitric oxide (NO) and peroxynitrite ($ONOO^-$) at rates which exceed the capacity of natural antioxidant and anti-nitrification defense mechanisms to detoxify these toxic products. This review is focusing on the role of oxidative and nitrosavtive stress in cerebral ischemic reperfusion injury by discussing the concepts, the mechanisms, and the pharmacological approaches of ROS and RNS modulation for preventing cerebral ischemic reperfusion injury.

Keywords Cerebral ischemic reperfusion injury · Oxidative stress

1   Introduction

Cerebral ischemic reperfusion injury is a heterogeneous phenomenon with a multi-factorial etiology [1]. It is characterized as a cascade of neurochemical processes evolving in time and space after restriction or sudden interruption of cerebral blood flow [2]. Insults of ischemic reperfusion trigger various stress responses caused by numerous molecular processes in the brain [2]. Oxidative and nitrosative stress have been identified playing significant roles in ischemic cell death within the penumbra. These stresses are partly downstream consequences of excitotoxicity, stemming

J. Ma · Z. Liu · Z. Shi (✉)
Department of Neurosurgery, Sun Yat-sen Memorial Hospital of Sun Yat-sen University, Guangzhou, China
e-mail: shizhs@mail.sysu.edu.cn

© Springer International Publishing AG, part of Springer Nature 2018
W. Jiang et al. (eds.), *Cerebral Ischemic Reperfusion Injuries (CIRI)*, Springer Series in Translational Stroke Research, https://doi.org/10.1007/978-3-319-90194-7_7

101

from an increase in secondary messenger systems paired to the free radicals generated form enzymes [3]. After cerebral ischemic reperfusion injury, favoured by a calcium overload in the cells, there is an increased production of reactive oxygen species (ROS) and reactive nitrogen species (RNS), including superoxide anion ($O_2\cdot-$), hydrogen peroxide (H_2O_2), hydroxyl radical (OH^-), nitric oxide ($NO\cdot$), or peroxynitrite ($OONO^-$) in the mitochondria of the penumbral tissue [4, 5]. Since the ischemic brain is highly susceptible to oxidative and nitrosative damage, the prevention of the effects of ROS and RNS is a potential therapeutic strategy for the cerebral ischemic reperfusion injury [4, 6].

2   Oxidative Stress in Cerebral Ischemic Reperfusion Injury

2.1   Reactive Oxygen Species

ROS is a collective term that defines the oxygen-derived small molecules that immediately react with various chemical substances and includes the superoxide anion ($O_2\cdot^-$), hydrogen peroxide (H_2O_2), and the hydroxyl radical ($HO\cdot$) [7, 8]. The generation of ROS is ubiquitous in many cellular respiration and metabolic processes. An electron is transferred to oxygen in the process of ROS generation, forming super-oxides ($O_2\cdot^-$). The process occurs mainly in mitochondria of cells and ROS is produced by peroxisomes and variety of enzymes (cyclooxygenase, lipoxygenases, xanthine oxidase, oxide synthase, P450 cytochromes, and enzymes in the mitochondrial election transfer chain), and NADPH is the dominant donor of electron in the reaction catalyzed by NADPH oxidases [9, 10].

Initially, ROS has been thought as critical mediators regulating the homeostasis and inflammatory responses of cells [11]. According to the "free radicals theory of aging" proposed by Harman in 1956, free radicals could be produced during ordinary cellular respiration and could contribute to tissue injury [12]. Mann [13] has found that hyperoxia and H_2O_2 have toxic effect on central nervous system in 1946. Moreover, ROS also operates as intracellular signaling molecules that regulates various intracellular pathways mediating cellar growth, differentiation, proliferation, apoptosis, migration, contraction, and cytoskeletal regulation [9].

Recently, the discovery of NADPH oxidases (NOX), which generates ROS and consume oxygen, put a challenge on the former paradigm. Then a new paradigm has been raised that regulated ROS generation plays a key role in physiology and cell signaling, while only excessive levels of ROS are injurious. This theory was verified by the evidences from most organ systems, including the central nervous system [14, 15]. Therefore, normal ROS generation regulating the physiological cellular processes, may be regard to as "redox regulation" while uncontrolled ROS generation causing tissue injury may be related to as "oxidative stress" [7].

2.2 The Oxidative Stress Hypothesis

Oxidative stress can be described as a relative overload of reactive oxygen species resulting from inordinate ROS production and/or injured ROS degradation [16]. It is characterized by a transform from a cellular environment in which the metabolism and generation of oxidant substances are balanced to one of the raised levels of the same molecules [10]. The unbalanced relationship between oxidant generation mechanism and anti-oxidant production mechanism are resulted from an overproduction of ROS, such as superoxide (O_2^-), or a decrease in the removal of ROS. The main anti-oxidants contain catalase, glutathione peroxidases, and superoxide dismutases (SOD). After the balance between oxidant generation system and oxidant defense system had been disturbed by pathological condition, the surplus of ROS becomes adverse and plays a key role in various acute and chronic diseases that affect the vasculature [10, 17]. It also has important impacts on the innate immune response, cellular signaling, vascular tone, oxygen sensing, and angiogenesis [18].

The hypothesis that ROS are involved in cerebral ischemic reperfusion injury can be dated back to the 1970s [19]. Moreover, many lines of evidence have shown that oxidative stress is the central element of the pathological development after cerebral ischemic reperfusion injury [20–23]. In addition to other consequences, one of the significant effects of oxidative stress is injured signaling of the vasoprotective molecule nitric oxide (NO), leading to changes in vascular structure and function [10]. After an ischemic injury, microglial cells and recruited phagocytes synthesized higher NOX2, and an increased expression of NOX4 could be found in neuronal cells and brain microvascular endothelial cells [24–26].

Because of low oxidative defense capacity and high oxygen consumption, our brain is especially sensitive to oxidative stress [27]. In cerebral ischemic reperfusion injury, neuronal damage and cellular energy failure is caused by oxygen deficiency and ischemia-derived nutrient, and damages driven by oxidative stress could be even aggravated by reperfusion because the freshly arriving oxygen will enhance new ROS production [16, 28]. Therefore, using antioxidant therapies to scavenge ROS, which is based on the oxidative stress hypothesis, has been performed in clinical trials recently. Unfortunately, these studies failed to perform a significant improvement in stroke patents even though preclinical results are promising [29, 30]. However, the failure of antioxidant trials, especially in stroke patients, does not prove that ROS is irrelevant to ischemia-reperfusion injury [31]. On the contrary, the reasons of this failure was that focus on single ROS species has limited relevance because of the effects of interactions and secondary reactions among other ROS species [32]. Thus, there are urgent demands to find alternative approaches that relevant molecular sources of oxidative stress are more focused on and to seek a much deeper understanding of the underlying mechanisms of the oxidative stress.

2.3 Main Sources of Oxidative Stress in Cerebral Ischemic Reperfusion Injury

Activity of the mitochondrial respiratory chain is one of the major cellular source of ROS generation. During the process of the adenosine triphosphate (ATP) and water generation, when mitochondria consumes an oxygen, unpaired electrons escape from the respiratory complexes I and III. Then oxygen radicals are formed by this escape of electrons causing the damage of biological molecules [7, 33]. Because the consumption of oxygen in brain accounts for 20% of total body oxygen consumption, mitochondria are thought as the basic source of ROS in brain aging and necrotic and apoptotic cell death [34–36].

In addition to the mitochondrial election transport chain, numerous enzymes within the vasculature is another source of ROS. These include cyclooxygenases [37, 38], uncoupled nitric oxide synthase (NOS) [39–41], xanthine oxidase [37, 42, 43], lipoxygenases, peroxisome, heme oxygenase, and monoamine oxidases, among others [44]. These enzymes generate ROS when the protein is in a dysfunctional state or as a by-product of normal enzyme activity [7, 10, 17]. By contrast, the NADPH oxidases from the family of ROS-producing enzymes, their fundamental function is known for generating ROS [45]. It is the source of deliberate ROS production required for normal signaling and for the excessive ROS associated with oxidative stress [10, 45]. Indeed, the process of generation of ROS by NADPH oxidases contains a cascade of events including feed-forward mechanisms and recruitment of other enzymes that promote further oxidative stress [46, 47].

2.4 NADPH Oxidase and Expression of NADPH in the Central Nervous System

NADPH oxidase is a kind of multicomponent complexes which is containing a catalytic NOX subunit that transfers electrons form NADPH to oxygen to produce ROS [48]. These subunits include another membrane protein, cytosolic protein, and P22phox. The activity of NADPH oxidase was discovered initially in studies of neutrophils, where NOX2 plays a role in immunological host defense [49, 50]. Five NOX isoforms have been identified in most mammals: NOX1 to NOX5, and two additional proteins, dual oxidase (DUOX)1 and DUOX2, have been found that contain an oxidase domain. Unlike other NOX isoforms, NOX3 may be irrelevant to cerebral ischemic reperfusion injury because the main expression of it was found in the inner ear but not in the vasculature [16, 51]. NOX1, NOX2, NOX4, NOX5 are all expressed in the vasculature, and it should be noted that NOX5 is not expressed in rodents [26, 32, 51, 52]. In the cerebrovascular system, evidence has shown that NOX1, NOX2, and NOX4 catalytic subunits are expressed both in rat basilar artery and cerebral arteries [52, 53].

The specific cellular localization of different NOX isoforms has been studied widely over the past decade. Several studies have revealed that NOX1, NOX2, and NOX4 are expressed in endothelial cells, and the expression of NOX2 and NOX4 have been identified in smooth muscle cells, and NOX2 and NOX4 are expressed in adventitia [10, 45]. Moreover, evidence has also shown that NOX1, NOX2, and NOX4 are expressed in neurons and that lower expression of NOX1 and NOX2 but higher expression of NOX4 can be detected in astrocyte [54, 55]. Furthermore, it has also been documented NOX2 is mainly expressed in endothelia and adventitial cells of cerebral arteries as well as in the activated state of microglial cells [56–60]. In addition to NOX2, NOX1 and NOX4 protein is also expressed in microglial cells [61, 62].

On the contrary, there has been little interest in the specific subcellular localization of NADPH oxidases in cerebral vessels. NOX1 has been shown to be expressed in the plasma of membrane of the vascular smooth muscle cells [63]. NOX2 has shown to be expressed in the cytoplasm, lipid raft rich fraction, and the plasma membrane of the endothelial cells [56, 64]. NOX3 has been found in the specific cellular membraneous compartments [65]. NOX4 has mostly been found in the endoplasmic reticulum and the nuclear envelope as well as mitochondria [66–69]. DUOX enzymes has shown to be expressed at the apical membranes of epithelial cells [70].

2.5   The Role of NADPH Oxidase in Cerebral Ischemic Reperfusion Injury

NADPH oxidase plays an important role in the cerebral ischemic reperfusion injury. Evidence based on protein and mRNA level have confirmed that NOX2 and NOX4 largely contribute to the pathologic development in ischemic stroke [71–73]. After an ischemic stroke, expression of NOX2 and NOX4 protein are up-regulated from 24 to 72 h within the microglia, neurons and endothelia cells respectively [26, 74]. This up-regulation reveals the detrimental effects of NOX2 and NOX4 activation after ischemic stroke; therefore, pave the way for treatments that are targeting the NOX2 and NOX4 isoforms pharmacologically. Moreover, evidence from NOX2 knockout mice has shown a strong neurological protection after ischemic stroke by improving the blood-brain barrier (BBB) permeability and reducing the post-stroke brain swelling [25, 75]. Similarly, this neurological protection effect has also found in mice deficient for NOX4 [71]. This study supposes that NOX4-mediated oxidative stress result in neuronal injury via leakage of neuronal apoptosis and blood-brain barriers, and experimental data has shown that deficiency of NOX4 can reduce post-stroke mortality and improve neurological functions [71]. In contrast to NOX2 and NOX4, the role of NOX1 in ischemic stroke remains controversial. Several studies have pointed out deficiency of NOX1 has no impact on infract volume, neurological outcome and post-stroke ROS generation [71, 76]. However, neurological protective effect of NOX1 deficiency was identified in another study [77]. The reason of these divergent results might stem from different experimental protocols. Nevertheless, pharmacological approaches to inhibit NOX activity, specifically using NOX4 inhibitors, have become a new treatment strategy for ischemic stroke.

2.6   Pharmacological Approaches to Inhibit NOX Activity

2.6.1   Antioxidants Therapies

Before the emersion of NOX-based strategies, the administration of antioxidant molecules was the very first promising therapeutic approach targeting oxidative stress in the central nervous system. However, clinical studies have shown that this approach failed to reduce the oxidative injury because rather than block the whole process of ROS generation, antioxidants could only capture ROS that have already produced [78, 79]. Indeed, the varied origin of ROS, the time when ROS was released, and the complicated interaction between ROS and physiological cell function also contribute to this failure [9, 79]. Moreover, another limitation of antioxidants therapies is associated with their bioavailability. Several cross-sectional studies have confirmed that the dose of 100 mg/day of vitamin C rarely reach to potentially therapeutic levels in plasma concentrations [80–82].

2.6.2   Non-specific NOX Inhibitors

Apocynin and diphenylene iodonium (DPI) have been used as non-specific NOX inhibitors widely, and both of them can be useful when their mechanisms of action and limitation are well noticed.

Apocynin is a polyphenolic molecule and characterized by its anti-inflammatory functions. The mechanisms of apocynin action is supposed to block the assembly of NOX2 and p47^{phox}, and has to be activated by myeloperoxidases [16]. It is most widely used for in vivo studies because a remarkable inhibition of oxidative stress could be triggered by relatively low amounts of apocynin [83]. However, the limitation of apocynin is only extremely high concentrations of it can inhibit ROS production in vitro because of its antioxidant capacity and inhibition of Rho kinase [84, 85].

Diphenylene iodonium (DPI) is a potent non-specific inhibitor of the NADPH oxidase which inhibits unspecific electron transfer chains by formatting an iodonium radical that block the electron transporter irreversibly [86–88]. The mechanisms of DPI action involve NOX2 inhibition, leukocyte recruitment, inflammatory microglial response, and MMP activity within the ischemic brain [89–92]. DPI has been shown neuroprotective effects in vivo when it was administered after ischemic stroke; however, using DPI in clinic is limited because it is highly insoluble and toxic [90, 93].

2.6.3 Specific NOX Inhibitors

In contrast to using antioxidants and non-specific NOX inhibitors, specific NOX inhibitors, which can inhibit the relevant sources of ROS in different pathologies, are in dire need. Several researcher groups have already started embarking on the production of specific NOX inhibitors [94].

The triazolo pyrimidine VAS2870, 3-benzyl-7-(2-benzoxazolyl) thio-1,2,3-triazolo (4,5-d) pyrimidine, is a more specific NOX inhibitor that has already shown its strong protective effects in preclinical stroke studies [71, 95, 96]. VAS2870 and its derivate VAS3947 was thought to likely inhibit all NOX isoforms by inhibiting assembly or conformation changes to active NOX complexes [66, 93, 97]. It can inhibit NADPH oxidase activity in platelet-derived growth factor (PDGF)-stimulated primary rat aortic vascular smooth muscle cell and oxLDL-exposed human endothelial cell at 10 μM and inhibit the H_2O_2 generation around wound margin without apparent toxicity in zebrafish larvae [95–97]. It should be noted that the therapeutically relevant time window of intrathecal treatment with VAS2870 is 2 h after cerebral ischemic reperfusion injury [74].

The pyrazolopyridine derivate GKT136901, 2-(2-chlorophenyl)-4-methyl-5-(pyridin-2-ylmethyl)-1H-pyrazolo [4,3-c] pyridine-3,5(2H,5H)-dione, and its close analogue GKT137831 are recently introduced as a dual NOX1/4 inhibitor [98]. They can inhibit NADPH oxidase activity, TGF-β1/2 and fibronectin induction, and p38 MAP kinase activation in mouse proximal tublar cells at 10 μM and inhibit intracellular ROS formation, thrombin-induced CD44, and HAS2 protein and mRNA levels in human aortic smooth muscle cells at 30 μM [99, 100]. Moreover, their oral bioavailability and little impact on plasma total triglyceride or cholesterol levels and body weight make these compounds the most suitable for a clinical translation [100].

In addition to VAS2870 and GKT136901, many novel inhibitors, such as Ebselen, S17834, M171, Fulvene-5 were introduced recently by their specific pharmacological functions [101–104]. However, although the first and promising data in vivo of them have been published, more detailed understanding about their functions is still needed.

3 Nitrosative Stress in Cerebral Ischemic Reperfusion Injury

3.1 Reactive Nitrogen Species

Reactive nitrogen species (RNS) refers a family of free radicals derived from nitric oxide (·NO) and superoxide ($O_2^{·-}$) generated via the enzymatic activity of inducible nitric aynthase (NOS) and NADPH oxidase respectively [105]. The main reactive nitrogen species are comprised of nitric oxide (NO), peroxynitrite ($ONOO^-$), dinitrogen trioxide (N_2O_3), and nitrosonium ion (NO^+) [106, 107]. In most case, NO and $ONOO^-$ are the crucial species of RNS during cerebral ischemic reperfusion injury because NO is generated simultaneously with $O_2^{·-}$ and rapidly reacts with $O_2^{·-}$ to produce $ONOO^-$, and both NO and $ONOO^-$ contribute to brain damage and BBB breakdown by inducing the influx of substances into brain parenchyma and mediating the degradation of tight juctions in the BBB [107].

3.2   Main Sources of Nitric Oxide

Nitric oxide (NO) is the signaling molecule primitively known as endothelium-derived relaxing factor (EDRF) regulating relaxation of blood vessels [108]. It is a small gaseous molecule synthesised from L-arginine, which controls vascular tone and neurotransmission, induces protein post-translational modification, and regulates mRNA translation and gene transcription [109]. Nitric oxide is synthesized by three isoforms of the enzyme NOS: endothelial nitric oxide synthase (eNOS), neuronal nitric oxide synthase (nNOS), and inducible nitric oxide synthase (iNOS). NO derived from eNOS has physiological functions including preserving and maintaining the brain's microcirculation, reducing smooth muscle proliferation, and inhibiting platelet aggregation [110, 111]. NO derived from nNOS acts as a neurotransmitter that plays a role in neuronal plasticity, transmission of pain signals, memory formation, neurotransmitter release, and regulation of central nervous system blood flow [110, 112]. NO derived from iNOS also has multiple functions including contribution to the neurotoxic actions after ischemic stroke and traumatic brain injury and regulation of cerebral blood flow [113–115]. In contrast to eNOS and nNOS, which are calcium-dependent and produce nanomolar levels of NO, iNOS is calcium-independent and produces micromolar levels of NO [116]. The physiological concentration of NO derived from eNOS is vital to communication of neuron, synaptic transmission, regulation of vascular tone, inflammatory responses, platelet aggregation while the high concentration of NO generated from iNOS and nNOS is detrimental to the ischemic brain [107, 117–119].

3.3   Roles of RNS in Cerebral Ischemic Reperfusion Injury

NO and $ONOO^-$ are two common species of RNS that are representative in cerebral ischemic reperfusion injury. In the early phase of ischemic stroke, transient shortage of the blood supply results in the increased eNOS activity that generates low concentration of NO to protect the brain vasculature [120]. At the same time, energy crisis caused by ischemia induces the glutamate accumulation and triggers the activation of calcium channels that can stimulate nNOS to produce NO [121, 122]. In the early stage of reperfusion, there is a transient raise of stable NO metabolites, and the up-regulated expression of iNOS lead to excessive NO generation [121, 122]. Moreover, there are two stage of NO generation after ischemic stroke, and both of stages are correlated with increased iNOS and nNOS respectively [123]. The first stage of NO generation was after 1 h of ischemic stroke while the second stage occurred at 24–48 h of reperfusion after 1 h of ischemic stroke [123]. This increased generation of NO, which was derived from nNOS and iNOS, are neurotoxic and contribute to cell death and BBB disruption. In contrast to the high concentration of NO from iNOS and eNOS, the low concentration of NO generated from eNOS exerts neuroprotective effects. Evidence has shown that infarction volumes of eNOS knockout mice is significantly larger than those of wild-type mice after ischemic stroke [124]. Similarly, using medications, which could increase eNOS activity, has shown the same results in ischemic animal model [124]. The mechanisms of neuroprotective effects by these medications refer to the reduction of thrombosis formation, the enhancement of vasorelaxation, elevation of cerebral blood flow, the suppression of NMDA receptor activation, and the improvement of vasorelaxation [124–126].

Likewise, the overproduction of peroxynitrite also plays a detrimental role in neurons and endothelial cells during ischemic reperfusion injury [127–129]. In physiological status, the reaction between NO and O_2^- that generates peroxynitrite remains at a diffusion controlled level [130]. During both the ischemic stage and the reperfusion stage, the generation of $ONOO^-$ is extremely increased because of the dramatically rapid generation of NO. Coincidently, evidence from blood samples of ischemic stroke has discovered that the increase of $ONOO^-$ concentration occurs at 24 and 48 h after ischemic stroke [131, 132]. Because the penetrating capacity of $ONOO^-$ across lipid bilayers is 400 times higher than its parent radical superoxide anions approximately, $ONOO^-$ is far more neurotoxic than NO. Indeed, the mechanisms of cytotoxic effects of $ONOO^-$ include lipid membrane peroxidation, protein tyrosine nitration, induction of mitochondrial dysfunction, DNA breakage caused by PARP activation, enzymatic activity inhibition, signal transduction dysfunction, and cytoskeletal disruption by altering protein structure and dysfunction [107, 133–136]. In addition to its neurotoxic effect, the overproduction of $ONOO^-$ also leads to the BBB breakdown by mediating the activation of MMPs, the degradation of the tight junction proteins, and the rearrangement of the tight junction proteins [137–141].

3.4 Pharmacological Approaches to Regulate RNS Activity

Because RNS play crucial roles in regulating different functions both in physiological and pathological states, RNS could be potential drug targets for the treatment of ischemic stroke. Pharmacological approaches by targeting RNS has been well developed in NO-based therapeutic strategies; however, because of the technical limitation of detection, drug development by targeting $ONOO^-$ is much slower [107].

The fundamental aim of NO-based therapeutic strategies is to establish balanced concentration of NO by increasing level of NO derived from eNOS and reducing the cytotoxic level of NO derived from iNOS and nNOS. There are three approaches could regulate the NO activity.

3.4.1 Increasing eNOS Activity

One of the promising agents, which could enhance the eNOS activity, is statins. It can improve expression of eNOS via LDL-dependent and independent pathways [142, 143]. Researches in experimental animal models have shown that statins can reduce both the edema formation and infraction volume after ischemic stroke [144, 145]. Deficiency of the protective effect of statin in eNOS knockout mice has confirmed that the protective effects of stain are eNOS-dependent [146, 147]. Moreover, based on its beneficial effects, supported by abundant preclinical and clinical studies, statins are one of the recommended drug to prevent stroke [148, 149]. Furthermore, the antioxidant characteristics of statins also contribute to the reduction of oxidative stress in brain after ischemic stroke [150]. However, it has been reported that using stain may lead to increased risk of hemorrhagic stroke and higher incidence of infection [151–153]. Thus, when perform stain treatment, the negative side effects of stain should not be ignored because of its multiple pharmacological activities.

3.4.2 nNOS and iNOS Inhibition

L-NAME, a non-selective NOS inhibitor, has been proven to be beneficial for cerebral ischemic stroke injury in experimental mouse models including prevention of BBB breakdown, reduction of infarction volume, and improvement of the recovery of neurological functions [154, 155]. Moreover, because L-NAME can increase the eNOS activity, it also can exert its protective effects for ischemic stroke via eNOS activation.

Delta-(S-methylisothioureido)-L-norvaline (L-MIN), a specific nNOS inhibitor, has been shown to be able to reduce infraction volume in animal stroke models [156]. Other typical experimental inhibitors of nNOS, such as ARL-17477, tirilazad, 7-nitroindazole, BN80933, and PPBP were also reported to decrease the neurological deficits and infarct volume of animal ischemic stroke models [157–161].

1400 W and aminoguanidine, specific iNOS inhibitors, have been shown that can reduce the infraction volume and attenuate ischemic brain injury [113]. The use of 1400 W and aminoguanidine to inhibit iNOS activity has been proposed to be a valuable approach for human stroke because the expression of iNOS in human brain has been identified after ischemic stroke and the time window for administration of iNOS ihhibitors is longer than other treatments [162].

3.4.3 Increasing Substrates of NO Production

NO donors and substrates of eNOS has been utilized to ameliorate the prognosis of the patients suffering with ischemic stroke for a long time. L-Arginine, a NO precursor, has shown to be able to decrease the infract volume, to enhance the blood flow, and to improve the neurological function in rat ischemic stroke models [163, 164]. However, the clinical application of the L-arginine is limited because of the risk of reducing the blood flow to the ischemic penumbra [165].

4   Conclusion

Stroke is the second leading cause of death and long-term disability of adults in human disease [166]. However, the accumulation of failure of neuroprotective drugs in the clinical trials for patients with ischemic stroke indicates that new pharmacological strategies are needed both in the preclinical and clinic field. Cerebral ischemic reperfusion injury is a deleterious, but salvageable, aggravation of an ischemic injury after reperfusion [1]. During the reperfusion, ROS and RNS, which are very important mediators of BBB breakdown, neurotoxicity, cell death, and tissue damage, are mainly produced in the ischemic penumbra [167, 168]. Therefore, the prevention of the effects of ROS and RNS might be a potential therapeutic strategy for the cerebral ischemic reperfusion injury. Indeed, although most drugs remain at the preclinical stage, some of candidates have already been employed in clinical trials. After gained more understanding about the mechanisms of ROS and RNS in cerebral ischemia, it will be possible to develop novel pharmacological strategies for cerebral ischemic reperfusion injury.FundingThis work was supported by National Natural Science Foundation of China (8171001013).

References

1. Pan J, Konstas AA, Bateman B, Ortolano GA, Pile-Spellman J. Reperfusion injury following cerebral ischemia: pathophysiology, MR imaging, and potential therapies. Neuroradiology. 2007;49(2):93–102.
2. Chomova M, Zitnanova I. Look into brain energy crisis and membrane pathophysiology in ischemia and reperfusion. Stress. 2016;19(4):341–8.
3. Chamorro A, Dirnagl U, Urra X, Planas AM. Neuroprotection in acute stroke: targeting excitotoxicity, oxidative and nitrosative stress, and inflammation. Lancet Neurol. 2016;15(8):869–81.
4. Fukuyama N, Takizawa S, Ishida H, Hoshiai K, Shinohara Y, Nakazawa H. Peroxynitrite formation in focal cerebral ischemia-reperfusion in rats occurs predominantly in the peri-infarct region. J Cereb Blood Flow Metab. 1998;18(2):123–9.
5. Pacher P, Beckman JS, Liaudet L. Nitric oxide and peroxynitrite in health and disease. Physiol Rev. 2007;87(1):315–424.
6. Deng ZF, Rui Q, Yin X, Liu HQ, Tian Y. In vivo detection of superoxide anion in bean sprout based on ZnO nanodisks with facilitated activity for direct electron transfer of superoxide dismutase. Anal Chem. 2008;80(15):5839–46.
7. Nayernia Z, Jaquet V, Krause KH. New insights on NOX enzymes in the central nervous system. Antioxid Redox Signal. 2014;20(17):2815–37.
8. D'Autreaux B, Toledano MB. ROS as signalling molecules: mechanisms that generate specificity in ROS homeostasis. Nat Rev Mol Cell Biol. 2007;8(10):813–24.
9. Carbone F, Teixeira PC, Braunersreuther V, Mach F, Vuilleumier N, Montecucco F. Pathophysiology and treatments of oxidative injury in ischemic stroke: focus on the phagocytic NADPH oxidase 2. Antioxid Redox Signal. 2015;23(5):460–89.
10. De Silva TM, Faraci FM. Effects of angiotensin II on the cerebral circulation: role of oxidative stress. Front Physiol. 2012;3:484.
11. Kvietys PR, Granger DN. Role of reactive oxygen and nitrogen species in the vascular responses to inflammation. Free Radic Biol Med. 2012;52(3):556–92.

12. Harman D. Aging: a theory based on free radical and radiation chemistry. J Gerontol. 1956;11(3):298–300.
13. Mann PJ, Quastel JH. Toxic effects of oxygen and of hydrogen peroxide on brain metabolism. Biochem J. 1946;40(1):139–44.
14. Burgoyne JR, Mongue-Din H, Eaton P, Shah AM. Redox signaling in cardiac physiology and pathology. Circ Res. 2012;111(8):1091–106.
15. Milton VJ, Sweeney ST. Oxidative stress in synapse development and function. Dev Neurobiol. 2012;72(1):100–10.
16. Radermacher KA, Wingler K, Langhauser F, Altenhofer S, Kleikers P, Hermans JJR, de Angelis MH, Kleinschnitz C, Schmidt HHHW. Neuroprotection after stroke by targeting NOX4 as a source of oxidative stress. Antioxid Redox Sign. 2013;18(12):1418–27.
17. Radermacher KA, Wingler K, Langhauser F, Altenhofer S, Kleikers P, Hermans JJ, Hrabe de Angelis M, Kleinschnitz C, Schmidt HH. Neuroprotection after stroke by targeting NOX4 as a source of oxidative stress. Antioxid Redox Signal. 2013;18(12):1418–27.
18. Halliwell B. Phagocyte-derived reactive species: salvation or suicide? Trends Biochem Sci. 2006;31(9):509–15.
19. Bagenholm R, Nilsson UA, Gotborg CW, Kjellmer I. Free radicals are formed in the brain of fetal sheep during reperfusion after cerebral ischemia. Pediatr Res. 1998;43(2):271–5.
20. Chan PH. Reactive oxygen radicals in signaling and damage in the ischemic brain. J Cereb Blood Flow Metab. 2001;21(1):2–14.
21. Geng X, Li F, Yip J, Peng C, Elmadhoun O, Shen J, Ji X, Ding Y. Neuroprotection by chlorpromazine and promethazine in severe transient and permanent ischemic stroke. Mol Neurobiol. 2017;54(10):8140–50.
22. Cai L, Thibodeau A, Peng C, Ji X, Rastogi R, Xin R, Singh S, Geng X, Rafols JA, Ding Y. Combination therapy of normobaric oxygen with hypothermia or ethanol modulates pyruvate dehydrogenase complex in thromboembolic cerebral ischemia. J Neurosci Res. 2016;94(8):749–58.
23. Jung YS, Lee SW, Park JH, Seo HB, Choi BT, Shin HK. Electroacupuncture preconditioning reduces ROS generation with NOX4 down-regulation and ameliorates blood-brain barrier disruption after ischemic stroke. J Biomed Sci. 2016;23:32.
24. Green SP, Cairns B, Rae J, Errett-Baroncini C, Hongo JA, Erickson RW, Curnutte JT. Induction of gp91-phox, a component of the phagocyte NADPH oxidase, in microglial cells during central nervous system inflammation. J Cereb Blood Flow Metab. 2001;21(4):374–84.
25. Kahles T, Luedike P, Endres M, Galla HJ, Steinmetz H, Busse R, Neumann-Haefelin T, Brandes RP. NADPH oxidase plays a central role in blood-brain barrier damage in experimental stroke. Stroke. 2007;38(11):3000–6.
26. Vallet P, Charnay Y, Steger K, Ogier-Denis E, Kovari E, Herrmann F, Michel JP, Szanto I. Neuronal expression of the NADPH oxidase NOX4, and its regulation in mouse experimental brain ischemia. Neuroscience. 2005;132(2):233–8.
27. Flamm ES, Demopoulos HB, Seligman ML, Poser RG, Ransohoff J. Free radicals in cerebral ischemia. Stroke. 1978;9(5):445–7.
28. Chan PH. Role of oxidants in ischemic brain damage. Stroke. 1996;27(6):1124–9.
29. Shuaib A, Lees KR, Lyden P, Grotta J, Davalos A, Davis SM, Diener HC, Ashwood T, Wasiewski WW, Emeribe U, et al. NXY-059 for the treatment of acute ischemic stroke. N Engl J Med. 2007;357(6):562–71.
30. Lees KR, Zivin JA, Ashwood T, Davalos A, Davis SM, Diener HC, Grotta J, Lyden P, Shuaib A, Hardemark HG, et al. NXY-059 for acute ischemic stroke. N Engl J Med. 2006;354(6):588–600.
31. Diener HC, Lees KR, Lyden P, Grotta J, Davalos A, Davis SM, Shuaib A, Ashwood T, Wasiewski W, Alderfer V, et al. NXY-059 for the treatment of acute stroke: pooled analysis of the SAINT I and II Trials. Stroke. 2008;39(6):1751–8.
32. Wingler K, Hermans JJ, Schiffers P, Moens A, Paul M, Schmidt HH. NOX1, 2, 4, 5: counting out oxidative stress. Br J Pharmacol. 2011;164(3):866–83.

33. Sas K, Robotka H, Toldi J, Vecsei L. Mitochondria, metabolic disturbances, oxidative stress and the kynurenine system, with focus on neurodegenerative disorders. J Neurol Sci. 2007;257(1–2):221–39.
34. Federico A, Cardaioli E, Da Pozzo P, Formichi P, Gallus GN, Radi E. Mitochondria, oxidative stress and neurodegeneration. J Neurol Sci. 2012;322(1–2):254–62.
35. Gomez-Cabrera MC, Sanchis-Gomar F, Garcia-Valles R, Pareja-Galeano H, Gambini J, Borras C, Vina J. Mitochondria as sources and targets of damage in cellular aging. Clin Chem Lab Med. 2012;50(8):1287–95.
36. Beal MF. Energetics in the pathogenesis of neurodegenerative diseases. Trends Neurosci. 2000;23(7):298–304.
37. Didion SP, Hathaway CA, Faraci FM. Superoxide levels and function of cerebral blood vessels after inhibition of CuZn-SOD. Am J Phys Heart Circ Phys. 2001;281(4):H1697–703.
38. Niwa K, Haensel C, Ross ME, Iadecola C. Cyclooxygenase-1 participates in selected vasodilator responses of the cerebral circulation. Circ Res. 2001;88(6):600–8.
39. Vasquez-Vivar J, Martasek P, Hogg N, Masters BS, Pritchard KA Jr, Kalyanaraman B. Endothelial nitric oxide synthase-dependent superoxide generation from adriamycin. Biochemistry. 1997;36(38):11293–7.
40. Santhanam AV, d'Uscio LV, Smith LA, Katusic ZS. Uncoupling of eNOS causes superoxide anion production and impairs NO signaling in the cerebral microvessels of hph-1 mice. J Neurochem. 2012;122(6):1211–8.
41. Narayanan D, Xi Q, Pfeffer LM, Jaggar JH. Mitochondria control functional CaV1.2 expression in smooth muscle cells of cerebral arteries. Circ Res. 2010;107(5):631–41.
42. Warner DS, Sheng H, Batinic-Haberle I. Oxidants, antioxidants and the ischemic brain. J Exp Biol. 2004;207(Pt 18):3221–31.
43. Kinugawa S, Huang H, Wang Z, Kaminski PM, Wolin MS, Hintze TH. A defect of neuronal nitric oxide synthase increases xanthine oxidase-derived superoxide anion and attenuates the control of myocardial oxygen consumption by nitric oxide derived from endothelial nitric oxide synthase. Circ Res. 2005;96(3):355–62.
44. Braunersreuther V, Jaquet V. Reactive oxygen species in myocardial reperfusion injury: from physiopathology to therapeutic approaches. Curr Pharm Biotechnol. 2012;13(1):97–114.
45. Drummond GR, Selemidis S, Griendling KK, Sobey CG. Combating oxidative stress in vascular disease: NADPH oxidases as therapeutic targets. Nat Rev Drug Discov. 2011;10(6):453–71.
46. Faraci FM, Lamping KG, Modrick ML, Ryan MJ, Sigmund CD, Didion SP. Cerebral vascular effects of angiotensin II: new insights from genetic models. J Cerebr Blood F Met. 2006;26(4):449–55.
47. Selemidis S, Sobey CG, Wingler K, Schmidt HHHW, Drummond GR. NADPH oxidases in the vasculature: molecular features, roles in disease and pharmacological inhibition. Pharmacol Ther. 2008;120(3):254–91.
48. Bedard K, Krause KH. The NOX family of ROS-generating NADPH oxidases: physiology and pathophysiology. Physiol Rev. 2007;87(1):245–313.
49. Schappi MG, Jaquet V, Belli DC, Krause KH. Hyperinflammation in chronic granulomatous disease and anti-inflammatory role of the phagocyte NADPH oxidase. Semin Immunopathol. 2008;30(3):255–71.
50. Miller AA, Drummond GR, Sobey CG. Novel isoforms of NADPH-oxidase in cerebral vascular control. Pharmacol Ther. 2006;111(3):928–48.
51. Infanger DW, Sharma RV, Davisson RL. NADPH oxidases of the brain: distribution, regulation, and function. Antioxid Redox Sign. 2006;8(9–10):1583–96.
52. Ago T, Kitazono T, Kuroda J, Kumai Y, Kamouchi M, Ooboshi H, Wakisaka M, Kawahara T, Rokutan K, Ibayashi S, et al. NAD(P)H oxidases in rat basilar arterial endothelial cells. Stroke. 2005;36(5):1040–6.
53. Paravicini TM, Chrissobolis S, Drummond GR, Sobey CG. Increased NADPH-oxidase activity and nox4 expression during chronic hypertension is associated with enhanced cerebral vasodilatation to NADPH in vivo. Stroke. 2004;35(2):584–9.

54. Sorce S, Krause KH. NOX enzymes in the central nervous system: from signaling to disease. Antioxid Redox Signal. 2009;11(10):2481–504.

55. Antony S, Wu Y, Hewitt SM, Anver MR, Butcher D, Jiang G, Meitzler JL, Liu H, Juhasz A, Lu J, et al. Characterization of NADPH oxidase 5 expression in human tumors and tumor cell lines with a novel mouse monoclonal antibody. Free Radic Biol Med. 2013;65:497–508.

56. Kazama K, Anrather J, Zhou P, Girouard H, Frys K, Milner TA, Iadecola C. Angiotensin II impairs neurovascular coupling in neocortex through NADPH oxidase-derived radicals. Circ Res. 2004;95(10):1019–26.

57. De Silva TM, Broughton BR, Drummond GR, Sobey CG, Miller AA. Gender influences cerebral vascular responses to angiotensin II through Nox2-derived reactive oxygen species. Stroke. 2009;40(4):1091–7.

58. Miller AA, Drummond GR, De Silva TM, Mast AE, Hickey H, Williams JP, Broughton BRS, Sobey CG. NADPH oxidase activity is higher in cerebral versus systemic arteries of four animal species: role of Nox2. Am J Physiol Heart C. 2009;296(1):H220–5.

59. Wu DC, Re DB, Nagai M, Ischiropoulo H, Przedborski S. The inflammatory NADPH oxidase enzyme modulates motor neuron degeneration in amyotrophic lateral sclerosis mice. Proc Natl Acad Sci U S A. 2006;103(32):12132–7.

60. Fischer MT, Sharma R, Lim JL, Haider L, Frischer JM, Drexhage J, Mahad D, Bradl M, van Horssen J, Lassmann H. NADPH oxidase expression in active multiple sclerosis lesions in relation to oxidative tissue damage and mitochondrial injury. Brain J Neurol. 2012;135:886–99.

61. Cheret C, Gervais A, Lelli A, Colin C, Amar L, Ravassard P, Mallet J, Cumano A, Krause KH, Mallat M. Neurotoxic activation of microglia is promoted by a Nox1-dependent NADPH oxidase. J Neurosci. 2008;28(46):12039–51.

62. Li B, Bedard K, Sorce S, Hinz B, Dubois-Dauphin M, Krause KH. NOX4 expression in human microglia leads to constitutive generation of reactive oxygen species and to constitutive IL-6 expression. J Innate Immun. 2009;1(6):570–81.

63. Hilenski LL, Clempus RE, Quinn MT, Lambeth JD, Griendling KK. Distinct subcellular localizations of Nox1 and Nox4 in vascular smooth muscle cells. Arterioscl Throm Vas. 2004;24(4):677–83.

64. Zhang AY, Yi F, Zhang G, Gulbins E, Li PL. Lipid raft clustering and redox signaling platform formation in coronary arterial endothelial cells. Hypertension. 2006;47(1):74–80.

65. Gianni D, Diaz B, Taulet N, Fowler B, Courtneidge SA, Bokoch GM. Novel p47(phox)-related organizers regulate localized NADPH oxidase 1 (Nox1) activity. Sci Signal. 2009;2(88):ra54.

66. Altenhofer S, Kleikers PWM, Radermacher KA, Scheurer P, Hermans JJR, Schiffers P, Ho HD, Wingler K, Schmidt HHHW. The NOX toolbox: validating the role of NADPH oxidases in physiology and disease. Cell Mol Life Sci. 2012;69(14):2327–43.

67. Chen K, Kirber MT, Xiao H, Yang Y, Keaney JF. Regulation of ROS signal transduction by NADPH oxidase 4 localization. J Cell Biol. 2008;181(7):1129–39.

68. Wu RF, Ma ZY, Liu Z, Terada LS. Nox4-derived H2O2 mediates endoplasmic reticulum signaling through local Ras activation. Mol Cell Biol. 2010;30(14):3553–68.

69. Boucherie C, Schafer S, Lavand'homme P, Maloteaux JM, Hermans E. Chimerization of astroglial population in the lumbar spinal cord after mesenchymal stem cell transplantation prolongs survival in a rat model of amyotrophic lateral sclerosis. J Neurosci Res. 2009;87(9):2034–46.

70. Koziel R, Pircher H, Kratochwil M, Lener B, Hermann M, Dencher NA, Jansen-Durr P. Mitochondrial respiratory chain complex I is inactivated by NADPH oxidase Nox4. Biochem J. 2013;452:231–9.

71. Kleinschnitz C, Grund H, Wingler K, Armitage ME, Jones E, Mittal M, Barit D, Schwarz T, Geis C, Kraft P, et al. Post-stroke inhibition of induced NADPH oxidase type 4 prevents oxidative stress and neurodegeneration. PLoS Biol. 2010;8(9):e1000479.

72. McCann SK, Dusting GJ, Roulston CL. Early increase of Nox4 NADPH oxidase and super-oxide generation following endothelin-1-induced stroke in conscious rats. J Neurosci Res. 2008;86(11):2524–34.

73. Miller AA, Drummond GR, Schmidt HHHW, Sobey CG. NADPH oxidase activity and function are profoundly greater in cerebral versus systemic arteries. Circ Res. 2005;97(10):1055–62.
74. Tuo YH, Liu Z, Chen JW, Wang QY, Li SL, Li MC, Dai G, Wang JS, Zhang YL, Feng L, Shi ZS. NADPH oxidase inhibitor improves outcome of mechanical reperfusion by suppressing hemorrhagic transformation. J Neurointerv Surg. 2017;9(5):492–8.
75. Chen H, Song YS, Chan PH. Inhibition of NADPH oxidase is neuroprotective after ischemia-reperfusion. J Cereb Blood Flow Metab. 2009;29(7):1262–72.
76. Jackman KA, Miller AA, De Silva TM, Crack PJ, Drummond GR, Sobey CG. Reduction of cerebral infarct volume by apocynin requires pretreatment and is absent in Nox2-deficient mice. Br J Pharmacol. 2009;156(4):680–8.
77. Kahles T, Kohnen A, Heumueller S, Rappert A, Bechmann I, Liebner S, Wittko IM, Neumann-Haefelin T, Steinmetz H, Schroeder K, et al. NADPH oxidase Nox1 contributes to ischemic injury in experimental stroke in mice. Neurobiol Dis. 2010;40(1):185–92.
78. Sutherland BA, Minnerup J, Balami JS, Arba F, Buchan AM, Kleinschnitz C. Neuroprotection for ischaemic stroke: translation from the bench to the bedside. Int J Stroke. 2012;7(5):407–18.
79. Miller ER, Pastor-Barriuso R, Dalal D, Riemersma RA, Appel LJ, Guallar E. Meta-analysis: high-dosage vitamin E supplementation may increase all-cause mortality. Ann Intern Med. 2005;142(1):37–46.
80. Kubota Y, Iso H, Date C, Kikuchi S, Watanabe Y, Wada Y, Inaba Y, Tamakoshi A, Group JS. Dietary intakes of antioxidant vitamins and mortality from cardiovascular disease: the Japan Collaborative Cohort Study (JACC) study. Stroke. 2011;42(6):1665–72.
81. Myint PK, Luben RN, Welch AA, Bingham SA, Wareham NJ, Khaw KT. Plasma vitamin C concentrations predict risk of incident stroke over 10 y in 20 649 participants of the European Prospective Investigation into Cancer Norfolk prospective population study. Am J Clin Nutr. 2008;87(1):64–9.
82. Yokoyama T, Date C, Kokubo Y, Yoshiike N, Matsumura Y, Tanaka H. Serum vitamin C concentration was inversely associated with subsequent 20-year incidence of stroke in a Japanese rural community – the Shibata study. Stroke. 2000;31(10):2287–94.
83. Schiavone S, Sorce S, Dubois-Dauphin M, Jaquet V, Colaianna M, Zotti M, Cuomo V, Trabace L, Krause KH. Involvement of NOX2 in the development of behavioral and pathologic alterations in isolated rats. Biol Psychiatry. 2009;66(4):384–92.
84. Heumuller SWS, Barbosa-Sicard E, Schmidt HH, Busse R, Schroder K, Brandes RP. Apocynin is not an inhibitor of vascular NADPH oxidases but an antioxidant. Hypertension. 2008;51:211–7.
85. Schluter T, Steinbach AC, Steffen A, Rettig R, Grisk O. Apocynin-induced vasodilation involves Rho kinase inhibition but not NADPH oxidase inhibition. Cardiovasc Res. 2008;80(2):271–9.
86. Bardutzky J, Meng X, Bouley J, Duong TQ, Ratan R, Fisher M. Effects of intravenous dimethyl sulfoxide on ischemia evolution in a rat permanent occlusion model. J Cereb Blood Flow Metab. 2005;25(8):968–77.
87. Shimizu S, Simon RP, Graham SH. Dimethylsulfoxide (DMSO) treatment reduces infarction volume after permanent focal cerebral ischemia in rats. Neurosci Lett. 1997;239(2–3):125–7.
88. O'Donnell BV, Tew DG, Jones OT, England PJ. Studies on the inhibitory mechanism of iodonium compounds with special reference to neutrophil NADPH oxidase. Biochem J. 1993;290(Pt 1):41–9.
89. Ishikawa M, Cooper D, Arumugam TV, Zhang TH, Nanda A, Granger DN. Platelet-leukocyte-endothelial cell interactions after middle cerebral artery occlusion and reperfusion. J Cereb Blood Flow Metab. 2004;24(8):907–15.
90. Nagel S, Genius J, Heiland S, Horstmann S, Gardner H, Wagner S. Diphenyleneiodonium and dimethylsulfoxide for treatment of reperfusion injury in cerebral ischemia of the rat. Brain Res. 2007;1132(1):210–7.

91. Wang TG, Qin L, Liu B, Liu YX, Wilson B, Eling TE, Langenbach R, Taniura S, Hong JS. Role of reactive oxygen species in LPS-induced production of prostaglandin E-2 in microglia. J Neurochem. 2004;88(4):939–47.

92. Nagel S, Hadley G, Pfleger K, Grond-Ginsbach C, Buchan AM, Wagner S, Papadakis M. Suppression of the inflammatory response by diphenyleneiodonium after transient focal cerebral ischemia. J Neurochem. 2012;123:98–107.

93. Wind S, Beuerlein K, Eucker T, Muller H, Scheurer P, Armitage ME, Ho H, Schmidt HHHW, Wingler K. Comparative pharmacology of chemically distinct NADPH oxidase inhibitors. Br J Pharmacol. 2010;161(4):885–98.

94. Jaquet V, Scapozza L, Clark RA, Krause KH, Lambeth JD. Small-molecule NOX inhibitors: ROS-generating NADPH oxidases as therapeutic targets. Antioxid Redox Sign. 2009;11(10):2535–52.

95. Stielow C, Catar RA, Muller G, Wingler K, Scheurer P, Schmidt HHHW, Morawietz H. Novel Nox inhibitor of oxLDL-induced reactive oxygen species formation in human endothelial cells. Biochem Bioph Res Co. 2006;344(1):200–5.

96. Niethammer P, Grabher C, Look AT, Mitchison TJ. A tissue-scale gradient of hydrogen peroxide mediates rapid wound detection in zebrafish. Nature. 2009;459(7249):996–U123.

97. ten Freyhaus H, Huntgeburth M, Wingler K, Schnitker J, Baumer AT, Vander M, Bekhite MM, Wartenberg M, Sauer H, Rosenkranz S. Novel Nox inhibitor VAS2870 attenuates PDGF-dependent smooth muscle cell chemotaxis, but not proliferation. Cardiovasc Res. 2006;71(2):331–41.

98. Galli S, Arciuch VGA, Poderoso C, Converso DP, Zhou QQ, Joffe EBD, Cadenas E, Boczkowski J, Carreras MC, Poderoso JJ. Tumor cell phenotype is sustained by selective MAPK oxidation in mitochondria. PLoS One. 2008;3(6):e2379.

99. Sedeek M, Callera G, Montezano A, Gutsol A, Heitz F, Szyndralewiez C, Page P, Kennedy CRJ, Burns KD, Touyz RM, et al. Critical role of Nox4-based NADPH oxidase in glucose-induced oxidative stress in the kidney: implications in type 2 diabetic nephropathy. Am J Physiol Renal. 2010;299(6):F1348–58.

100. Vendrov AE, Madamanchi NR, Niu XL, Molnar KC, Runge M, Szyndralewiez C, Page P, Runge MS. NADPH oxidases regulate CD44 and hyaluronic acid expression in thrombin-treated vascular smooth muscle cells and in atherosclerosis. J Biol Chem. 2010;285(34):26545–57.

101. Smith SM, Min J, Ganesh T, Diebold B, Kawahara T, Zhu Y, McCoy J, Sun A, Snyder JP, Fu H, et al. Ebselen and congeners inhibit NADPH oxidase 2-dependent superoxide generation by interrupting the binding of regulatory subunits. Chem Biol. 2012;19(6):752–63.

102. Verbeuren TJ, Bouskela E, Cohen RA, Vanhoutte PM. Regulation of adhesion molecules: a new target for the treatment of chronic venous insufficiency. Microcirculation. 2000;7(6 Pt 2):S41–8.

103. Gianni D, Taulet N, Zhang H, DerMardirossian C, Kister J, Martinez L, Roush WR, Brown SJ, Bokoch GM, Rosen H. A novel and specific NADPH oxidase-1 (Nox1) small-molecule inhibitor blocks the formation of functional invadopodia in human colon cancer cells. ACS Chem Biol. 2010;5(10):981–93.

104. Bhandarkar SS, Jaconi M, Fried LE, Bonner MY, Lefkove B, Govindarajan B, Perry BN, Parhar R, Mackelfresh J, Sohn A, et al. Fulvene-5 potently inhibits NADPH oxidase 4 and blocks the growth of endothelial tumors in mice. J Clin Invest. 2009;119(8):2359–65.

105. Daiber A, Frein D, Namgaladze D, Ullrich V. Oxidation and nitrosation in the nitrogen monoxide/superoxide system. J Biol Chem. 2002;277(14):11882–8.

106. Cipak Gasparovic A, Zarkovic N, Zarkovic K, Semen K, Kaminskyy D, Yelisyeyeva O, Bottari SP. Biomarkers of oxidative and nitro-oxidative stress: conventional and novel approaches. Br J Pharmacol. 2017;174(12):1771–83.

107. Chen XM, Chen HS, Xu MJ, Shen JG. Targeting reactive nitrogen species: a promising therapeutic strategy for cerebral ischemia-reperfusion injury. Acta Pharmacol Sin. 2013;34(1):67–77.

108. Furchgott RF, Zawadzki JV. The obligatory role of endothelial cells in the relaxation of arterial smooth muscle by acetylcholine. Nature. 1980;288(5789):373–6.
109. Moncada S, Palmer RM, Higgs EA. Nitric oxide: physiology, pathophysiology, and pharmacology. Pharmacol Rev. 1991;43(2):109–42.
110. Toda N, Ayajiki K, Okamura T. Cerebral blood flow regulation by nitric oxide: recent advances. Pharmacol Rev. 2009;61(1):62–97.
111. Broos K, Feys HB, De Meyer SF, Vanhoorelbeke K, Deckmyn H. Platelets at work in primary hemostasis. Blood Rev. 2011;25(4):155–67.
112. Schuman EM, Madison DV. Nitric oxide and synaptic function. Annu Rev Neurosci. 1994;17:153–83.
113. Iadecola C, Zhang F, Xu X. Inhibition of inducible nitric oxide synthase ameliorates cerebral ischemic damage. Am J Phys. 1995;268(1 Pt 2):R286–92.
114. Iadecola C, Zhang F, Casey R, Nagayama M, Ross ME. Delayed reduction of ischemic brain injury and neurological deficits in mice lacking the inducible nitric oxide synthase gene. J Neurosci. 1997;17(23):9157–64.
115. Clark RS, Kochanek PM, Obrist WD, Wong HR, Billiar TR, Wisniewski SR, Marion DW. Cerebrospinal fluid and plasma nitrite and nitrate concentrations after head injury in humans. Crit Care Med. 1996;24(7):1243–51.
116. Pautz A, Art J, Hahn S, Nowag S, Voss C, Kleinert H. Regulation of the expression of inducible nitric oxide synthase. Nitric Oxide. 2010;23(2):75–93.
117. Lundberg JO, Weitzberg E, Gladwin MT. The nitrate-nitrite-nitric oxide pathway in physiology and therapeutics. Nat Rev Drug Discov. 2008;7(2):156–67.
118. Kiss JP, Vizi ES. Nitric oxide: a novel link between synaptic and nonsynaptic transmission. Trends Neurosci. 2001;24(4):211–5.
119. Forstermann U. Nitric oxide and oxidative stress in vascular disease. Pflug Arch Eur J Phy. 2010;459(6):923–39.
120. Bolanos JP, Almeida A. Roles of nitric oxide in brain hypoxia-ischemia. Biochim Biophys Acta. 1999;1411(2–3):415–36.
121. Grandati M, Verrecchia C, Revaud ML, Allix M, Boulu RG, Plotkine M. Calcium-independent NO-synthase activity and nitrites/nitrates production in transient focal cerebral ischaemia in mice. Br J Pharmacol. 1997;122(4):625–30.
122. Iadecola C, Xu X, Zhang F, el-Fakahany EE, Ross ME. Marked induction of calcium-independent nitric oxide synthase activity after focal cerebral ischemia. J Cereb Blood Flow Metab. 1995;15(1):52–9.
123. Shen JG, Ma S, Chan PS, Lee W, Fung PCW, Cheung RTF, Tong Y, Liu KJ. Nitric oxide down-regulates caveolin-1 expression in rat brains during focal cerebral ischemia and reperfusion injury. J Neurochem. 2006;96(4):1078–89.
124. Huang Z, Huang PL, Ma J, Meng W, Ayata C, Fishman MC, Moskowitz MA. Enlarged infarcts in endothelial nitric oxide synthase knockout mice are attenuated by nitro-L-arginine. J Cereb Blood Flow Metab. 1996;16(5):981–7.
125. Rudic RD, Sessa WC. Nitric oxide in endothelial dysfunction and vascular remodeling: clinical correlates and experimental links. Am J Hum Genet. 1999;64(3):673–7.
126. Iadecola C. Bright and dark sides of nitric oxide in ischemic brain injury. Trends Neurosci. 1997;20(3):132–9.
127. Fabian RH, DeWitt DS, Kent TA. In vivo detection of superoxide anion production by the brain using a cytochrome c electrode. J Cereb Blood Flow Metab. 1995;15(2):242–7.
128. Peters O, Back T, Lindauer U, Busch C, Megow D, Dreier J, Dirnagl U. Increased formation of reactive oxygen species after permanent and reversible middle cerebral artery occlusion in the rat. J Cereb Blood Flow Metab. 1998;18(2):196–205.
129. Kim GW, Kondo T, Noshita N, Chan PH. Manganese superoxide dismutase deficiency exacerbates cerebral infarction after focal cerebral ischemia/reperfusion in mice: implications for the production and role of superoxide radicals. Stroke. 2002;33(3):809–15.

130. Peter B, Van Waarde MA, Vissink A, s-Gravenmade EJ, Konings AW. The role of secretory granules in radiation-induced dysfunction of rat salivary glands. Radiat Res. 1995;141(2):176–82.
131. Al-Nimer MS, Al-Mahdawi AM, Sakeni RA. Assessment of nitrosative oxidative stress in patients with middle cerebral artery occlusion. Neurosciences. 2007;12(1):31–4.
132. Nanetti L, Taffi R, Vignini A, Moroni C, Raffaelli F, Bacchetti T, Silvestrini M, Provinciali L, Mazzanti L. Reactive oxygen species plasmatic levels in ischemic stroke. Mol Cell Biochem. 2007;303(1–2):19–25.
133. Moro MA, Almeida A, Bolanos JP, Lizasoain I. Mitochondrial respiratory chain and free radical generation in stroke. Free Radical Biol Med. 2005;39(10):1291–304.
134. Greenacre SA, Ischiropoulos H. Tyrosine nitration: localisation, quantification, consequences for protein function and signal transduction. Free Radic Res. 2001;34(6):541–81.
135. Suzuki M, Tabuchi M, Ikeda M, Tomita T. Concurrent formation of peroxynitrite with the expression of inducible nitric oxide synthase in the brain during middle cerebral artery occlusion and reperfusion in rats. Brain Res. 2002;951(1):113–20.
136. Schopfer FJ, Baker PR, Freeman BA. NO-dependent protein nitration: a cell signaling event or an oxidative inflammatory response? Trends Biochem Sci. 2003;28(12):646–54.
137. Viappiani S, Nicolescu AC, Holt A, Sawicki G, Crawford BD, Leon H, van Mulligen T, Schulz R. Activation and modulation of 72 kDa matrix metalloproteinase-2 by peroxynitrite and glutathione. Biochem Pharmacol. 2009;77(5):826–34.
138. Rajagopalan S, Meng XP, Ramasamy S, Harrison DG, Galis ZS. Reactive oxygen species produced by macrophage-derived foam cells regulate the activity of vascular matrix metalloproteinases in vitro. Implications for atherosclerotic plaque stability. J Clin Invest. 1996;98(11):2572–9.
139. Migita K, Maeda Y, Abiru S, Komori A, Yokoyama T, Takii Y, Nakamura M, Yatsuhashi H, Eguchi K, Ishibashi H. Peroxynitrite-mediated matrix metalloproteinase-2 activation in human hepatic stellate cells. FEBS Lett. 2005;579(14):3119–25.
140. Donnini S, Monti M, Roncone R, Morbidelli L, Rocchigiani M, Oliviero S, Casella L, Giachetti A, Schulz R, Ziche M. Peroxynitrite inactivates human-tissue inhibitor of metalloproteinase-4. FEBS Lett. 2008;582(7):1135–40.
141. Tan KH, Harrington S, Purcell WM, Hurst RD. Peroxynitrite mediates nitric oxide-induced blood-brain barrier damage. Neurochem Res. 2004;29(3):579–87.
142. Laufs U, Liao JK. Post-transcriptional regulation of endothelial nitric oxide synthase mRNA stability by Rho GTPase. J Biol Chem. 1998;273(37):24266–71.
143. Laufs U. Beyond lipid-lowering: effects of statins on endothelial nitric oxide. Eur J Clin Pharmacol. 2003;58(11):719–31.
144. Sironi L, Cimino M, Guerrini U, Calvio AM, Lodetti B, Asdente M, Balduini W, Paoletti R, Tremoli E. Treatment with statins after induction of focal ischemia in rats reduces the extent of brain damage. Arterioscler Thromb Vasc Biol. 2003;23(2):322–7.
145. Prinz V, Laufs U, Gertz K, Kronenberg G, Balkaya M, Leithner C, Lindauer U, Endres M. Intravenous rosuvastatin for acute stroke treatment: an animal study. Stroke. 2008;39(2):433–8.
146. Endres M, Laufs U, Huang Z, Nakamura T, Huang P, Moskowitz MA, Liao JK. Stroke protection by 3-hydroxy-3-methylglutaryl (HMG)-CoA reductase inhibitors mediated by endothelial nitric oxide synthase. Proc Natl Acad Sci U S A. 1998;95(15):8880–5.
147. Di Napoli M, Papa F. Inflammation, statins, and outcome after ischemic stroke. Stroke. 2001;32(10):2446–7.
148. Heart Protection Study Collaborative G. MRC/BHF Heart Protection Study of cholesterol lowering with simvastatin in 20,536 high-risk individuals: a randomised placebo-controlled trial. Lancet. 2002;360(9326):7–22.
149. Adams HP Jr, del Zoppo G, Alberts MJ, Bhatt DL, Brass L, Furlan A, Grubb RL, Higashida RT, Jauch EC, Kidwell C, et al. Guidelines for the early management of adults with ischemic stroke: a guideline from the American Heart Association/American Stroke Association Stroke Council, Clinical Cardiology Council, Cardiovascular Radiology and Intervention Council,

and the Atherosclerotic Peripheral Vascular Disease and Quality of Care Outcomes in Research Interdisciplinary Working Groups: the American Academy of Neurology affirms the value of this guideline as an educational tool for neurologists. Stroke. 2007;38(5):1655–711.

150. Vaughan CJ, Delanty N. Neuroprotective properties of statins in cerebral ischemia and stroke. Stroke. 1999;30(9):1969–73.

151. Collins R, Armitage J, Parish S, Sleight P, Peto R, Heart Protection Study Collaborative G. Effects of cholesterol-lowering with simvastatin on stroke and other major vascular events in 20536 people with cerebrovascular disease or other high-risk conditions. Lancet. 2004;363(9411):757–67.

152. Vergouwen MD, de Haan RJ, Vermeulen M, Roos YB. Statin treatment and the occurrence of hemorrhagic stroke in patients with a history of cerebrovascular disease. Stroke. 2008;39(2):497–502.

153. Becker K, Tanzi P, Kalil A, Shibata D, Cain K. Early statin use is associated with increased risk of infection after stroke. J Stroke Cerebrovasc. 2013;22(1):66–71.

154. Ding-Zhou L, Marchand-Verrecchia C, Croci N, Plotkine M, Margaill I. L-NAME reduces infarction, neurological deficit and blood-brain barrier disruption following cerebral ischemia in mice. Eur J Pharmacol. 2002;457(2–3):137–46.

155. Nakagawa H, Ikota N, Ozawa T, Kotake Y. Dose- and time-dependence of radiation-induced nitric oxide formation in mice as quantified with electron paramagnetic resonance. Nitric Oxide Biol Ch. 2001;5(1):47–52.

156. Nagafuji T, Sugiyama M, Muto A, Makino T, Miyauchi T, Nabata H. The neuroprotective effect of a potent and selective inhibitor of type I NOS (L-MIN) in a rat model of focal cerebral ischaemia. Neuroreport. 1995;6(11):1541–5.

157. Palmer C, Roberts RL. Reduction in hypoxic-ischemic brain swelling following delayed inhibition of inducible nitric oxide synthase in 7 day old rats. Pediatr Res. 2002;51(4):446a.

158. Yoshida T, Limmroth V, Irikura K, Moskowitz MA. The NOS inhibitor, 7-nitroindazole, decreases focal infarct volume but not the response to topical acetylcholine in pial vessels. J Cereb Blood Flow Metab. 1994;14(6):924–9.

159. del Pilar Fernandez Rodriguez M, Belmonte A, Meizoso MJ, Garcia-Novio M, Garcia-Iglesias E. Effect of tirilazad on brain nitric oxide synthase activity during cerebral ischemia in rats. Pharmacology. 1997;54(2):108–12.

160. Chabrier PE, Auguet M, Spinnewyn B, Auvin S, Cornet S, Demerle-Pallardy C, Guilmard-Favre C, Marin JG, Pignol B, Gillard-Roubert V, et al. BN 80933, a dual inhibitor of neuronal nitric oxide synthase and lipid peroxidation: a promising neuroprotective strategy. Proc Natl Acad Sci U S A. 1999;96(19):10824–9.

161. Goyagi T, Goto S, Bhardwaj A, Dawson VL, Hurn PD, Kirsch JR. Neuroprotective effect of sigma(1)-receptor ligand 4-phenyl-1-(4-phenylbutyl) piperidine (PPBP) is linked to reduced neuronal nitric oxide production. Stroke. 2001;32(7):1613–20.

162. Forster C, Clark HB, Ross ME, Iadecola C. Inducible nitric oxide synthase expression in human cerebral infarcts. Acta Neuropathol. 1999;97(3):215–20.

163. Morikawa E, Moskowitz MA, Huang Z, Yoshida T, Irikura K, Dalkara T. L-Arginine infusion promotes nitric oxide-dependent vasodilation, increases regional cerebral blood flow, and reduces infarction volume in the rat. Stroke. 1994;25(2):429–35.

164. Willmot M, Gray L, Gibson C, Murphy S, Bath PM. A systematic review of nitric oxide donors and L-arginine in experimental stroke; effects on infarct size and cerebral blood flow. Nitric Oxide. 2005;12(3):141–9.

165. Bath PM, Willmot M, Leonardi-Bee J, Bath FJ. Nitric oxide donors (nitrates), L-arginine, or nitric oxide synthase inhibitors for acute stroke. Cochrane Database Syst Rev. 2002;4:CD000398.

166. Heron M. Deaths: leading causes for 2007. Natl Vital Stat Rep. 2011;59(8):1–95.

167. Kunz A, Park L, Abe T, Gallo EF, Anrather J, Zhou P, Iadecola C. Neurovascular protection by ischemic tolerance: role of nitric oxide and reactive oxygen species. J Neurosci. 2007;27(27):7083–93.

168. Maneen MJ, Cipolla MJ. Peroxynitrite diminishes myogenic tone in cerebral arteries: role of nitrotyrosine and F-actin. Am J Phys Heart Circ Phys. 2007;292(2):H1042–50.

Chapter 8
Extracellular Matrix in Stroke

Yao Yao

Abstract The extracellular matrix (ECM) is a non-cellular structure found in all tissues. It undergoes tightly controlled remodeling in both physiological (e.g., during development and upon adaption to physiological needs) and pathological conditions, and exerts diverse and important functions. After stroke, the structure of ECM is disrupted and the expression of ECM proteins is altered. The significance of these changes, however, remains largely unknown, mainly due to the intrinsic complexity of the ECM. In this chapter, I first introduce the three types of ECM in the brain with a focus on the basement membranes (BMs). Next, BM assembly and function are briefly summarized. Third, BM changes in stroke are discussed in detail. Furthermore, important questions that need to be answered in future studies are described.

Keywords Extracellular matrix · Basement membrane · Stroke · Blood brain barrier

1 Extracellular Matrix

The extracellular matrix (ECM) is a unique voluminous structure found in almost every organ. In the brain, it constitutes approximately 10–20% of brain volume [1, 2]. The ECM consists of a large group of proteins with highly conserved structural domains, including collagen, laminin, nidogen, and many other glycoproteins. In addition to providing structural support for tissue integrity and elasticity, the ECM

Y. Yao (✉)
Department of Pharmaceutical and Biomedical Sciences, University of Georgia,
Athens, GA, USA
e-mail: yyao@uga.edu

© Springer International Publishing AG, part of Springer Nature 2018
W. Jiang et al. (eds.), *Cerebral Ischemic Reperfusion Injuries (CIRI)*, Springer Series in Translational Stroke Research, https://doi.org/10.1007/978-3-319-90194-7_8

121

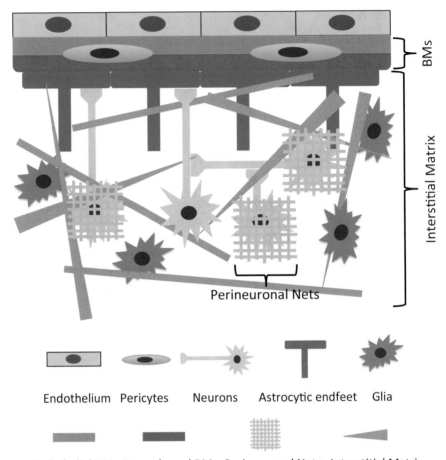

Fig. 8.1 Diagram of three major forms of ECM in the brain. The BMs are located in the abluminal side of endothelial cells. Note the two layers of BMs: endothelial BM (green) and parenchymal BM (red). The interstitial matrix (orange) is widely distributed in the brain parenchyma. The perineuronal nets (light purple) are primarily found in the cell bodies and dendrites of neurons. *ECM* extracellular matrix, *BM* basement membrane

also undergoes constant remodeling to regulate tissue homeostasis [3, 4]. The functional significance of the ECM is reflected by the large variety of deficits/phenotypes observed in mice and/or humans with mutations in genes encoding ECM components [5–7]. Based on its biochemical composition and structure, the ECM is broadly divided into three forms: interstitial matrix, perineuronal nets, and basement membranes (BMs) (Fig. 8.1).

1.1 Interstitial Matrix

The interstitial matrix is a loose network mainly composed of fibrillar collagens, including collagen I, II, III, V, XI, XXIV and XXVII, among which collagen I is the major form. It also contains non-fibrillar collagens, including fibronectin and chondroitin sulfate proteoglycans [8, 9]. The interstitial matrix is diffusely distributed between cells in the parenchyma and functions to provide scaffolding for tissues [8, 10].

1.2 Perineuronal Nets

The perineuronal nets are mesh-like ECM aggregates surrounding neuronal cell bodies and dendrites [11]. They are mainly composed of lecticans, hyaluronan, tenascin and link proteins [11, 12]. It has been shown that synaptic contacts are absent in neuronal surfaces covered by perineuronal nets [12], suggesting that perineuronal nets inhibit synaptic formation. In addition, there is evidence showing that perineuronal nets function to maintain synaptic plasticity and preserve neuronal health [9, 13, 14].

1.3 BM

The BM is an amorphous layer (50–100 nm in thickness) of highly cross-linked ECM proteins [15–20]. Biochemically, the BM is composed of four major types of proteins: collagen IV, laminin, nidogen, and perlecan [21–24]. Some minor components, including fibulins, osteonectin and netrin-4, are also found in the BM [25–32]. Anatomically, the BM is located on the abluminal side of endothelial cells [22, 24, 25, 33, 34], at the interface between brain parenchyma and the circulation system in the CNS. This close relationship with brain vessels makes the BM an immediate target during stroke. Therefore, we focus our discussion on the BM hereafter.

2 BM Components

At the blood-brain barrier (BBB), two BMs are found: a parenchymal BM and an endothelial BM, which are separated by pericytes [35–37] (Fig. 8.2). In physiological conditions, these two BMs are indistinguishable and appear as one in regions without pericytes [35, 36]. In pathological conditions, especially inflammation, the parenchymal and endothelial BMs can be separated at postcapillary venules by

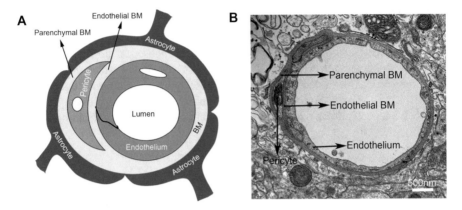

Fig. 8.2 The BMs at the BBB. (**a**) Diagram illustration of the parenchymal and endothelial BMs at the BBB. Astrocytes are shown in blue, pericytes in orange, and endothelial cells in green. The BMs are highlighted in yellow. (**b**) A representative electron microscopy image from a 2-month-old mouse brain showing the BMs at the BBB. Note that the parenchymal and endothelial BMs are separated by pericytes and indistinguishable in regions without pericytes. Scale bar represents 500 nm. *BM* basement membrane, *BBB* blood-brain barrier

infiltrating leukocytes [24, 38], forming perivascular cuffs. Most BM components are ubiquitously expressed at the parenchymal and endothelial BMs. Laminin, however, shows differential expression in these two BMs [36] (see below for details). Below we discuss the four major BM components individually.

2.1 Collagen IV

Collagen IV is the most abundant component of the BMs. It is a trimeric protein containing three α chains (Fig. 8.3). Currently, six collagen IV α chains (COL4A1–6) have been identified [39–44]. Each α chain is composed of three domains: an N-terminal 7S domain, a middle triple-helical domain, and a C-terminal globular non-collagenous-1 (NC1) domain [22, 45] (Fig. 8.3a). Collagen IV network formation involves the following steps [22, 46, 47]. First, the NC1 domain initiates the assembly of three α chains, eventually forming a trimer (protomer). Next, two protomers dimerize via their NC1 domains, generating a NC1 hexamer (Fig. 8.3c). Third, four protomers associate via their N-terminal 7S domains to form a spider-shaped tetramer (Fig. 8.3c). These dimers and tetramers further assemble into sheet-like suprastructure (Fig. 8.3c), forming collagen IV network.

At the BBB, collagen IV is secreted by endothelial cells, pericytes, and astrocytes [4]. Collagen IV has been shown to increase the transendothelial electrical resistance (TEER) of brain capillary endothelial cells *in vitro*, suggesting an important role of collagen IV in maintaining BBB integrity [48]. To further investigate the biological function of collagen IV, mice deficient in COL4A1/2, two of the most frequently used α chains that make the major collagen IV isoform $\alpha 1(IV)_2 \alpha 2(IV)$, were gener-

Fig. 8.3 Collagen IV
network formation. (a)
Structural illustration of a
collagen IV monomer (α
chain), which contains a
7S domain, a triple helix
domain, and an NC1
domain. (b) Structural
illustration of a collagen
IV trimer (protomer),
containing three collagen
IV monomers. (c)
Sheet-like collagen IV
network formed by
dimerization and
tetramerization of
protomers through their
NC1 and 7S domains,
respectively

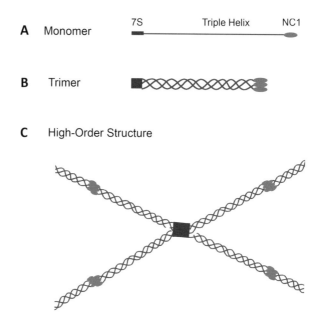

ated [49]. BMs were detected in the mutants at early developmental stages (up to E9.5), but showed defects at E10.5–11.5, leading to lethality [49]. These data suggest that collagen IV is required for the maintenance of BM integrity and function, but dispensable for the initial assembly of BM during early development. Furthermore, a series of mutations in COL4A1/2 have been generated in mice [50, 51] and identified in humans [51, 52]. These mutations are associated with disrupted vascular integrity, stress-induced hemorrhage, and adult-onset stroke in both mice and humans [50–52].

2.2 Laminin

Laminin is a group of cross-shaped heterotrimeric glycoproteins composed of one α, one β, and one γ polypeptide chains [7, 36, 53–55]. So far, five α, four β, and three γ variants have been identified [56]. Each laminin chain consists of various globular (LN, L4, and LF) domains, rod-like repeats called epidermal growth factor-like (LE) domains, and a coiled-coil domain [57]. The diagram of the structures of all 12 laminin polypeptide chains is shown in Fig. 8.4a. Compared to β and γ chains, α chains have five homologous globular (LG) domains in the C-terminus, which are generally involved in receptor-engaging. The α, β, and γ chains bind to each other via their coiled-coil domains, forming a large (400–900 kDa) laminin molecule with one long arm and two or three short arms [56, 57] (Fig. 8.4b). Various combinations of these chains generate a large number of different laminin molecules [57, 58]. It should be noted, however, that not all possible combinations have been experimentally

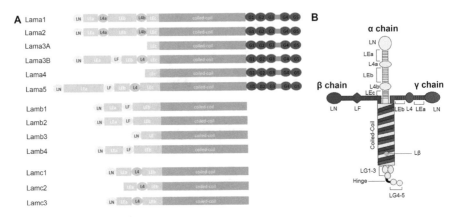

Fig. 8.4 Structures of individual laminin polypeptide chains and a fully assembled laminin molecule. (**a**) Structural illustration of each laminin polypeptide chain. These chains contain various globular domains, rod-like LE domains, and a coiled-coil domain. Note that the C-terminal globular domains are unique to laminin α chains. (**b**) Structural illustration of a fully assembled laminin molecule composed of α, β, γ chains

isolated. In mammals, 16 laminin molecules have been identified so far, and four novel combinations have been proposed based on *in vivo* and *in vitro* studies [58]. All these laminin molecules are summarized in Table 8.1. Like collagen IV, laminin molecules are able to interact with each other, forming sheet-like structures. This intermolecular self-assembly occurs at the N-terminus of each laminin polypeptide chain via non-covalent bonds [59].

These laminin molecules were originally named using numbers based on the order of their discovery, such as Laminin-1, Laminin-2, and so on [60]. Although simple to use, this system fails to provide information regarding the trimeric composition of laminins. Therefore, an alternative nomenclature system that indicates each individual polypeptide chain was developed, and is widely used nowadays. In this new system, laminins are named by Greek letters representing polypeptide chains followed by numbers indicating the variants [57]. For instance: laminin-1 is written as laminin-$\alpha1\beta1\gamma1$ or simply abbreviated as laminin-111. A side-by-side comparison of these two nomenclature systems is summarized in Table 8.1.

At the BBB, laminin is expressed by endothelial cells, astrocytes, and pericytes [7, 36]. Interestingly, these cells express different laminin isoforms. Specifically, endothelial cells predominantly synthesize laminin-411 and -511 [35, 61]; astrocytes mainly make laminin-211 [35, 62]; and pericytes produce α4- and α5-containing laminins [63]. This unique expression pattern allows differential distribution of laminins in the parenchymal and endothelial BMs: the former contains laminin-211, whereas the latter is rich in laminin-411 and -511 [36].

Laminin plays many important roles at the BBB. First, laminin actively regulates the barrier function and maintains BBB integrity under physiological conditions. It has been reported that laminin is able to increase the TEER of brain capillary endothelial cells *in vitro* [48]. Consistent with this study, research from our laboratory shows

Table 8.1 Currently identified laminin molecules and their names

Original name	Alternative name
Laminin-1	Laminin-α1β1γ1 (Laminin-111)
Laminin-2	Laminin-α2β1γ1 (Laminin-211)
Laminin-3	Laminin-α1β2γ1 (Laminin-121)
Laminin-4	Laminin-α2β2γ1 (Laminin-221)
Laminin-5	Laminin-α3Aβ3γ2 (Laminin-3A32)
Laminin-5B	Laminin-α3Bβ3γ2 (Laminin-3B32)
Laminin-6	Laminin-α3β1γ1 (Laminin-311)
Laminin-7	Laminin-α3β2γ1 (Laminin-321)
Laminin-8	Laminin-α4β1γ1 (Laminin-411)
Laminin-9	Laminin-α4β2γ1 (Laminin-421)
Laminin-10	Laminin-α5β1γ1 (Laminin-511)
Laminin-11	Laminin-α5β2γ1 (Laminin-521)
Laminin-12	Laminin-α2β1γ3 (Laminin-213)
Laminin-13	Laminin-α3β2γ3 (Laminin-323)
Laminin-14	Laminin-α4β2γ3 (Laminin-423)
Laminin-15	Laminin-α5β2γ3 (Laminin-523)
–	Laminin-α2β1γ2 (Laminin-212)
–	Laminin-α2β2γ2 (Laminin-222)
–	Laminin-α3β3γ3 (Laminin-333)
–	Laminin-α5β2γ2 (Laminin-522)

that loss of astrocytic laminin leads to age-dependent BBB disruption in deep brain regions [64]; whereas loss of pericytic laminin results in a mild BBB breakdown phenotype, possibly due to compensation by endothelial laminin [63]. Additionally, lack of laminin α4 results in hemorrhage in embryonic and perinatal stages [65]. This hemorrhagic phenotype, however, is absent in adulthood probably due to compensatory expression of laminin α5 [61, 66]. Second, accumulating evidence suggests that laminin is essential for BM assembly. Genetic ablation of laminin α1 led to loss of BMs and embryonic lethality at approximately embryonic day (E) 6.5–7 [67, 68]. Similar results are found in mice lacking laminin β1 [67] or γ1 [69, 70]. These data strongly indicate an indispensable role of laminin in BM assembly. Third, laminin also participates in a large variety of signaling pathways and contributes to many other important functions, including embryonic development, neuromuscular junction maturation, and blood pressure regulation [7]. For a systemic review of the loss-of-function studies on laminin, the readers are referred to ref. [7].

2.3 Nidogen

Nidogen, also called entactin, is a glycoprotein expressed ubiquitously in the BMs. In mammals, two nidogen genes (nidogen-1 and nidogen-2) are identified. Structurally, nidogens have two N-terminal (G1 and G2) and one C-terminal (G3)

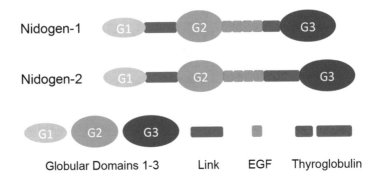

Fig. 8.5 Structural illustration of mouse nidogen-1 and -2. Both nidogens contain two N-terminal globular domains (G1 and G2) connected by a link, four rod-like EGF domains, a thyroglobulin domain, and a C-terminal globular domain (G3)

globular domains, which are connected by rod-like domains mainly composed of epidermal growth factor repeats (Fig. 8.5). Unlike collagen IV and laminin, nidogens are unable to self-assemble and form sheet-like supramolecules. Biochemical studies have shown that the G3 domain binds to laminin γ1 at high affinity, whereas the G2 domain binds to collagen IV and perlecan. These data suggest that nidogens function as a linker to connect collagen IV and laminin networks. Surprisingly, genetic deletion of nidogen-1 only causes a mild phenotype, and most BMs except those in brain capillaries are unaffected [71–74]. Like nidogen-1 null mice, nidogen-2 null mice have a normal phenotype and intact BMs [75]. Additionally, nidogen-2 is redistributed and up-regulated in nidogen-1 null mice [72, 76], although nidogen-1 expression is unchanged in nidogen-2 null mice [75]. These results suggest that nidogen-1 and nidogen-2 may compensate for each other's loss in these single mutants. Consistent with this hypothesis, double mutant mice lacking both nidogen isoforms show severe defects in BM assembly and die shortly after birth [77, 78], suggesting a critical role of nidogens in BM assembly. It should be noted, however, that ultrastructurally normal BMs are found in the kidney [77] and skin [79] in these double mutants, suggesting that the requirement for nidogens in BM assembly is tissue-specific.

2.4 Perlecan

Perlecan, also called heparin sulfate proteoglycan 2 (HSPG2), is a large (~470 kDa) multi-domain protein expressed in most BMs throughout the body [80]. Perlecan contains five domains (domain I–V) [81–83] (Fig. 8.6): Domain I has three glycosaminoglycan attachment sites and a sea urchin sperm protein-enterokinase-argin (SEA) motif; Domain II contains four low-density lipoprotein (LDL) receptor class A repeats and an immunoglobulin (Ig) domain; Domain III has multiple laminin B and laminin EGF domains; Domain IV has a series of Ig domains; and Domain V

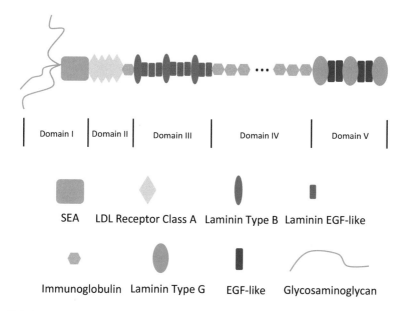

Fig. 8.6 Structural illustration of perlecan. Perlecan is a five-domain protein that contains three glycosaminoglycan chains at its N-terminus and multiple functional subdomains. The black lines underneath the structural diagram indicate the regions of these five domains (I–V)

contains many laminin G and EGF domains. Unlike collagen IV and laminin, perlecan is unable to form supramolecular sheet-like structures [81].

The various domains and motifs enable perlecan to interact with a large number of molecules [81, 84–91], including heparin-binding growth factors, other ECM proteins, and extracellular domains of cell surface components. It should be noted that perlecan, especially its domains IV and V, can be recognized and cleaved by many proteases [81, 82, 92–99], including thrombin, plasmin, and collagenase. It has been shown that protease cleavage of perlecan generates many bioactive fragments [81, 83, 92, 93, 95, 98, 99], suggesting that perlecan may have multiple important functions. Consistent with this speculation, perlecan null mice show massive developmental abnormalities in many organs [100–102]. Interestingly, the formation of BMs is not affected in these mutants, although BMs deteriorate in regions with increased mechanical stress [101]. These data suggest that perlecan, like collagen IV, is dispensable for BM formation, but required for BM maintenance.

3   BM Assembly

Although many efforts have been devoted to BM assembly research, the exact mechanism remains not fully understood. It is generally accepted that the assembly process involves three steps [22, 26, 49, 103–109]. First, cell-surface receptors (e.g.

integrins and dystroglycans), via interacting with laminin, facilitate the deposition of laminin network. Next, collagen IV network is connected to laminin network on cell surface via nidogen bridging, forming a 3D scaffold. This scaffold then interacts with other BM components, including perlecan, and eventually forms a fully functional BM.

4 BM Function

The BMs play important roles in physiological conditions, including providing structural support for cells, separating tissue compartments, and modulating molecular signaling [36, 110]. In the brain, the BMs actively participate in the regulation of BBB integrity. First, as the non-cellular component of the BBB located between endothelial cells and astrocytes [111, 112], the BMs serve as a physical barrier between the circulation system and the CNS. Electron microscopy studies showed that leukocytes were frequently detected between the endothelial cells and the underlying BMs [113–115], suggesting that crossing the BMs is the rate-limiting step in leukocyte infiltration. Consistent with this observation, it has been demonstrated that leukocytes take about 5 min to cross endothelial monolayer, but 20–30 min to penetrate the underlying BM [116–119]. Additionally, it has been reported that T cells spend 9–10 min crossing the endothelium, but they are retained in the outer surface of vessels for approximately 30 min [120]. Second, BM components (ECM proteins) are able to regulate/maintain BBB integrity via signaling to various cell types that they have direct contact with, including endothelial cells, pericytes and astrocytes. In fact, various receptors for these ECM proteins, including integrins and dystroglycan, have been identified in these cells [4, 55]. The biological functions of ECM-receptor interaction have been investigated extensively. Accumulating evidence shows that knockout/blockage of ECM receptors or their respective ligands results in similar phenotypes [21, 55], strongly indicating that ECM proteins exert these important functions by binding to their receptors.

5 BM in Stroke

Stroke is the fifth leading cause of death and the leading cause of serious long-term disability in the United States [121]. It is broadly categorized into ischemic stroke and hemorrhagic stroke, based on the nature of injury.

Accumulating evidence shows that ischemic stroke induces dissolution of the BMs. For example, BM dissolution was observed soon after the onset of ischemia in two different stroke models [122]. In addition, changes in BM components were detected as early as 10 min after reperfusion in the middle cerebral artery occlusion (MCAO) model [123]. At the ultrastructural level, the well-defined electron-dense BMs become fuzzy and faint after ischemia [124, 125]. Biochemically, ischemia significantly reduces the levels of various BM components [122, 126], including collagen IV and laminin (see below for details). In addition to ischemic stroke, loss of BM has also been reported in hemorrhagic stroke. For example, BM degradation and BBB breakdown were found in rats 24–72 h after subarachnoid hemorrhage [127]. Furthermore, intracerebral injection of collagenase, an enzyme that degrades collagen, is widely used as a hemorrhage model [128–130]. Additionally, various mutations in COL4A1/2 genes, which encode collagen IV, lead to intracerebral hemorrhage [51, 52, 131].

Stroke induced decrease of major BM components could be due to either reduced protein synthesis or enhanced protein degradation. Given that: (1) the loss of BM components can be detected as early as 1–3 h after ischemia [122, 123], and (2) these ECM proteins have a long turnover rate [132], it is logical to hypothesize that increased degradation is responsible for the reduced BM components after stroke. Consistent with this hypothesis, the activities of various proteolytic enzymes have been found increased after stroke. For instance, the activities of tissue-type [133] and urokinase-type [134–136] plasminogen activators, which are able to degrade collagen IV and laminin [26, 137–139], are significantly increased after MCAO. In addition, matrix metalloproteinases, a large family of proteinases that cleave various ECM proteins [138, 140], are also up-regulated after stroke [141–149]. Another type of proteolytic enzymes is cathepsins, which digest perlecan and many other cellular components [150]. It has been shown that the enzymatic activities and expression levels of cathepsin B and L are increased early after ischemia [150, 151]. Below, we discuss the changes of four major BM components in stroke individually.

5.1 Collagen IV

Collagen IV, found in almost all BMs, plays an important role in vascular integrity and stroke pathology. There is evidence showing that collagen IV level is reduced after ischemic stroke. For example, decreased expression of collagen IV was found in both thromboembolic and MCAO stroke models [138, 152]. Similar results were reported at 3 and 24 h after focal ischemia [138, 153]. Furthermore, ischemic stroke was detected in patients with mutations in COL4A1 gene [154–156], which encodes collagen IV α1 chain. These data suggest that collagen IV may be involved in the pathogenesis of ischemic stroke. In contrast to the reduced level of collagen IV in brain ischemic injury, increased collagen IV expression was found in the spinal cord 24 h after ischemia-reperfusion injury [157], suggesting that the regulation of collagen after ischemia is organ-specific.

In addition to ischemic stroke, collagen IV also plays a critical role in the pathogenesis of hemorrhagic stroke. First, severe collagen IV degradation was found in hemorrhagic regions in ischemic brain [158]. Next, mutations in COL4A1 gene have also been shown to cause perinatal cerebral hemorrhage [159]. Although about 50% of these mutants die within 1 day after birth [159], all those that develop into adulthood have hemorrhagic stroke [160]. In addition, many other mutations in COL4A1 gene result in various degrees of intracerebral hemorrhage in adult mice [51, 52] (Fig. 8.7). Like COL4A1, a G-to-D mutation in COL4A2, which encodes collagen IV α2 chain, also leads to intracerebral hemorrhage [131]. Mutations in COL4A1/2 and the severity of hemorrhage are summarized in Fig. 8.7. Consistent with animal work, mutations in COL4A1/2 have been identified in human patients with intracerebral hemorrhages [51, 131, 154, 160–173], suggesting a crucial role of collagen IV in the pathogenesis of intracerebral hemorrhage. Further mechanistic studies demonstrate that mutant collagen IV proteins are stored inside cells at the expense of secretion [50, 131, 159], and that intracellular accumulation of mutant proteins in endothelial cells and pericytes is a key triggering factor of intracerebral hemorrhage [52].

5.2   Laminin

Laminin changes after ischemic stroke have been controversial. On one hand, there is evidence showing that laminin level is reduced after ischemic stroke. For example, using an ischemia-reperfusion model in baboons, researchers have shown that both the number of laminin-positive vessels and laminin expression are gradually and continuously decreased after reperfusion [150, 174]. Additionally, substantial diminish of laminin expression was also observed in Mongolian gerbils up to 72 h of reperfusion after ischemia [175]. Interestingly, the reduction of laminin expression correlates with the up-regulation of various proteases [150, 174, 175], suggesting that ischemia-induced increase of proteases may be responsible for the degradation of laminin and other ECM proteins. Furthermore, similar changes in laminin and protease expression were also found in stroke patients [176].

Laminin expression, on the other hand, has also been reported to be up-regulated after ischemic stroke. It has been shown that laminin expression is increased in the injury core within the first 24 h after ischemia [177, 178]. By 2–3 days after ischemia, reactive astrocytes strongly express laminin, forming a laminin-positive barrier that separates the ischemic regions from healthy tissue [177, 178]. Additionally, Ji and Tsirka demonstrated a transient up-regulation of laminin expression in mice after MCAO [179]. They further showed that the increase of laminin was blocked by COX-2 inhibitor celecoxib or in COX-2$^{-/-}$ mice [179], suggesting that the transient up-regulation of laminin after MCAO is COX-2 dependent.

This discrepancy on laminin expression after ischemic stroke may be explained by differences in animal models/antibodies/time points or distinct fixation methods. It has been reported that fixation method heavily affects the immunohistochemical staining of laminin on brain tissue [180]. For example, heavy formaldehyde fixation masks vascular BM laminin antigen, whereas mild-moderate formaldehyde fixation favors or reveals vascular BM laminin antigen [180]. Therefore, it is important to indicate the fixation protocol used for laminin immunostaining in future studies.

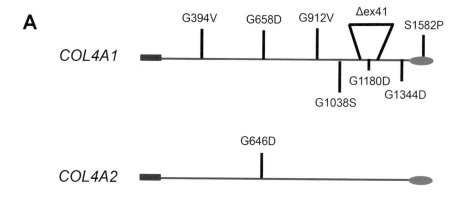

Fig. 8.7 Collagen IV mutations and the hemorrhagic phenotypes. (**a**) Diagram illustration of various mutations and their locations in COL4A1/2 genes. (**b**) Comparison of the severity of hemorrhage caused by these mutations in COL4A1/2

Unlike ischemic stroke, a causative effect of loss of laminin in intracerebral hemorrhage has been reported. Using a conditional knockout mouse line, we have shown that astrocytic laminin is essential for smooth muscle cell maturation and loss of astrocytic laminin leads to age-dependent and region-specific intracerebral hemorrhage [181]. In addition, micro-hemorrhages were also found in mice deficient in pericytic laminin [63]. It should be noted that only approximately 11% of the mutants demonstrated micro-hemorrhages [63], suggesting an incomplete penetrance of this phenotype. This incomplete penetrance may be due to the mixed genetic background. Furthermore, a significantly lower level of laminin expression was observed in hemorrhagic regions compared to non-hemorrhagic areas in ischemic brains [182]. These studies reveal a strong correlation between laminin level and intracerebral hemorrhage, indicating a critical role of laminin in maintaining the integrity of blood vessels in the brain.

5.3 Nidogen

Nidogen changes in ischemic stroke are largely unknown. It has been reported that nidogen expression in human brain endothelial cells increases over time under nitric oxide-induced oxidative stress [183], which mimics one of the pathological changes in ischemic stroke, suggesting that nidogen may be increased after ischemic stroke. This result, however, needs to be validated *in vivo* using ischemic models.

Like in ischemic stroke, nidogen expression in intracerebral hemorrhage remains unclear. In an *in vitro* study, fragmentations of nidogen-1 and -2 have been detected in astrocyte medium after exposure to plasma kallikrein [184], which replicates the pathology of intracerebral hemorrhage, suggesting that nidogens may be degraded after intracerebral hemorrhage. No *in vivo* studies, however, are currently available. Future studies should focus on understanding how ischemia and intracerebral hemorrhage affect nidogen expression.

5.4 Perlecan

Perlecan has been suggested as the most sensitive ECM protein to proteolysis in the ischemic core [150]. Immunohistochemical analysis revealed a 43–63% decrease of perlecan expression within a few hours after MCAO [150]. There is evidence demonstrating that, rather than being degraded, perlecan is cleaved into various bioactive fragments after stroke. For example, perlecan domain V (termed DV hereafter) is persistently up-regulated in rodents in both transient MCAO [98] and photothrombotic stroke models [185]. Consistent with this finding, enhanced DV level was observed in human brains around regions of ischemic injury [186]. In addition, the C-terminal fragment of DV—laminin-like globular domain 3 (termed LG3 hereafter) was also found increased after ischemic stroke in mice and rats [99]. The same result was replicated in the oxygen/glucose-deprivation model [99]. Further examination of the biological functions of these perlecan fragments has shown that they play multiple beneficial roles after ischemic stroke [98, 99, 187], including reducing ischemic lesion and preventing neuronal death, inducing angiogenesis without disrupting BBB integrity, improving motor function, and modulating astrogliosis. These data suggest that perlecan is rapidly processed after ischemic injury and its proteolytic fragments (DV and LG3) actively regulate the pathogenesis of ischemic stroke, promoting recovery.

Currently, there are no data available on perlecan expression in intracerebral hemorrhage. Future studies should focus on determining how perlecan expression changes in hemorrhagic stroke and elucidating the therapeutic potential of DV and LG3 in this disease.

6   Future Directions

The ECM/BM has been understudied owing to its intrinsic complexity. With the advancements in molecular biology and genetic techniques, significant progress has been made in the field of ECM research. For example, the generation of various knockout and/or conditional knockout mice has allowed investigation of the biological functions of many BM components. Many important questions, however, remain to be answered. These key questions include: what are the temporal and spatial expression profiles of ECM proteins in physiological and pathological conditions? How do ischemia and hemorrhage affect the expression of each individual ECM components? What are the molecular mechanisms underlying stroke-induced ECM alterations? What are the biological functions of proteolytic fragments of ECM proteins after stroke? Can exogenous ECM proteins prevent and/or ameliorate pathological changes in stroke? Elucidating these questions will substantially enrich our knowledge in ECM biology and stroke pathology, and may lead to innovative therapies for stroke.

Funding Information This work was supported by AHA Scientist Development Grant (16SDG29320001).

References

1. Bignami A, Hosley M, Dahl D. Hyaluronic acid and hyaluronic acid-binding proteins in brain extracellular matrix. Anat Embryol (Berl). 1993;188:419–33.
2. Cragg B. Brain extracellular space fixed for electron microscopy. Neurosci Lett. 1979;15:301–6.
3. Hynes RO. The extracellular matrix: not just pretty fibrils. Science. 2009;326:1216–9.
4. Baeten KM, Akassoglou K. Extracellular matrix and matrix receptors in blood-brain barrier formation and stroke. Dev Neurobiol. 2011;71:1018–39.
5. Jarvelainen H, Sainio A, Koulu M, Wight TN, Penttinen R. Extracellular matrix molecules: potential targets in pharmacotherapy. Pharmacol Rev. 2009;61:198–223.
6. Bateman JF, Boot-Handford RP, Lamande SR. Genetic diseases of connective tissues: cellular and extracellular effects of ecm mutations. Nat Rev Genet. 2009;10:173–83.
7. Yao Y. Laminin: loss-of-function studies. Cell Mol Life Sci. 2017;74(6):1095–115.
8. Rauch U. Brain matrix: structure, turnover and necessity. Biochem Soc Trans. 2007;35:656–60.
9. Lau LW, Cua R, Keough MB, Haylock-Jacobs S, Yong VW. Pathophysiology of the brain extracellular matrix: a new target for remyelination. Nat Rev Neurosci. 2013;14:722–9.
10. Bonnans C, Chou J, Werb Z. Remodelling the extracellular matrix in development and disease. Nat Rev Mol Cell Biol. 2014;15:786–801.
11. Kwok JC, Dick G, Wang D, Fawcett JW. Extracellular matrix and perineuronal nets in cns repair. Dev Neurobiol. 2011;71:1073–89.

12. Bonneh-Barkay D, Wiley CA. Brain extracellular matrix in neurodegeneration. Brain Pathol. 2009;19:573–85.
13. Yamaguchi Y. Lecticans: organizers of the brain extracellular matrix. Cell Mol Life Sci. 2000;57:276–89.
14. Dityatev A, Schachner M. Extracellular matrix molecules and synaptic plasticity. Nat Rev Neurosci. 2003;4:456–68.
15. Vracko R. Significance of basal lamina for regeneration of injured lung. Virchows Arch A Pathol Pathol Anat. 1972;355:264–74.
16. Vracko R, Benditt EP. Basal lamina: the scaffold for orderly cell replacement. Observations on regeneration of injured skeletal muscle fibers and capillaries. J Cell Biol. 1972;55:406–19.
17. Vracko R, Benditt EP. Capillary basal lamina thickening. Its relationship to endothelial cell death and replacement. J Cell Biol. 1970;47:281–5.
18. Vracko R. Basal lamina scaffold-anatomy and significance for maintenance of orderly tissue structure. Am J Pathol. 1974;77:314–46.
19. Vracko R, Strandness DE Jr. Basal lamina of abdominal skeletal muscle capillaries in diabetics and nondiabetics. Circulation. 1967;35:690–700.
20. Ruben GC, Yurchenco PD. High resolution platinum-carbon replication of freeze-dried basement membrane. Microsc Res Tech. 1994;28:13–28.
21. Yurchenco PD, Amenta PS, Patton BL. Basement membrane assembly, stability and activities observed through a developmental lens. Matrix Biol. 2004;22:521–38.
22. Kalluri R. Basement membranes: structure, assembly and role in tumour angiogenesis. Nat Rev Cancer. 2003;3:422–33.
23. Sorokin L. The impact of the extracellular matrix on inflammation. Nat Rev Immunol. 2010;10:712–23.
24. Hallmann R, Zhang X, Di Russo J, Li L, Song J, Hannocks MJ, et al. The regulation of immune cell trafficking by the extracellular matrix. Curr Opin Cell Biol. 2015;36:54–61.
25. Paulsson M. Basement membrane proteins: Structure, assembly, and cellular interactions. Crit Rev Biochem Mol Biol. 1992;27:93–127.
26. Yurchenco PD, Schittny JC. Molecular architecture of basement membranes. FASEB J. 1990;4:1577–90.
27. Schittny JC, Yurchenco PD. Basement membranes: molecular organization and function in development and disease. Curr Opin Cell Biol. 1989;1:983–8.
28. Erickson AC, Couchman JR. Still more complexity in mammalian basement membranes. J Histochem Cytochem. 2000;48:1291–306.
29. Myers JC, Dion AS, Abraham V, Amenta PS. Type xv collagen exhibits a widespread distribution in human tissues but a distinct localization in basement membrane zones. Cell Tissue Res. 1996;286:493–505.
30. Myers JC, Li D, Bageris A, Abraham V, Dion AS, Amenta PS. Biochemical and immunohistochemical characterization of human type xix defines a novel class of basement membrane zone collagens. Am J Pathol. 1997;151:1729–40.
31. Saarela J, Rehn M, Oikarinen A, Autio-Harmainen H, Pihlajaniemi T. The short and long forms of type xviii collagen show clear tissue specificities in their expression and location in basement membrane zones in humans. Am J Pathol. 1998;153:611–26.
32. Tomono Y, Naito I, Ando K, Yonezawa T, Sado Y, Hirakawa S, et al. Epitope-defined monoclonal antibodies against multiplexin collagens demonstrate that type xv and xviii collagens are expressed in specialized basement membranes. Cell Struct Funct. 2002;27:9–20.
33. Yurchenco PD. Basement membranes: cell scaffoldings and signaling platforms. Cold Spring Harb Perspect Biol. 2011;3:pii: a004911.
34. Martin GR, Timpl R. Laminin and other basement membrane components. Annu Rev Cell Biol. 1987;3:57–85.
35. Sixt M, Engelhardt B, Pausch F, Hallmann R, Wendler O, Sorokin LM. Endothelial cell laminin isoforms, laminins 8 and 10, play decisive roles in t cell recruitment across the blood-brain barrier in experimental autoimmune encephalomyelitis. J Cell Biol. 2001;153:933–46.

36. Hallmann R, Horn N, Selg M, Wendler O, Pausch F, Sorokin LM. Expression and function of laminins in the embryonic and mature vasculature. Physiol Rev. 2005;85:979–1000.
37. Owens T, Bechmann I, Engelhardt B. Perivascular spaces and the two steps to neuroinflammation. J Neuropathol Exp Neurol. 2008;67:1113–21.
38. van Horssen J, Bo L, Vos CM, Virtanen I, de Vries HE. Basement membrane proteins in multiple sclerosis-associated inflammatory cuffs: potential role in influx and transport of leukocytes. J Neuropathol Exp Neurol. 2005;64:722–9.
39. Hudson BG, Reeders ST, Tryggvason K. Type iv collagen: structure, gene organization, and role in human diseases. Molecular basis of goodpasture and alport syndromes and diffuse leiomyomatosis. J Biol Chem. 1993;268:26033–6.
40. Kalluri R. Discovery of type iv collagen non-collagenous domains as novel integrin ligands and endogenous inhibitors of angiogenesis. Cold Spring Harb Symp Quant Biol. 2002;67:255–66.
41. Filie JD, Burbelo PD, Kozak CA. Genetic mapping of the alpha 1 and alpha 2 (iv) collagen genes to mouse chromosome 8. Mamm Genome. 1995;6:487.
42. Sugimoto M, Oohashi T, Ninomiya Y. The genes col4a5 and col4a6, coding for basement membrane collagen chains alpha 5(iv) and alpha 6(iv), are located head-to-head in close proximity on human chromosome xq22 and col4a6 is transcribed from two alternative promoters. Proc Natl Acad Sci U S A. 1994;91:11679–83.
43. Soininen R, Huotari M, Hostikka SL, Prockop DJ, Tryggvason K. The structural genes for alpha 1 and alpha 2 chains of human type iv collagen are divergently encoded on opposite DNA strands and have an overlapping promoter region. J Biol Chem. 1988;263:17217–20.
44. Momota R, Sugimoto M, Oohashi T, Kigasawa K, Yoshioka H, Ninomiya Y. Two genes, col4a3 and col4a4 coding for the human alpha3(iv) and alpha4(iv) collagen chains are arranged head-to-head on chromosome 2q36. FEBS Lett. 1998;424:11–6.
45. Ortega N, Werb Z. New functional roles for non-collagenous domains of basement membrane collagens. J Cell Sci. 2002;115:4201–14.
46. Khoshnoodi J, Pedchenko V, Hudson BG. Mammalian collagen iv. Microsc Res Tech. 2008;71:357–70.
47. Vanacore RM, Shanmugasundararaj S, Friedman DB, Bondar O, Hudson BG, Sundaramoorthy M. The alpha1.Alpha2 network of collagen iv. Reinforced stabilization of the noncollagenous domain-1 by noncovalent forces and the absence of met-lys cross-links. J Biol Chem. 2004;279:44723–30.
48. Tilling T, Korte D, Hoheisel D, Galla HJ. Basement membrane proteins influence brain capillary endothelial barrier function in vitro. J Neurochem. 1998;71:1151–7.
49. Poschl E, Schlotzer-Schrehardt U, Brachvogel B, Saito K, Ninomiya Y, Mayer U. Collagen iv is essential for basement membrane stability but dispensable for initiation of its assembly during early development. Development. 2004;131:1619–28.
50. Kuo DS, Labelle-Dumais C, Mao M, Jeanne M, Kauffman WB, Allen J, et al. Allelic heterogeneity contributes to variability in ocular dysgenesis, myopathy and brain malformations caused by col4a1 and col4a2 mutations. Hum Mol Genet. 2014;23:1709–22.
51. Favor J, Gloeckner CJ, Janik D, Klempt M, Neuhauser-Klaus A, Pretsch W, et al. Type iv procollagen missense mutations associated with defects of the eye, vascular stability, the brain, kidney function and embryonic or postnatal viability in the mouse, mus musculus: an extension of the col4a1 allelic series and the identification of the first two col4a2 mutant alleles. Genetics. 2007;175:725–36.
52. Jeanne M, Jorgensen J, Gould DB. Molecular and genetic analyses of collagen type iv mutant mouse models of spontaneous intracerebral hemorrhage identify mechanisms for stroke prevention. Circulation. 2015;131:1555–65.
53. Miner JH, Yurchenco PD. Laminin functions in tissue morphogenesis. Annu Rev Cell Dev Biol. 2004;20:255–84.
54. Domogatskaya A, Rodin S, Tryggvason K. Functional diversity of laminins. Annu Rev Cell Dev Biol. 2012;28:523–53.

55. Colognato H, Yurchenco PD. Form and function: the laminin family of heterotrimers. Dev Dyn. 2000;218:213–34.
56. Aumailley M. The laminin family. Cell Adhes Migr. 2013;7:48–55.
57. Aumailley M, Bruckner-Tuderman L, Carter WG, Deutzmann R, Edgar D, Ekblom P, et al. A simplified laminin nomenclature. Matrix Biol. 2005;24:326–32.
58. Durbeej M. Laminins. Cell Tissue Res. 2010;339:259–68.
59. McKee KK, Harrison D, Capizzi S, Yurchenco PD. Role of laminin terminal globular domains in basement membrane assembly. J Biol Chem. 2007;282:21437–47.
60. Burgeson RE, Chiquet M, Deutzmann R, Ekblom P, Engel J, Kleinman H, et al. A new nomenclature for the laminins. Matrix Biol. 1994;14:209–11.
61. Sorokin LM, Pausch F, Frieser M, Kroger S, Ohage E, Deutzmann R. Developmental regulation of the laminin alpha5 chain suggests a role in epithelial and endothelial cell maturation. Dev Biol. 1997;189:285–300.
62. Jucker M, Tian M, Norton DD, Sherman C, Kusiak JW. Laminin alpha 2 is a component of brain capillary basement membrane: reduced expression in dystrophic dy mice. Neuroscience. 1996;71:1153–61.
63. Gautam J, Zhang X, Yao Y. The role of pericytic laminin in blood brain barrier integrity maintenance. Sci Rep. 2016;6:36450.
64. Yao Y, Chen ZL, Norris EH, Strickland S. Astrocytic laminin regulates pericyte differentiation and maintains blood brain barrier integrity. Nat Commun. 2014;5:3413.
65. Thyboll J, Kortesmaa J, Cao R, Soininen R, Wang L, Iivanainen A, et al. Deletion of the laminin alpha4 chain leads to impaired microvessel maturation. Mol Cell Biol. 2002;22:1194–202.
66. Patton BL, Miner JH, Chiu AY, Sanes JR. Distribution and function of laminins in the neuromuscular system of developing, adult, and mutant mice. J Cell Biol. 1997;139:1507–21.
67. Miner JH, Li C, Mudd JL, Go G, Sutherland AE. Compositional and structural requirements for laminin and basement membranes during mouse embryo implantation and gastrulation. Development. 2004;131:2247–56.
68. Alpy F, Jivkov I, Sorokin L, Klein A, Arnold C, Huss Y, et al. Generation of a conditionally null allele of the laminin alpha1 gene. Genesis. 2005;43:59–70.
69. Smyth N, Vatansever HS, Murray P, Meyer M, Frie C, Paulsson M, et al. Absence of basement membranes after targeting the lamc1 gene results in embryonic lethality due to failure of endoderm differentiation. J Cell Biol. 1999;144:151–60.
70. Murray P, Edgar D. Regulation of programmed cell death by basement membranes in embryonic development. J Cell Biol. 2000;150:1215–21.
71. Kang SH, Kramer JM. Nidogen is nonessential and not required for normal type iv collagen localization in caenorhabditis elegans. Mol Biol Cell. 2000;11:3911–23.
72. Murshed M, Smyth N, Miosge N, Karolat J, Krieg T, Paulsson M, et al. The absence of nidogen 1 does not affect murine basement membrane formation. Mol Cell Biol. 2000;20:7007–12.
73. Dong L, Chen Y, Lewis M, Hsieh JC, Reing J, Chaillet JR, et al. Neurologic defects and selective disruption of basement membranes in mice lacking entactin-1/nidogen-1. Lab Investig. 2002;82:1617–30.
74. May CA. Distribution of nidogen in the murine eye and ocular phenotype of the nidogen-1 knockout mouse. ISRN Ophthalmol. 2012;2012:378641.
75. Schymeinsky J, Nedbal S, Miosge N, Poschl E, Rao C, Beier DR, et al. Gene structure and functional analysis of the mouse nidogen-2 gene: Nidogen-2 is not essential for basement membrane formation in mice. Mol Cell Biol. 2002;22:6820–30.
76. Miosge N, Sasaki T, Timpl R. Evidence of nidogen-2 compensation for nidogen-1 deficiency in transgenic mice. Matrix Biol. 2002;21:611–21.
77. Bader BL, Smyth N, Nedbal S, Miosge N, Baranowsky A, Mokkapati S, et al. Compound genetic ablation of nidogen 1 and 2 causes basement membrane defects and perinatal lethality in mice. Mol Cell Biol. 2005;25:6846–56.
78. Bose K, Nischt R, Page A, Bader BL, Paulsson M, Smyth N. Loss of nidogen-1 and -2 results in syndactyly and changes in limb development. J Biol Chem. 2006;281:39620–9.

79. Mokkapati S, Baranowsky A, Mirancea N, Smyth N, Breitkreutz D, Nischt R. Basement membranes in skin are differently affected by lack of nidogen 1 and 2. J Invest Dermatol. 2008;128:2259–67.

80. Knox SM, Whitelock JM. Perlecan: how does one molecule do so many things? Cell Mol Life Sci. 2006;63:2435–45.

81. Farach-Carson MC, Carson DD. Perlecan—a multifunctional extracellular proteoglycan scaffold. Glycobiology. 2007;17:897–905.

82. Whitelock JM, Melrose J, Iozzo RV. Diverse cell signaling events modulated by perlecan. Biochemistry. 2008;47:11174–83.

83. Roberts J, Kahle MP, Bix GJ. Perlecan and the blood-brain barrier: beneficial proteolysis? Front Pharmacol. 2012;3:155.

84. Costell M, Sasaki T, Mann K, Yamada Y, Timpl R. Structural characterization of recombinant domain ii of the basement membrane proteoglycan perlecan. FEBS Lett. 1996;396:127–31.

85. Crossin KL, Krushel LA. Cellular signaling by neural cell adhesion molecules of the immunoglobulin superfamily. Dev Dyn. 2000;218:260–79.

86. Dolan M, Horchar T, Rigatti B, Hassell JR. Identification of sites in domain i of perlecan that regulate heparan sulfate synthesis. J Biol Chem. 1997;272:4316–22.

87. Fjeldstad K, Kolset SO. Decreasing the metastatic potential in cancers—targeting the heparan sulfate proteoglycans. Curr Drug Targets. 2005;6:665–82.

88. Hopf M, Gohring W, Mann K, Timpl R. Mapping of binding sites for nidogens, fibulin-2, fibronectin and heparin to different ig modules of perlecan. J Mol Biol. 2001;311:529–41.

89. Jiang X, Couchman JR. Perlecan and tumor angiogenesis. J Histochem Cytochem. 2003;51:1393–410.

90. Merz DC, Alves G, Kawano T, Zheng H, Culotti JG. Unc-52/perlecan affects gonadal leader cell migrations in C. elegans hermaphrodites through alterations in growth factor signaling. Dev Biol. 2003;256:173–86.

91. Yang Y. Wnts and wing: Wnt signaling in vertebrate limb development and musculoskeletal morphogenesis. Birth Defects Res C Embryo Today. 2003;69:305–17.

92. Gonzalez EM, Reed CC, Bix G, Fu J, Zhang Y, Gopalakrishnan B, et al. Bmp-1/tolloid-like metalloproteases process endorepellin, the angiostatic c-terminal fragment of perlecan. J Biol Chem. 2005;280:7080–7.

93. Saini MG, Bix GJ. Oxygen-glucose deprivation (ogd) and interleukin-1 (il-1) differentially modulate cathepsin b/l mediated generation of neuroprotective perlecan lg3 by neurons. Brain Res. 2012;1438:65–74.

94. Whitelock JM, Murdoch AD, Iozzo RV, Underwood PA. The degradation of human endothelial cell-derived perlecan and release of bound basic fibroblast growth factor by stromelysin, collagenase, plasmin, and heparanases. J Biol Chem. 1996;271:10079–86.

95. Bix G, Fu J, Gonzalez EM, Macro L, Barker A, Campbell S, et al. Endorepellin causes endothelial cell disassembly of actin cytoskeleton and focal adhesions through alpha2beta1 integrin. J Cell Biol. 2004;166:97–109.

96. Irving-Rodgers HF, Catanzariti KD, Aspden WJ, D'Occhio MJ, Rodgers RJ. Remodeling of extracellular matrix at ovulation of the bovine ovarian follicle. Mol Reprod Dev. 2006;73:1292–302.

97. Thadikkaran L, Crettaz D, Siegenthaler MA, Gallot D, Sapin V, Iozzo RV, et al. The role of proteomics in the assessment of premature rupture of fetal membranes. Clin Chim Acta. 2005;360:27–36.

98. Lee B, Clarke D, Al Ahmad A, Kahle M, Parham C, Auckland L, et al. Perlecan domain v is neuroprotective and proangiogenic following ischemic stroke in rodents. J Clin Invest. 2011;121:3005–23.

99. Saini MG, Pinteaux E, Lee B, Bix GJ. Oxygen-glucose deprivation and interleukin-1alpha trigger the release of perlecan lg3 by cells of neurovascular unit. J Neurochem. 2011;119:760–71.

100. Arikawa-Hirasawa E, Watanabe H, Takami H, Hassell JR, Yamada Y. Perlecan is essential for cartilage and cephalic development. Nat Genet. 1999;23:354–8.

101. Costell M, Gustafsson E, Aszodi A, Morgelin M, Bloch W, Hunziker E, et al. Perlecan maintains the integrity of cartilage and some basement membranes. J Cell Biol. 1999;147:1109–22.
102. Rossi M, Morita H, Sormunen R, Airenne S, Kreivi M, Wang L, et al. Heparan sulfate chains of perlecan are indispensable in the lens capsule but not in the kidney. EMBO J. 2003;22:236–45.
103. Mouw JK, Ou G, Weaver VM. Extracellular matrix assembly: a multiscale deconstruction. Nat Rev Mol Cell Biol. 2014;15:771–85.
104. Takagi J, Yang Y, Liu JH, Wang JH, Springer TA. Complex between nidogen and laminin fragments reveals a paradigmatic beta-propeller interface. Nature. 2003;424:969–74.
105. Willem M, Miosge N, Halfter W, Smyth N, Jannetti I, Burghart E, et al. Specific ablation of the nidogen-binding site in the laminin gamma1 chain interferes with kidney and lung development. Development. 2002;129:2711–22.
106. Aumailley M, Pesch M, Tunggal L, Gaill F, Fassler R. Altered synthesis of laminin 1 and absence of basement membrane component deposition in (beta)1 integrin-deficient embryoid bodies. J Cell Sci. 2000;113(Pt 2):259–68.
107. Tsiper MV, Yurchenco PD. Laminin assembles into separate basement membrane and fibrillar matrices in schwann cells. J Cell Sci. 2002;115:1005–15.
108. Charonis AS, Tsilibary EC, Yurchenco PD, Furthmayr H. Binding of laminin to type iv collagen: a morphological study. J Cell Biol. 1985;100:1848–53.
109. Sasaki T, Forsberg E, Bloch W, Addicks K, Fassler R, Timpl R. Deficiency of beta 1 integrins in teratoma interferes with basement membrane assembly and laminin-1 expression. Exp Cell Res. 1998;238:70–81.
110. Di Russo J, Hannocks MJ, Luik AL, Song J, Zhang X, Yousif L, et al. Vascular laminins in physiology and pathology. In: Matrix Biol, vol. 57-58; 2017. p. 140–8.
111. Engelhardt B, Ransohoff RM. Capture, crawl, cross: the T cell code to breach the blood-brain barriers. Trends Immunol. 2012;33:579–89.
112. Zlokovic BV. The blood-brain barrier in health and chronic neurodegenerative disorders. Neuron. 2008;57:178–201.
113. Furie MB, Naprstek BL, Silverstein SC. Migration of neutrophils across monolayers of cultured microvascular endothelial cells. An in vitro model of leucocyte extravasation. J Cell Sci. 1987;88(Pt 2):161–75.
114. Hurley JV. An electron microscopic study of leucocytic emigration and vascular permeability in rat skin. Aust J Exp Biol Med Sci. 1963;41:171–86.
115. Marchesi VT, Florey HW. Electron micrographic observations on the emigration of leucocytes. Q J Exp Physiol Cogn Med Sci. 1960;45:343–8.
116. Ohashi KL, Tung DK, Wilson J, Zweifach BW, Schmid-Schonbein GW. Transvascular and interstitial migration of neutrophils in rat mesentery. Microcirculation. 1996;3:199–210.
117. Hoshi O, Ushiki T. Neutrophil extravasation in rat mesenteric venules induced by the chemotactic peptide n-formyl-methionyl-luecylphenylalanine (fmlp), with special attention to a barrier function of the vascular basal lamina for neutrophil migration. Arch Histol Cytol. 2004;67:107–14.
118. Yadav R, Larbi KY, Young RE, Nourshargh S. Migration of leukocytes through the vessel wall and beyond. Thromb Haemost. 2003;90:598–606.
119. Bixel MG, Petri B, Khandoga AG, Khandoga A, Wolburg-Buchholz K, Wolburg H, et al. A cd99-related antigen on endothelial cells mediates neutrophil but not lymphocyte extravasation in vivo. Blood. 2007;109:5327–36.
120. Bartholomaus I, Kawakami N, Odoardi F, Schlager C, Miljkovic D, Ellwart JW, et al. Effector t cell interactions with meningeal vascular structures in nascent autoimmune cns lesions. Nature. 2009;462:94–8.
121. Writing Group M, Mozaffarian D, Benjamin EJ, Go AS, Arnett DK, Blaha MJ, et al. Heart disease and stroke statistics-2016 update: a report from the american heart association. Circulation. 2016;133:e38–60.
122. Hamann GF, Burggraf D, Martens HK, Liebetrau M, Jager G, Wunderlich N, et al. Mild to moderate hypothermia prevents microvascular basal lamina antigen loss in experimental focal cerebral ischemia. Stroke. 2004;35:764–9.

123. Yepes M, Sandkvist M, Wong MK, Coleman TA, Smith E, Cohan SL, et al. Neuroserpin reduces cerebral infarct volume and protects neurons from ischemia-induced apoptosis. Blood. 2000;96:569–76.
124. Nahirney PC, Reeson P, Brown CE. Ultrastructural analysis of blood-brain barrier breakdown in the peri-infarct zone in young adult and aged mice. J Cereb Blood Flow Metab. 2016;36:413–25.
125. Kwon I, Kim EH, del Zoppo GJ, Heo JH. Ultrastructural and temporal changes of the microvascular basement membrane and astrocyte interface following focal cerebral ischemia. J Neurosci Res. 2009;87:668–76.
126. Wang CX, Shuaib A. Critical role of microvasculature basal lamina in ischemic brain injury. Prog Neurobiol. 2007;83:140–8.
127. Scholler K, Trinkl A, Klopotowski M, Thal SC, Plesnila N, Trabold R, et al. Characterization of microvascular basal lamina damage and blood-brain barrier dysfunction following subarachnoid hemorrhage in rats. Brain Res. 2007;1142:237–46.
128. Wang J, Tsirka SE. Tuftsin fragment 1–3 is beneficial when delivered after the induction of intracerebral hemorrhage. Stroke. 2005;36:613–8.
129. Wang J, Rogove AD, Tsirka AE, Tsirka SE. Protective role of tuftsin fragment 1–3 in an animal model of intracerebral hemorrhage. Ann Neurol. 2003;54:655–64.
130. Wang J, Tsirka SE. Contribution of extracellular proteolysis and microglia to intracerebral hemorrhage. Neurocrit Care. 2005;3:77–85.
131. Jeanne M, Labelle-Dumais C, Jorgensen J, Kauffman WB, Mancini GM, Favor J, et al. Col4a2 mutations impair col4a1 and col4a2 secretion and cause hemorrhagic stroke. Am J Hum Genet. 2012;90:91–101.
132. Hynes RO. Stretching the boundaries of extracellular matrix research. Nat Rev Mol Cell Biol. 2014;15:761–3.
133. Yepes M, Sandkvist M, Moore EG, Bugge TH, Strickland DK, Lawrence DA. Tissue-type plasminogen activator induces opening of the blood-brain barrier via the ldl receptor-related protein. J Clin Invest. 2003;112:1533–40.
134. Ahn MY, Zhang ZG, Tsang W, Chopp M. Endogenous plasminogen activator expression after embolic focal cerebral ischemia in mice. Brain Res. 1999;837:169–76.
135. Hosomi N, Lucero J, Heo JH, Koziol JA, Copeland BR, del Zoppo GJ. Rapid differential endogenous plasminogen activator expression after acute middle cerebral artery occlusion. Stroke. 2001;32:1341–8.
136. Wang CX, Yang T, Shuaib A. Clot fragments formed from original thrombus obstruct downstream arteries in the ischemic injured brain. Microcirculation. 2006;13:229–36.
137. Liotta LA, Goldfarb RH, Brundage R, Siegal GP, Terranova V, Garbisa S. Effect of plasminogen activator (urokinase), plasmin, and thrombin on glycoprotein and collagenous components of basement membrane. Cancer Res. 1981;41:4629–36.
138. Hamann GF, Liebetrau M, Martens H, Burggraf D, Kloss CU, Bultemeier G, et al. Microvascular basal lamina injury after experimental focal cerebral ischemia and reperfusion in the rat. J Cereb Blood Flow Metab. 2002;22:526–33.
139. Lijnen HR, Van Hoef B, Lupu F, Moons L, Carmeliet P, Collen D. Function of the plasminogen/plasmin and matrix metalloproteinase systems after vascular injury in mice with targeted inactivation of fibrinolytic system genes. Arterioscler Thromb Vasc Biol. 1998;18:1035–45.
140. Cunningham LA, Wetzel M, Rosenberg GA. Multiple roles for mmps and timps in cerebral ischemia. Glia. 2005;50:329–39.
141. Fujimura M, Gasche Y, Morita-Fujimura Y, Massengale J, Kawase M, Chan PH. Early appearance of activated matrix metalloproteinase-9 and blood-brain barrier disruption in mice after focal cerebral ischemia and reperfusion. Brain Res. 1999;842:92–100.
142. Gasche Y, Fujimura M, Morita-Fujimura Y, Copin JC, Kawase M, Massengale J, et al. Early appearance of activated matrix metalloproteinase-9 after focal cerebral ischemia in mice: a possible role in blood-brain barrier dysfunction. J Cereb Blood Flow Metab. 1999;19:1020–8.
143. Heo JH, Lucero J, Abumiya T, Koziol JA, Copeland BR, del Zoppo GJ. Matrix metalloproteinases increase very early during experimental focal cerebral ischemia. J Cereb Blood Flow Metab. 1999;19:624–33.

144. Romanic AM, White RF, Arleth AJ, Ohlstein EH, Barone FC. Matrix metalloproteinase expression increases after cerebral focal ischemia in rats: Inhibition of matrix metalloproteinase-9 reduces infarct size. Stroke. 1998;29:1020–30.
145. Petty MA, Wettstein JG. Elements of cerebral microvascular ischaemia. Brain Res Brain Res Rev. 2001;36:23–34.
146. Liu XS, Zhang ZG, Zhang L, Morris DC, Kapke A, Lu M, et al. Atorvastatin downregulates tissue plasminogen activator-aggravated genes mediating coagulation and vascular permeability in single cerebral endothelial cells captured by laser microdissection. J Cereb Blood Flow Metab. 2006;26:787–96.
147. Rosenberg GA. Matrix metalloproteinases in neuroinflammation. Glia. 2002;39:279–91.
148. Rosenberg GA, Estrada EY, Dencoff JE. Matrix metalloproteinases and timps are associated with blood-brain barrier opening after reperfusion in rat brain. Stroke. 1998;29:2189–95.
149. Rosenberg GA, Navratil M, Barone F, Feuerstein G. Proteolytic cascade enzymes increase in focal cerebral ischemia in rat. J Cereb Blood Flow Metab. 1996;16:360–6.
150. Fukuda S, Fini CA, Mabuchi T, Koziol JA, Eggleston LL Jr, del Zoppo GJ. Focal cerebral ischemia induces active proteases that degrade microvascular matrix. Stroke. 2004;35:998–1004.
151. Seyfried D, Han Y, Zheng Z, Day N, Moin K, Rempel S, et al. Cathepsin b and middle cerebral artery occlusion in the rat. J Neurosurg. 1997;87:716–23.
152. Vosko MR, Busch E, Burggraf D, Bultemeier G, Hamann GF. Microvascular basal lamina damage in thromboembolic stroke in a rat model. Neurosci Lett. 2003;353:217–20.
153. Trinkl A, Vosko MR, Wunderlich N, Dichgans M, Hamann GF. Pravastatin reduces microvascular basal lamina damage following focal cerebral ischemia and reperfusion. Eur J Neurosci. 2006;24:520–6.
154. Breedveld G, de Coo IF, Lequin MH, Arts WF, Heutink P, Gould DB, et al. Novel mutations in three families confirm a major role of col4a1 in hereditary porencephaly. J Med Genet. 2006;43:490–5.
155. Sibon I, Coupry I, Menegon P, Bouchet JP, Gorry P, Burgelin I, et al. Col4a1 mutation in axenfeld-rieger anomaly with leukoencephalopathy and stroke. Ann Neurol. 2007;62:177–84.
156. Plaisier E, Gribouval O, Alamowitch S, Mougenot B, Prost C, Verpont MC, et al. Col4a1 mutations and hereditary angiopathy, nephropathy, aneurysms, and muscle cramps. N Engl J Med. 2007;357:2687–95.
157. Anik I, Kokturk S, Genc H, Cabuk B, Koc K, Yavuz S, et al. Immunohistochemical analysis of timp-2 and collagen types i and iv in experimental spinal cord ischemia-reperfusion injury in rats. J Spinal Cord Med. 2011;34:257–64.
158. Rosell A, Cuadrado E, Ortega-Aznar A, Hernandez-Guillamon M, Lo EH, Montaner J. Mmp-9-positive neutrophil infiltration is associated to blood-brain barrier breakdown and basal lamina type iv collagen degradation during hemorrhagic transformation after human ischemic stroke. Stroke. 2008;39:1121–6.
159. Gould DB, Phalan FC, Breedveld GJ, van Mil SE, Smith RS, Schimenti JC, et al. Mutations in col4a1 cause perinatal cerebral hemorrhage and porencephaly. Science. 2005;308:1167–71.
160. Gould DB, Phalan FC, van Mil SE, Sundberg JP, Vahedi K, Massin P, et al. Role of col4a1 in small-vessel disease and hemorrhagic stroke. N Engl J Med. 2006;354:1489–96.
161. de Vries LS, Mancini GM. Intracerebral hemorrhage and col4a1 and col4a2 mutations, from fetal life into adulthood. Ann Neurol. 2012;71:439–41.
162. Gunda B, Mine M, Kovacs T, Hornyak C, Bereczki D, Varallyay G, et al. Col4a2 mutation causing adult onset recurrent intracerebral hemorrhage and leukoencephalopathy. J Neurol. 2014;261:500–3.
163. Corlobe A, Tournier-Lasserve E, Mine M, Menjot de Champfleur N, Carra Dalliere C, Ayrignac X, et al. Col4a1 mutation revealed by an isolated brain hemorrhage. Cerebrovasc Dis. 2013;35:593–4.
164. de Vries LS, Koopman C, Groenendaal F, Van Schooneveld M, Verheijen FW, Verbeek E, et al. Col4a1 mutation in two preterm siblings with antenatal onset of parenchymal hemorrhage. Ann Neurol. 2009;65:12–8.

165. Shah S, Kumar Y, McLean B, Churchill A, Stoodley N, Rankin J, et al. A dominantly inherited mutation in collagen iv a1 (col4a1) causing childhood onset stroke without porencephaly. Eur J Paediatr Neurol. 2010;14:182–7.
166. Lichtenbelt KD, Pistorius LR, De Tollenaer SM, Mancini GM, De Vries LS. Prenatal genetic confirmation of a col4a1 mutation presenting with sonographic fetal intracranial hemorrhage. Ultrasound Obstet Gynecol. 2012;39:726–7.
167. Garel C, Rosenblatt J, Moutard ML, Heron D, Gelot A, Gonzales M, et al. Fetal intracerebral hemorrhage and col4a1 mutation: promise and uncertainty. Ultrasound Obstet Gynecol. 2013;41:228–30.
168. Colin E, Sentilhes L, Sarfati A, Mine M, Guichet A, Ploton C, et al. Fetal intracerebral hemorrhage and cataract: Think col4a1. J Perinatol. 2014;34:75–7.
169. Weng YC, Sonni A, Labelle-Dumais C, de Leau M, Kauffman WB, Jeanne M, et al. Col4a1 mutations in patients with sporadic late-onset intracerebral hemorrhage. Ann Neurol. 2012;71:470–7.
170. van der Knaap MS, Smit LM, Barkhof F, Pijnenburg YA, Zweegman S, Niessen HW, et al. Neonatal porencephaly and adult stroke related to mutations in collagen iv a1. Ann Neurol. 2006;59:504–11.
171. Vahedi K, Massin P, Guichard JP, Miocque S, Polivka M, Goutieres F, et al. Hereditary infantile hemiparesis, retinal arteriolar tortuosity, and leukoencephalopathy. Neurology. 2003;60:57–63.
172. Bilguvar K, DiLuna ML, Bizzarro MJ, Bayri Y, Schneider KC, Lifton RP, et al. Col4a1 mutation in preterm intraventricular hemorrhage. J Pediatr. 2009;155:743–5.
173. Lanfranconi S, Markus HS. Col4a1 mutations as a monogenic cause of cerebral small vessel disease: a systematic review. Stroke. 2010;41:e513–8.
174. Hamann GF, Okada Y, Fitridge R, del Zoppo GJ. Microvascular basal lamina antigens disappear during cerebral ischemia and reperfusion. Stroke. 1995;26:2120–6.
175. Zalewska T, Ziemka-Nalecz M, Sarnowska A, Domanska-Janik K. Transient forebrain ischemia modulates signal transduction from extracellular matrix in gerbil hippocampus. Brain Res. 2003;977:62–9.
176. Horstmann S, Kalb P, Koziol J, Gardner H, Wagner S. Profiles of matrix metalloproteinases, their inhibitors, and laminin in stroke patients: influence of different therapies. Stroke. 2003;34:2165–70.
177. Liesi P, Kaakkola S, Dahl D, Vaheri A. Laminin is induced in astrocytes of adult brain by injury. EMBO J. 1984;3:683–6.
178. Jucker M, Bialobok P, Kleinman HK, Walker LC, Hagg T, Ingram DK. Laminin-like and laminin-binding protein-like immunoreactive astrocytes in rat hippocampus after transient ischemia. Antibody to laminin-binding protein is a sensitive marker of neural injury and degeneration. Ann N Y Acad Sci. 1993;679:245–52.
179. Ji K, Tsirka SE. Inflammation modulates expression of laminin in the central nervous system following ischemic injury. J Neuroinflammation. 2012;9:159.
180. Jucker M, Bialobok P, Hagg T, Ingram DK. Laminin immunohistochemistry in brain is dependent on method of tissue fixation. Brain Res. 1992;586:166–70.
181. Chen ZL, Yao Y, Norris EH, Kruyer A, Jno-Charles O, Akhmerov A, et al. Ablation of astrocytic laminin impairs vascular smooth muscle cell function and leads to hemorrhagic stroke. J Cell Biol. 2013;202:381–95.
182. Hamann GF, Okada Y, del Zoppo GJ. Hemorrhagic transformation and microvascular integrity during focal cerebral ischemia/reperfusion. J Cereb Blood Flow Metab. 1996;16:1373–8.
183. Ning M, Sarracino DA, Kho AT, Guo S, Lee SR, Krastins B, et al. Proteomic temporal profile of human brain endothelium after oxidative stress. Stroke. 2011;42:37–43.
184. Liu J, Gao BB, Feener EP. Proteomic identification of novel plasma kallikrein substrates in the astrocyte secretome. Transl Stroke Res. 2010;1:276–86.
185. Bix GJ, Gowing EK, Clarkson AN. Perlecan domain v is neuroprotective and affords functional improvement in a photothrombotic stroke model in young and aged mice. Transl Stroke Res. 2013;4:515–23.

186. Kahle MP, Lee B, Pourmohamad T, Cunningham A, Su H, Kim H, et al. Perlecan domain v is upregulated in human brain arteriovenous malformation and could mediate the vascular endothelial growth factor effect in lesional tissue. Neuroreport. 2012;23:627–30.
187. Al-Ahmad AJ, Lee B, Saini M, Bix GJ. Perlecan domain v modulates astrogliosis in vitro and after focal cerebral ischemia through multiple receptors and increased nerve growth factor release. Glia. 2011;59:1822–40.

Chapter 9
Inflammation and Ischemic Stroke

Junwei Hao, Kai Zheng, and Heng Zhao

Abstract Stroke is the leading cause of death and disability worldwide (Shan and Guo, BMC Neurol 17:33, 2017). Neuroinflammation plays a significant role in the pathogenesis of stroke. In this chapter, we will first review the initial factors that trigger neuroinflammation in the ischemic brain. We will then summarize the main molecules involved in neuroinflammation after stroke as well as the major inflammatory cells derived from brain resident cells and circulating blood. In addition, we will discuss the relationship between post-ischemic inflammation and brain repairs. Lastly, anti-inflammatory therapies will be summarized. The aim of this chapter is not to meticulously review all of the abovementioned aspects, but to provide an overview of the essential components to understand neuroinflammation after stroke.

Keywords Ischemic stroke · Inflammatory cell · Blood vessels · Inflammatory mediator

1 Introduction

Neuroinflammation after ischemic stroke may destroy, or protect and repair brain tissues [1]. Inflammatory response after ischemic stroke involves the activation of brain-resident cells and the infiltration of peripheral blood cells into the brain parenchyma [2]. Initially, ischemic stroke results in neuronal necrosis and the production of reactive oxygen species (ROS), which triggers microglia activities. This leads to a series of inflammatory reactions, and promotes the secretion of inflammatory cytokines and chemokines cytokines, such as interleukins 1 (IL-1) and tumor necrosis factor α (TNF-α). These inflammatory mediators up-regulate the protein expressions of cerebrovascular wall attached molecules (such as P-select

J. Hao (✉) · K. Zheng
Tianjin Neurological Institute, Tianjin Medical University General Hospital, Tianjin, China
e-mail: hjw@tmu.edu.cn

H. Zhao
Department of Neurosurgery, School of Medicine, Stanford University, Stanford, CA, USA

element and E-select element), which are also released into the peripheral blood to promote the activation of peripheral leukocytes. The expressions of adhesion molecules and chemotactic factors prompt the peripheral white blood cells to be attracted to the vascular endothelium and to infiltrate the brain parenchyma [3]. As a result, both the activated microglia/macrophages and the infiltrated leukocytes secrete cytotoxic inflammatory mediators, such as matrix metalloproteinases (MMPs), nitric oxide (NO), cytokines and ROS, enhance brain edema, promoting hemorrhagic transformation, and aggravating neuronal death.

We will discuss the major cascades of inflammatory responses involved in cerebral ischemia and brain injury. We will first introduce the factors that initiate or trigger neuroinflammation after stroke, and the associated cell signaling pathways during neuroinflammation. We will then summarize the major inflammatory products and molecules that modulate neuroinflammation. Thereafter, we will discuss the inflammatory cells which participate in brain injury after stroke, and the relationship between neuroinflammation and brain repair will be clarified. Lastly, we will discuss the current experimental evidence for anti-inflammatory therapies against stroke.

2 Neuroinflammatory Triggers, Molecules, and the Associated Cell Signaling Pathways After Stroke

The injured brain cells induce various inflammatory mediators and damage-associated molecular patterns (DAMPs), which enhance the recruitment of peripheral circulating inflammatory cells, making them more efficient participants in promoting inflammation [4]. After stroke, the lack of adequate energy delivery to the ischemic core causes a disruption of the ATP-dependent ionic gradient maintenance across the neuronal membrane, resulting in cellular swelling and organelle ruptures [5], increased intracellular calcium levels, excitotoxicity, and reactive oxygen species (ROS) production [6]. Following the initial ischemic injury, the complement system is activated, the blood-brain barrier (BBB) is interrupted, and matrix metalloproteases (MMPs) and adhesion molecules are induced [7]. These activities, along with the initial brain injury, promote DAMPs production in the ischemic brain [8], which stimulate intracellular and extracellular pattern recognition receptors (PRRs), initiating neuroinflammation. In this section, we will summarize the major DAMP and PRR proteins, and the associated cell signaling pathways.

2.1 DAMPs

DAMPsare released from stressed or damaged cells. Various substances have been found to be DAMPs, including purines (ATP and UTP), mRNA, hyaluronic acid, heat shock proteins (HSPs), high-mobility group box 1 (HMGB1) protein, and the peroxiredoxin family. For instance, purines (ATP and UTP) are released from the injured brain cells and their receptors, P2X and P2Y, function as alerting signals in the CNS. ATP also activates inflammasomes, which are large multimolecular complexes that control the activity of the proteolytic enzyme caspase-1, that cleaves pro-IL-1β to an active 17 kDa form. These DAMPs are similar to pathogen associated molecular patterns (PAMPs) expressed on the surface of gram negative bacteria [9], which are recognized by the Toll-like receptor (TLRs). DAMPS are also named as danger signals or alarmins, which induce chemotaxis and interact with the receptors on antigen-presenting cells, dendritic cells and macrophages. DAMPs also induce the activation of other PRRs, such as the receptor for advanced glycation end products (RAGE) and c-type lectin receptors [10].

HMGB1, also known as amphoterin or HMG1, is normally localized in the cell nuclei in normal brain cells, but in necrotic cells, it translocates into the cytosolic compartment, and then released into the extracellular space after stroke. It can be secreted by activated monocytes, macrophage, myeloid dendritic cells, and natural killer cells (NK cells), in response to endotoxin and other inflammatory stimuli. HMGB1 promotes necrosis and the influx of damaging inflammatory cells after ischemic stroke. The mechanism of inflammation and damage is by binding to TLR4, which mediates HMGB1-dependent activation of macrophage cytokine release. HMGB1 also activates inflammatory cells through the multiple surface receptors including TLR2, TLR4, and RAGE. RAGE expression is usually low under normal conditions, but increases to high levels upon the occurrence of inflammation [11]. TLR2 and TLR4 have been shown to be involved in infectious diseases. The expression levels of TLR4 have been found to be considerably high in brain regions that lack a tight BBB (blood-brain barrier), such as the circumventricular organs and choroid plexus [12]. The activation of NF-κB and MAP kinase usually follows the activation process of HMGB1 receptors. Recent studies indicate that a blockade of either RAGE or TLR4 contributes to the reduction of cytokine and nitric oxide production and decrease of inflammation, which indicates that HMGB1 potently participates in t inflammation induction [13].

2.2 Pattern-Recognition Receptor (PRR)

The major PRRs in the ischemic brain include Toll-like receptors (TLRs) and RAGE. Toll-like receptors TLRs) play an important role in post-ischemic inflammation [14]. They are expressed in macrophages and dendritic cells that recognize structurally conserved molecules derived from microbes. TLRs play key roles in proinflammatory cytokine expressions, leading to inflammatory responses and brain tissue damage. The activation of microglial cells during ischemic stroke has a close relationship with the significant induction of several TLRs, especially TLR2 and TLR4. In addition, TLRs are widely expressed on granulocytic cells and their activation is pivotal for the recruitment of bone-borne cells to the ischemic area.

The activation of TLRs initiates signaling events that stimulate the production of proinflammatory cytokines and chemokines, leading to the recruitment of leukocytes, especially tneutrophils and macrophages. The production of ROS and nitric oxide derivatives leads to the recruitment of leukocytes to the ischemic area, which subsequently exaggerate the initial damage. Recognition of TLRs on DCs enhances antigen presentations to the adaptive immunity elements. These events induce an early innate immune response, followed by the delayed T-cell activation phase. In fact, T cells play a pivotal role in the pathogenesis of ischemic stroke, depending on their cytokine production.

RAGE is a multi-ligand receptor, consisting of three Ig-domains (V1, C1 and C2), the transmembrane domain, and the cytosolic tail, required for RAGE-dependent cell signaling. The distal Ig-domain (V1) is implicated in the binding of HMGB1 and other RAGE ligands. HMGB1/RAGE regulates the cytoskeleton and the migratory responses of cells through the small GTPases, Cdc42 and Rac. Furthermore, HMGB1 binding to RAGE regulates migration through rapid integrin activation that requires the small GTPase, Rap1 [15]. Ligand binding to RAGE broadly regulates gene expression through the transcription factors CREB, SP1 and NF-κB. RAGE ligation activates the transcription of several plasma membrane proteins, such as the VCAM-1, ICAM-1 and the adhesion molecules, AMIGO. Additionally, RAGE signaling also plays a pivotal role in inducing the expression of intracellular proteins, including Bcl-2 and chromogranin B. RAGE not only plays a key role in regulating cell motility through the modulation of the cytoskeleton, HMGB1/RAGE has also been indicated in regulating integrin functions.

3   The Production of Inflammatory Molecules

The inflammatory reaction after stroke is associated with a variety of inflammatory mediators [16], such as cytokines, chemokines, and cell adhesion molecules, which will be reviewed in this section.

3.1   *Cytokines*

Cytokines are small proteins secreted by specific cell types, promoting interactions between immune cells, thus modulating inflammation. There are pro-inflammatory cytokines, including IL-1β, TNFα, and IL-6, and anti-inflammatory cytokines, including IL-4, IL-10, and TGFβ. These cytokines are secreted by various cells, including T cells, monocytes, and macrophages. Their functions are to promote or inhibit activities of these inflammatory cells.

IL-1β is one of the most important cytokines that affect brain injury and neuroinflammation after stroke. It is synthesized as a precursor protein with a small molecular mass of approximately 33 kD, mainly in monocytes and macrophages.

Upon stimulations, the precursor form of IL-1βhas minimal biological activity and becomes functional IL-1β by cleaving with the protease caspase-1; the activity of caspase-1 is modulated in inflammasome [17]. Thus, the mechanism of IL-1β production and caspase-1 activation regulated by inflammasome has received great attention. It is suggested that inflammasome is activated via hypoxia or ATP [17], and following the activation of caspase-1, generates active IL-1β. IL-1β is produced from monocytes and macrophages, which are activated by endogenous Toll-like receptor (TLR) ligands [18]. In contrast, gene KO of TLR2/4 results in reductions in IL-1β mRNA levels in infiltrating mononuclear cells in TLR2/4-double deficient mice after ischemic stroke.

IL-1β plays an important role in inducing the apoptosis of neuronal cells and contributes to the production of chemokine in astrocytes and microglia [19]. IL-1β is induced and expressed in the damaged brain tissue 30 min after ischemic stroke onset. IL-1β function loss is associated with infarct size reduction, which indirectly indicates that IL-1β might be a neurotoxic mediator [20].

Another important cytokine in stroke-induced inflammation is TNF-α, which is involved in the progress of acute pathological reaction of ischemic stroke, and contributes to the initiation of systemic inflammation [21]. Its expression initially increases at 1–3 h after the ischemic onset, and then reaches a second peak at 24–36 h. TNF-α protein expressions first occur in neurons, especially during the first hours after the ischemic stroke, then in microglia/macrophages, and in blood-borne inflammatory cells [22]. TNF-α also induces the expression of adhesion molecules via cerebral endothelial and glial cells; these adhesion molecules then stimulate the neutrophil accumulation and migration in the microvasculatures [23]. Additionally, TNF-α may also be implicated in BBB impairment, implement the phenotypic transformation of the endothelial cells to contribute to a pro-coagulant milieu in the cerebral microvessels, and activate glial cells to form scars as part of a repair and remodeling after ischemic tissue damage [24].

3.2 Chemokines

Chemokines are another set of cytokines. Their major function is to induce chemotaxis, i.e. they guide leukocyte migration to the inflammatory tissues. Various chemokines increase upon stroke onset. The migration of monocytes/macrophages requires the chemokine CCL2, which functions via the chemokine receptor CCR2, to attract monocytes/macrophages out of the bone marrow and into the ischemic brain. CCL2 is a small cytokine that belongs to the CC chemokine family. It is also known as monocyte chemoattractant protein 1 (MCP1) and small inducible cytokine A2 [25]. In addition to monocytes, CCL2 is also responsible for the recruitment of dendritic cells and memory T cells to the inflammation produced by ischemic insult [26]. Chemokines express in the infarct region. Intracerebroventricular injection (ICVI) of a chemokine-receptor antagonist (viral macrophage inflammatory protein-II) reduces infarct volume [27]. CCL2 expression is induced in response to ischemic insult, playing a key role in the recruitment of monocytes in ischemic tissue injury [28].

In contrast with CCL2, CCL5is another chemokine, which is responsible for the recruitment of T cells, basophils, and eosinophils, and plays an important role in recruiting leukocytes into ischemic lesions [29]. Aided by particular cytokines released by T cells, CCL5 also induces the proliferation and activation of certain natural-killer (NK) cells to form CHAK (CC-Chemokine-activated killer) cells [22]. CCL5 released from blood-borne cells is a pivotal mediator upon reperfusion. Recently, it is widely reported that CCL5 induced the recruitment of leukocyte and platelet adhesion in the cerebral microvasculature [30]. It has been suggested that CCL5 induces the migration of peripheral blood-borne cells across the impaired BBB by recognizing CCR1 and CCR5 receptors and adhesion molecules [31]. Additionally, CCL5 also increases cerebral damage through the secondary induction of other potent pro-inflammatory cytokines, such as IL-6.

3.3 MMPs

Another important group of molecules implicated in neuroinflammation are Matrix metalloproteinases (MMPs). MMPs, also known as matrixins, are calcium-dependent zinc-containing endopeptidases belonging to a larger family of proteases known as the metzincin superfamily [32]. Collectively, these enzymes are capable of degrading all kinds of extracellular matrix proteins, and can also process a number of bioactive molecules. They are known to be involved in the cleavage of cell surface receptors, the release of apoptotic ligands (such as the FAS ligand), and chemokine/cytokine inactivation [33]. MMPs are also thought to play a major role in cell behaviors such as cell proliferation, migration (adhesion/dispersion), differentiation, angiogenesis, apoptosis, and host defense. MMPs are essential neurotoxic mediators that promote BBB breakdown and post-ischemic inflammation. Functionally similar to IL-1β, MMPs induce apoptotic neuronal cell death by TNF-α and Fas ligands [34]. The neurotoxic function of MMP-9 is particularly established, given that infarct size is reduced in MMP-9-deficient mice compared to that in control mice.

3.4 Cell Adhesion Molecules

The adhesion, migration, and infiltration of blood leukocytes into the ischemic brain requires the involvement of cell adhesion molecules (CAMs), which are proteins that express on the inflammatory cell surface. They are involved in binding with other cells or with the extracellular matrix (ECM) in the process of cell adhesion. CAMs consist of three protein families: the integrins, the selectins, and Ig (immunoglobulin superfamily).

The integrin family includes heterodimeric membrane glycoproteins with an α and αβ subunit, which play an important role in ECM and cell-cell interactions [35]. Integrins are transmembrane receptors that stimulate cell-extracellular matrix (ECM) adhesion. Upon ligand binding, integrins are responsible for activating signal transduction pathways, which regulate cellular signals such as organization of the intracellular cytoskeleton, regulation of the cell cycle, and movement of new receptors to the cell membrane [36]. In the brain, integrins also play a pivotal role to unite the endothelial cells, astrocytes, and basal lamina that consist of the blood-brain barrier; consequently, they are extremely essential in maintaining the integrity of the cerebral microvasculature [37]. Another integrin, named α6β4 (CD104), interferes with the normal interactions between laminin-5 and astrocytes in the ECM upon ischemic insult. Above all, it is widely accepted that these integrins are involved in the progression of platelet activation [38], promoting adherence of neutrophils to endothelium and mediating cell-ECM interactions.

The selectins are a family of cell adhesion molecules that bind to sugar moieties or polymers [29], including P-, E-, and L-selectin. P-selectin expresses on platelets and endothelial cells, and its counter receptor on leukocytes consists of the oligosaccharide sialyl-Lewis X. The early re-localization of P-selectin facilitates the primary adhesion of leukocytes [39]. Nevertheless, it also plays a continuing role because its expression increases in post-ischemic cerebral vasculature.

E-selectin expresses on activated endothelium, in sequence after P-selectin under ischemic insult [40]. It has been suggested that transnasal E-selectin tolerization induces tolerance to a secondary ischemic insult in experimental stroke. Currently, a large number of studies are underway to advance these studies to clinical trials with the ultimate goal of using this form of immunomodulation for the secondary prevention of stroke [41].

L-selectin, which expresses on both endothelium and leukocytes, was originally identified as being a significant factor in homing in to the sites of infection and to peripheral lymph nodes. In addition to promoting neutrophil rolling [42], L-selectin also mediates neutrophil attachment to endothelial cells through interactions that are independent of integrin CD18 [43].

The last CAMs we will discuss are the Immunoglobulin superfamily. There are five major members of the immunoglobulin superfamily: Vascular cell adhesion molecule-1 (VCAM-1), Intercellular adhesion molecule-1(ICAM-1), Intercellular adhesion molecule-2 (ICAM-2), Mucosal vascular addressin cell adhesion molecule 1(CD146), and Platelet-endothelial cell adhesion molecule-1 (PECAM-1) [44] which are expressed on activated endothelial cells. Members of the immunoglobulin superfamily mediate a stronger binding of leukocytes to the vascular endothelium than the selectins under cerebral ischemia [45]. These cell adhesion molecules play an important role in the tight adhesion of leukocytes under ischemic insult, as well as the unique role of ICAM-2 in leukocyte-platelet interactions [46]. Therefore, treatments targeting the CAM family may interfere with the first step in the inflammatory pathway, resulting in the blocking of leukocyte transmigration into the ischemic lesions [47].

4 Inflammatory Cells and Ischemic Stroke

In the previous sections, we have discussed the initiation of post-ischemic inflammation and the molecules that modulate the inflammatory response. These inflammatory triggers and molecules modulate the central players of neuroinflammation, inflammatory cells, including both brain resident cells and peripheral blood leukocytes, which we will discuss in this section.

4.1 Brain Resident Cells in Ischemic Stroke

As we have discussed, the first immune cells activated in the ischemic lesion are the resident microglia in the brain. Microglia, the counterparts of monocytes/macrophages in the blood and other organs, are the main long-living resident immune cells in the central nervous system (CNS). Microglia have the ability to continuously sense and scan their environment for injuries and pathogens, then reacting to the damage-induced signals to protect against brain injury. Therefore, microglia maintain a stable chemical and physical microenvironment necessary for the CNS to conduct its normal functions [48]. When ischemic stroke occurs, microglia activate rapidly in response to ischemic injury [49]: their reactions reaching peaks at 2–3 days after stroke and lasting for several weeks. The role of microglia in ischemic stroke is a double-edged sword, as microglia produce inflammatory factors leading to cell damage and death. Microglia can also produce TGF-β1, which protects the central nervous system (CNS) [50].

Astrocytes are specialized and the most abundant cell type in the CNS, playing important roles in CNS disorders, including ischemic stroke [51]. Reactive astrogliosis is one of the important pathological features of ischemic stroke, which is accompanied by changes in morphology, proliferation, and gene expressions in the reactive astrocytes. Glial fibrillary acidic protein (GFAP) is overexpressed on the astrocytes after stroke [52], indicating the progress of astrocyte proliferation and differentiation. Astrocytes eventually fill up the empty space in the damaged brain to form a glial scar, replacing the CNS cells that cannot regenerate.

4.2 Peripheral Blood-Borne Cells in Ischemic Stroke

Many types of blood leukocytes participate in neuroinflammation and modulate brain injury after stroke, including neutrophils, monocytes, T cells, B cells, and NK cells. Neutrophils are the first leukocytes to infiltrate into the ischemic brain, acting from 30 min to a few hours after focal cerebral ischemia, then peaking from days 1–3, and rapidly decreasing thereafter [53]. Infiltrating neutrophils contribute to inflammation and brain injury by releasing pro-inflammatory factors, including

MMPs and inducible nitric oxide (iNOS) [17]. As neutrophils are detrimental to brain injury, ischemic brain injury might be decreased by an immuno-depletion of neutrophils or an antibody blockage of neutrophil infiltration [54].

A few hours after stroke onset, monocytes are recruited into the ischemic brain, participating in neuroinflammation. There are at least two types of blood monocytes, which can be divided into "pro-inflammatory" (Ly-6C^{high}/CCR2+) and "anti-inflammatory" (Ly-6C^{low}/CCR2−) subpopulations [55]. Monocyte/macrophage expressions can be divided into two types: inflammatory M1 and anti-inflammatory M2 macrophages [56]. M1 macrophages produce inflammatory responses by secreting inflammatory factors, such as IL-1β and TNF-α, and overexpressing iNOS, while M2 macrophages produce anti-inflammatory cytokines, such as IL-10 and TGF-β. As a result, M1 macrophages promote brain injury while M2 macrophages inhibit brain injury and promote brain recovery. Nevertheless, monocyte derived macrophages are not the only sources of M1 and M2 macrophages. Microglia are another major source of M1 and M2 macrophages, but the distinctive roles of microglia and monocytes in M1 and M2 polarization remain unknown.

Blood lymphocytes also play significant roles in neuroinflammation and brain injury [57]. Among the various types of lymphocytes, the significant role of T cells has been extensively studied [58], including CD4+, CD8+ T cells, γδT cells, and Treg cells in the pathogenesis of ischemic stroke [21]. The depletion of CD4 or CD8 T cells attenuates brain injury; γδT cells promote brain injury, while Treg activity attenuates brain injury and promotes functional recovery after stroke.

Lastly, dendritic cells (DC) play an important role in connecting innate and adaptive immunity. CD11c+ DC cells are increased in the ipsilateral hemisphere, peaking at 72 h after ischemic stroke. In addition, major histocompatibility complex (MHC)-II and co-stimulatory molecules are up-regulated in the brain resident and peripheral DC, which may be responsible for lymphocyte activation. Furthermore, the brain-resident DC might be involved in directing the early, local immune response and, later, participate in the recruitment of lymphocytes upon activation by INF-γ.

5 Post-ischemic Inflammation and Brain Repair

Neuroinflammation occurs after stroke in the ischemic brain, and is designed to remove necrotic tissues and promote lesion resolution. Generally, post-stroke neuro-restoration is achieved by enhancing neurogenesis, oligodendrogenesis and angio-genesis, which collaboratively contribute to neurological recovery [59]. Neurogenesis and oligodendrogenesis produce new parenchymal cells from neural stem cells which promote plasticity, restore neuronal signal transduction, and stimulate myelination [60]. Angiogenesis and arteriogenesis are the major forms of vascular remodeling, which contribute to increased cerebral blood flow (CBF). In addition, they generate trophic factors and proteases, which are crucial for restoration by helping to construct an environment for remyelination and neurite outgrowth [61].

5.1   The Removal of Necrotic Cells

The crucial process in brain repair mainly consists of the suppression of inflammation and clearing of dead cells [62], in which neutrophils and macrophages play critical roles. Macrophages include monocytes and microglia derived macrophages, which polarize into pro-inflammatory M1 and anti-inflammatory M2. The M2 macrophages are particularly important for inflammation resolution and brain repair. Nevertheless, how microglia and monocyte derived macrophages contribute to brain repair has not been distinguished. The phygocytotic mechanisms of macrophages after stroke are not completely understood. Nevertheless, the results of cultured microglia have indicated that ATP signaling through the G protein-coupled P2Y receptors results in rapid microglial membrane ruffling and whole-cell migration. ATP, ADP and UTP might be potent agonists for P2Y G protein-coupled receptors and P2X ligand-gated ion channels [63]. The microglia/macrophages are attracted by UTP and ATP through P2Y2 receptors. In addition, UDP acts on P2Y6 receptors to stimulate microglial phagocytosis. Furthermore, phosphatidylserine (PtdSer), which translocates to the outer leaflet of the plasma membrane of apoptotic cells, enables the apoptotic cells to be recognized by the phygocytotic cells and removed [64]. Other PtdSer binding proteins include MGF-E8 on microglia and TIM4 on macrophages, which also participate in the clearance of dead cells [65].

5.2   Oligodendrogenesis

Oligodendrocytes are quite sensitive to ischemic insult because white matter has lower blood flow than gray matter, and deep white matter has little collateral blood supply [66]. Post-ischemic neuroinflammation may have a detrimental effect on white matter cohesion by upregulating matrix MMPs, as both MMP-2 and MMP-9 exacerbate white matter lesions [67]. Oligodendrocyte progenitor cells (OPC) differentiate into mature oligodendrocytes under ischemic insult [68], and produce myelin sheaths, which wrap around axons, facilitating nerve conduction. Therefore, oligodendrogenesis plays an important role in behavioral and functional restoration after ischemic stroke [69].

Oligodendrogenesis after stroke occurs by recruiting resident OPCs from white and gray matter and generated by SVZ neural progenitor cells; these new oligodendrocytes become mature myelinating oligodendrocytes [70]. Such OPCs generated in the SVZ are observed in humans after demyelination [71]. These newly produced OPCs migrate to peri-infarct gray and white matter to participate in oligodendrogenesis. This OPCs migration is modulated by stromal-derived factor 1α (SDF-1α) and vascular endothelial growth factor (VEGF), secreted by activated cerebral endothelial cells in the ischemic boundary region [72]. In addition, glutamatergic signals resulting from damaged axons in the corpus callosum also induce the migration of OPCs from the SVZ to the peri-infarct areas.

5.3 Angiogenesis

As a crucial growth factor in post-ischemic angiogenesis, VEGF is generated by reactive astrocytes. Its action might need neutrophil MMPs, which indicates an association between inflammation and angiogenesis [73]. Angiogenesis usually occurs in the penumbra of the human ischemic brain hours after initial onset and continues to exist for several weeks [74]. The first step in angiogenesis is the nitric oxide (NO)-initiated vasodilation. Combining the vasodilation effect of NO with the increase in VEGF expression, which increases vascular permeability, allows the extravasation of plasma proteins that lay down a provisional scaffold for the migration of endothelial cells for vascular sprouting [75]. The second step starts with the dissociation of smooth muscle cells (SMCs) and loosening of the extracellular matrix that enwraps the mature vessels [76]. Angiopoietin-2 (Ang2), a Tie2 signaling inhibitor, may take part in stimulating the detachment of pericytes from endothelial cells, whereas the MMP family of proteinases decreases the matrix molecules, and then weakens vascular integrity [77]. The proliferating endothelial cells may migrate to distant sites as the sprouting path has been established. Consequently, the angiogenic process will be mediated vy a series of molecular signals such as VEGF, VEGF receptors, and placental growth factors [78]. Additionally, Ang1 will activate Tie2 receptors when the new blood vessel networks are formed [79], which will help to stabilize the networks initiated by VEGF.

6 Anti-Inflammatory Therapy in Ischemic Stroke

Since post-ischemic inflammation exacerbates brain injury after stroke, inflammation has been a target for stroke therapy. In this section, we will review some strategies targeted against pro-inflammatory responses.

6.1 Targeting Inflammasome Signaling Pathways

Inflammasome is a pivotal mediator in inflammation. This involves the activation of pro-caspase-1 into cleaved caspase-1 [80], which initiates and amplifies the generation of pro-inflammatory cytokines IL-1β and IL-18, resulting in apoptotic neuronal and glial cell death following ischemic stroke [81]. Therefore, targeting the upstream or downstream cascades of the inflammasome signaling pathways, including protein expressions in the pathway, as well as its activity and products, may provide avenues for developing therapeutics against ischemic stroke [82]. There are many potential targets involved in inflammasome signaling which can be explored: plasma membrane receptors/channels (i.e. P2X7 receptors, Pannexin 1 and K+ channels), cytokines (i.e. IL-1β and IL-18), inflammasome components (i.e. NLRPs, ASC and caspase-1), secondary messengers (i.e. ROS and PKR) [83], signaling pathways (i.e. NF-κB and MAPK) and cytokine receptors (i.e. IL-1R1and IL-18R) [84].

6.2   Targeting Neutrophil Recruitment

Clinical studies have tested three kinds of drugs or antibodies targeting neutrophil recruitment, which have potential values for ischemic stroke [85]: a humanized antibody against the CD11b/CD18, a mAb against ICAM-1, and the UK-279276. UK-279,276 is a recombinant glycoprotein, which inhibits neutrophil recruitment by selectively binding to the CD11b/CD18 integrin. Some studies have shown that these treatments are well tolerated. However, the resulting side effects, such as leukopenia and immunosuppression, make them ineffective for stroke treatment [86]. It seems that there is still a long way to go in translating neutrophil inhibitors for effective clinical use [87].

6.3   Immunosuppression Strategies

As previously discussed, T cells also participate in neuroinflammation and modulate brain injury after stroke. T cells consist of a number of subsets, including both pro-inflammatory and anti-inflammatory subsets. These subsets either promote or inhibit inflammation, leading either to brain injury exacerbation or attenuation. Nevertheless, the overall inhibition of T cell functions have been shown to attenuate brain injuries. T cell functions can be inhibited with the immunosuppressant drugs: FK-506, rapamycin, and cyclosporin A (CsA), combined with immunophilins, which are a high-affinity receptor proteins in the cytoplasm. This combination causes rotamase inhibition in T cells, thus interrupts cell activation. FK506 and CsA complexes, with their immunophilin receptors, have been indicated to impact a variety of cellular immune responses by suppressing Ca^{2+}-dependent serine/threonine phosphatase [88], calcineurin, and by mediating Ca^{2+} release via the ryanodine receptor. The rapamycin-FKBP12 complex suppresses the cell cycle process by mediating cell cycle kinases, including the PI3 kinase-like molecule known as 'target of rapamycin (TOR)' [89]. There are many other signaling pathways shown to be modulated by immunophilins such as, the interaction with heat-shock proteins, glucocorticoid receptor (s), nitric oxide synthase activity, the transforming growth factor receptor, and protein folding [90]; which have been involved in the immunomodulatory effects of immunophilin-binding immunosuppressants.

7 Summary

Neuroinflammation plays a crucial role in the pathogenesis of ischemic insult [91]. In this chapter, we have reviewed the initial trigger factors, summarized the molecules involved in post stroke neuroinflammation as well as the major inflammatory cells; the relationship between inflammation and brain repair, as well as possible anti-inflammatory therapies after stroke. Although there are many experimental results suggesting promising strategies to inhibit inflammation and attenuate brain injury, anti-inflammatory therapies for clinical translation have been unsuccessful. Therefore, there is still much to be done before pre-clinical trials can be translated for clinical use. As the role of inflammatory response upon ischemic insult is a double-edged sword, the challenge on how to curb its detrimental effects while promoting its beneficial roles for functional brain recovery remains.

Future research should synthesize each situation regarding pro- and anti-inflammatory responses under ischemic conditions, not by studying and analyzing them separately.

Many outstanding questions still remain: how to identify the exact dynamic balance between pro-and anti-inflammation generated during the different stages of ischemic stroke? Which pro-inflammatory or anti-inflammatory mediators should be targeted for therapy? How and when should the pro-inflammatory or anti-inflammatory pathways be activated or inhibited for treatment? Therefore, in order to design effective therapies, we must consider the dynamic balance between pro- and anti-inflammatory responses. Then we must identify the discrepancies between pre-clinical studies and clinical trials.

With a comprehensive understanding of the disease process, future studies will be able to identify a pipeline of new targets that control the cell signaling pathways and networks, rather than on a single mediator, and identify complementary holistic approaches to treat ischemic stroke.

References

1. Shan K, Guo W. Stroke caused by an inflammatory thrombus: a case report. BMC Neurol. 2017;17(1):33.
2. Silverman MG, et al. Association between lowering LDL-C and cardiovascular risk reduction among different therapeutic interventions: a systematic review and meta-analysis. JAMA. 2016;316(12):1289–97.
3. Bragg F, et al. Association between diabetes and cause-specific mortality in rural and urban areas of China. JAMA. 2017;317(3):280–9.
4. Fucikova J, et al. Calreticulin exposure by malignant blasts correlates with robust anticancer immunity and improved clinical outcome in AML patients. Blood. 2016;128(26):3113–24.
5. Choi HW, Klessig DF. DAMPs, MAMPs, and NAMPs in plant innate immunity. BMC Plant Biol. 2016;16(1):232.
6. Thoudam T, et al. Role of mitochondria-associated endoplasmic reticulum membrane in inflammation-mediated metabolic diseases. Mediat Inflamm. 2016;2016:1851420.

7. Santos SC, et al. Immunomodulation after ischemic stroke: potential mechanisms and implications for therapy. Crit Care. 2016;20(1):391.
8. Nakayama T. An inflammatory response is essential for the development of adaptive immunity-immunogenicity and immunotoxicity. Vaccine. 2016;34(47):5815–8.
9. Versluys M, Tarkowski LP, Van den Ende W. Fructans as DAMPs or MAMPs: evolutionary prospects, cross-tolerance, and multistress resistance potential. Front Plant Sci. 2016;7:2061.
10. Lu L, et al. Innate immune regulations and liver ischemia-reperfusion injury. Transplantation. 2016;100(12):2601–10.
11. Gougeon ML, et al. HMGB1/anti-HMGB1 antibodies define a molecular signature of early stages of HIV-associated neurocognitive disorders (HAND). Heliyon. 2017;3(2):e00245.
12. Wang Y, et al. Cigarette smoke attenuates phagocytic ability of macrophages through downregulating Milk fat globule-EGF factor 8 (MFG-E8) expressions. Sci Rep. 2017;7:42642.
13. Liu Y, et al. Blockade of HMGB1 preserves vascular homeostasis and improves blood perfusion in rats of acute limb ischemia/reperfusion. Microvasc Res. 2017;112:37–40.
14. Ji Y, et al. Temporal pattern of Toll-like receptor 9 upregulation in neurons and glial cells following cerebral ischemia reperfusion in mice. Int J Neurosci. 2016;126(3):269–77.
15. Olsson S, Jood K. Genetic variation in the receptor for advanced glycation end-products (RAGE) gene and ischaemic stroke. Eur J Neurol. 2013;20(6):991–3.
16. Vidale S, et al. Postischemic inflammation in acute stroke. J Clin Neurol. 2017;13(1):1–9.
17. Lee GA, et al. Interleukin 15 activates Akt to protect astrocytes from oxygen glucose deprivation-induced cell death. Cytokine. 2017;92:68–74.
18. Bronisz E, Kurkowska-Jastrzebska I. Matrix metalloproteinase 9 in epilepsy: the role of neuroinflammation in seizure development. Mediat Inflamm. 2016;2016:7369020.
19. Li N, et al. Bidirectional relationship of mast cells-neurovascular unit communication in neuroinflammation and its involvement in POCD. Behav Brain Res. 2017;322(Pt A):60–9.
20. Mijajlovic MD, et al. Post-stroke dementia – a comprehensive review. BMC Med. 2017;15(1):11.
21. Shukla V, et al. Cerebral ischemic damage in diabetes: an inflammatory perspective. J Neuroinflammation. 2017;14(1):21.
22. Zhang Y, et al. Effects of Shaoyao-Gancao decoction on infarcted cerebral cortical neurons: suppression of the inflammatory response following cerebral ischemia-reperfusion in a rat model. Biomed Res Int. 2016;2016:1859254.
23. Guo X, et al. miR-145 mediated the role of aspirin in resisting VSMCs proliferation and anti-inflammation through CD40. J Transl Med. 2016;14(1):211.
24. Kojima Y, et al. CD47-blocking antibodies restore phagocytosis and prevent atherosclerosis. Nature. 2016;536(7614):86–90.
25. Elkind MS, et al. The levels of inflammatory markers in the treatment of stroke study (LIMITS): inflammatory biomarkers as risk predictors after lacunar stroke. Int J Stroke. 2010;5(2):117–25.
26. Reaux-Le GA, et al. Current status of chemokines in the adult CNS. Prog Neurobiol. 2013;104:67–92.
27. Wacker BK, Perfater JL, Gidday JM. Hypoxic preconditioning induces stroke tolerance in mice via a cascading HIF, sphingosine kinase, and CCL2 signaling pathway. J Neurochem. 2012;123(6):954–62.
28. Kwon MJ, Yoon HJ, Kim BG. Regeneration-associated macrophages: a novel approach to boost intrinsic regenerative capacity for axon regeneration. Neural Regen Res. 2016;11(9):1368–71.
29. Zemer-Wassercug N, et al. The effect of dabigatran and rivaroxaban on platelet reactivity and inflammatory markers. J Thromb Thrombolysis. 2015;40(3):340–6.
30. Sajedi KM, et al. Correlation of early and late ejection fractions with CCL5 and CCL18 levels in acute anterior myocardial infarction. Iran J Immunol. 2016;13(2):100–13.
31. Rom S, et al. miR-98 and let-7g* protect the blood-brain barrier under neuroinflammatory conditions. J Cereb Blood Flow Metab. 2015;35(12):1957–65.
32. Wang XH, You YP. Epigallocatechin gallate extends therapeutic window of recombinant tissue plasminogen activator treatment for brain ischemic stroke: a randomized double-blind and placebo-controlled trial. Clin Neuropharmacol. 2017;40(1):24–8.

33. Zhang HT, et al. Early VEGF inhibition attenuates blood-brain barrier disruption in ischemic rat brains by regulating the expression of MMPs. Mol Med Rep. 2017;15(1):57–64.
34. Kanazawa M, et al. Therapeutic strategies to attenuate hemorrhagic transformation after tissue plasminogen activator treatment for acute ischemic stroke. J Atheroscler Thromb. 2017;24(3):240–53.
35. Fujioka T, et al. Beta1 integrin signaling promotes neuronal migration along vascular scaffolds in the post-stroke brain. EBioMedicine. 2017;16:195–203.
36. Rom S, et al. PARP inhibition in leukocytes diminishes inflammation via effects on integrins/cytoskeleton and protects the blood-brain barrier. J Neuroinflammation. 2016;13(1):254.
37. Xu XR, et al. Platelets and platelet adhesion molecules: novel mechanisms of thrombosis and anti-thrombotic therapies. Thromb J. 2016;14(Suppl 1):29.
38. Huang H, et al. Cerebral ischemia-induced angiogenesis is dependent on tumor necrosis factor receptor 1-mediated upregulation of alpha5beta1 and alphaVbeta3 integrins. J Neuroinflammation. 2016;13(1):227.
39. Zhao J, et al. Cinnamaldehyde inhibits inflammation and brain damage in a mouse model of permanent cerebral ischaemia. Br J Pharmacol. 2015;172(20):5009–23.
40. Wu N, et al. Association of inflammatory and hemostatic markers with stroke and thromboembolic events in atrial fibrillation: a systematic review and meta-analysis. Can J Cardiol. 2015;31(3):278–86.
41. Kurkowska-Jastrzebska I, et al. Carotid intima media thickness and blood biomarkers of atherosclerosis in patients after stroke or myocardial infarction. Croat Med J. 2016;57(6):548–57.
42. Pusch G, et al. Early dynamics of P-selectin and interleukin 6 predicts outcomes in ischemic stroke. J Stroke Cerebrovasc Dis. 2015;24(8):1938–47.
43. Yang S, et al. Biomarkers associated with ischemic stroke in diabetes mellitus patients. Cardiovasc Toxicol. 2016;16(3):213–22.
44. Guo M, et al. Polymorphisms in the receptor for advanced glycation end products gene are associated with susceptibility to drug-resistant epilepsy. Neurosci Lett. 2016;619:137–41.
45. Zhang D, et al. Up-regulation of VCAM1 relates to neuronal apoptosis after intracerebral hemorrhage in adult rats. Neurochem Res. 2015;40(5):1042–52.
46. Lu W, Bromley-Coolidge S, Li J. Regulation of GABAergic synapse development by postsynaptic membrane proteins. Brain Res Bull. 2017;129:30–42.
47. Riehl A, et al. The receptor RAGE: bridging inflammation and cancer. Cell Commun Signal. 2009;7:12.
48. Schofield ZV, et al. Neutrophils—a key component of ischemia-reperfusion injury. Shock. 2013;40(6):463–70.
49. Tabas I. 2016 Russell Ross memorial lecture in vascular biology: molecular-cellular mechanisms in the progression of atherosclerosis. Arterioscler Thromb Vasc Biol. 2017;37(2):183–9.
50. Altug CH, et al. Assessment of the relationship between serum vascular adhesion protein-1 (VAP-1) and severity of calcific aortic valve stenosis. J Heart Valve Dis. 2015;24(6):699–706.
51. Mandelbaum M, et al. A critical role for proinflammatory behavior of smooth muscle cells in hemodynamic initiation of intracranial aneurysm. PLoS One. 2013;8(9):e74357.
52. Tamma G, et al. Effect of roscovitine on intracellular calcium dynamics: differential enantioselective responses. Mol Pharm. 2013;10(12):4620–8.
53. Banjara M, Ghosh C. Sterile neuroinflammation and strategies for therapeutic intervention. Int J Inflamm. 2017;2017:8385961.
54. Courties G, Moskowitz MA, Nahrendorf M. The innate immune system after ischemic injury: lessons to be learned from the heart and brain. JAMA Neurol. 2014;71(2):233–6.
55. Pedersen DS, et al. Toxicological aspects of injectable gold-hyaluronan combination as a treatment for neuroinflammation. Histol Histopathol. 2014;29(4):447–56.
56. Zhang Y, et al. Treadmill exercise promotes neuroprotection against cerebral ischemia-reperfusion injury via downregulation of pro-inflammatory mediators. Neuropsychiatr Dis Treat. 2016;12:3161–73.
57. Perez-de-Puig I, et al. Neutrophil recruitment to the brain in mouse and human ischemic stroke. Acta Neuropathol. 2015;129(2):239–57.

58. Harmon EY, et al. Anti-inflammatory immune skewing is atheroprotective: Apoe−/- FcgammaRIIb−/− mice develop fibrous carotid plaques. J Am Heart Assoc. 2014;3(6):e001232.
59. Yu JH, et al. Induction of neurorestoration from endogenous stem cells. Cell Transplant. 2016;25(5):863–82.
60. Jarosiewicz B, et al. Virtual typing by people with tetraplegia using a self-calibrating intracortical brain-computer interface. Sci Transl Med. 2015;7(313):313ra179.
61. Kaya AH, Erdogan H, Tasdemiroglu E. Searching evidences of stroke in animal models: a review of discrepancies a review of discrepancies. Turk Neurosurg. 2017;27(2):167–73.
62. Rossi PJ, et al. Proceedings of the third annual deep brain stimulation think tank: a review of emerging issues and technologies. Front Neurosci. 2016;10:119.
63. Duricki DA, et al. Delayed intramuscular human neurotrophin-3 improves recovery in adult and elderly rats after stroke. Brain. 2016;139(Pt 1):259–75.
64. ElAli A, Jean LN. The role of monocytes in ischemic stroke pathobiology: new avenues to explore. Front Aging Neurosci. 2016;8:29.
65. Azad TD, Veeravagu A, Steinberg GK. Neurorestoration after stroke. Neurosurg Focus. 2016;40(5):E2.
66. Di Cesare F, et al. Phosphodiesterase-5 inhibitor PF-03049423 effect on stroke recovery: a double-blind, placebo-controlled randomized clinical trial. J Stroke Cerebrovasc Dis. 2016;25(3):642–9.
67. Liu Z, Chopp M. Astrocytes, therapeutic targets for neuroprotection and neurorestoration in ischemic stroke. Prog Neurobiol. 2016;144:103–20.
68. Popa-Wagner A, et al. Poststroke cell therapy of the aged brain. Neural Plast. 2015;2015:839638.
69. Wu X, et al. Long-term effectiveness of intensive therapy in chronic stroke. Neurorehabil Neural Repair. 2016;30(6):583–90.
70. Amar AP, Griffin JH, Zlokovic BV. Combined neurothrombectomy or thrombolysis with adjunctive delivery of 3K3A-activated protein C in acute ischemic stroke. Front Cell Neurosci. 2015;9:344.
71. Choi JC, et al. Effect of pre-stroke statin use on stroke severity and early functional recovery: a retrospective cohort study. BMC Neurol. 2015;15:120.
72. Wood H. Migraine: migraine is associated with increased risk of perioperative ischaemic stroke. Nat Rev Neurol. 2017;13(2):67.
73. Sullivan R, et al. A possible new focus for stroke treatment – migrating stem cells. Expert Opin Biol Ther. 2015;15(7):949–58.
74. Villapol S, et al. Neurorestoration after traumatic brain injury through angiotensin II receptor blockage. Brain. 2015;138(Pt 11):3299–315.
75. Kongbunkiat K, et al. Leukoaraiosis, intracerebral hemorrhage, and functional outcome after acute stroke thrombolysis. Neurology. 2017;88(7):638–45.
76. Sun L, et al. L-Serine treatment may improve neurorestoration of rats after permanent focal cerebral ischemia potentially through improvement of neurorepair. PLoS One. 2014;9(3):e93405.
77. Algra A, Wermer MJ. Stroke in 2016: stroke is treatable, but prevention is the key. Nat Rev Neurol. 2017;13(2):78–9.
78. Meimounn M, et al. Intensity in the neurorehabilitation of spastic paresis. Rev Neurol (Paris). 2015;171(2):130–40.
79. Ruscher K, Wieloch T. The involvement of the sigma-1 receptor in neurodegeneration and neurorestoration. J Pharmacol Sci. 2015;127(1):30–5.
80. Abeliovich A, Gitler AD. Defects in trafficking bridge Parkinson's disease pathology and genetics. Nature. 2016;539(7628):207–16.
81. Jackson JL, et al. Associations of 25-hydroxyvitamin D with markers of inflammation, insulin resistance and obesity in black and white community-dwelling adults. J Clin Transl Endocrinol. 2016;5:21–5.
82. Gogia S, Kaiser Y, Tawakol A. Imaging high-risk atherosclerotic plaques with PET. Curr Treat Options Cardiovasc Med. 2016;18(12):76.

83. Liu CL, Zhang K, Chen G. Hydrogen therapy: from mechanism to cerebral diseases. Med Gas Res. 2016;6(1):48–54.
84. Liang LJ, Yang JM, Jin XC. Cocktail treatment, a promising strategy to treat acute cerebral ischemic stroke? Med Gas Res. 2016;6(1):33–8.
85. Katayama Y, et al. Neuroprotective effects of clarithromycin against neuronal damage in cerebral ischemia and in cultured neuronal cells after oxygen-glucose deprivation. Life Sci. 2017;168:7–15.
86. Anrather J, Iadecola C. Inflammation and stroke: an overview. Neurotherapeutics. 2016;13(4):661–70.
87. Satani N, Savitz SI. Is immunomodulation a principal mechanism underlying how cell-based therapies enhance stroke recovery? Neurotherapeutics. 2016;13(4):775–82.
88. Vafadari B, Salamian A, Kaczmarek L. MMP-9 in translation: from molecule to brain physiology, pathology, and therapy. J Neurochem. 2016;139 Suppl 2:91–114.
89. Choi DH, Kang SH, Song H. Mean platelet volume: a potential biomarker of the risk and prognosis of heart disease. Korean J Intern Med. 2016;31(6):1009–17.
90. de Ramon L, et al. RNAi-based therapy in experimental ischemia-reperfusion injury. The new targets. Curr Pharm Des. 2016;22(30):4651–7.
91. Toraldo DM, et al. Statins may prevent atherosclerotic disease in OSA patients without co-morbidities? Curr Vasc Pharmacol. 2017;15(1):5–9.

Chapter 10
Cerebral Vascular Injury in Diabetic Ischemia and Reperfusion

Wenlu Li and Haibin Dai

Abstract About 30% of stroke patients are diabetic and more than 90% of them comprise type 2 diabetes (T2D). Diabetic stroke patients have higher mortality and worse neurological outcomes. Emerging clinical and experimental data suggest that blood-brain barrier (BBB) disruption, neuroinflammation, and stroke recovery impairment are exacerbated in diabetic patients. Hence, finding therapeutic approaches that can target these specific diabetic mechanisms in stroke is the thrust of the present translational study. Here, we summarize the ischemia-reperfusion injury in stroke, presenting the clinical evidence for involvement of hyperglycemia in severe damage of cerebral ischemia-reperfusion. We go on to consider the mechanisms that underlie such pathology, and highlight areas for future basic research and clinical studies into diabetic ischemia and reperfusion.

Keywords Vascular · Cerebral ischemia-reperfusion · Stroke · Diabetes

1 Ischemia-Reperfusion Injury in Stroke

Ischemia resulting from the restriction of blood supply to an organ is one of the most prevalent diseases globally. Ischemia is always followed by the restoration of vascular perfusion and reoxygenation, which is termed reperfusion. Loss of nutrition and oxygen constitutes a characteristic feature of ischemia. However, it has also been demonstrated that subsequent reperfusion can induce tissue damage and dysfunction [1]. Ischemia-reperfusion injury can occur as a complication of revascularization, such as thrombolytic therapy for acute ischemic stroke, intracranial stenting, carotid endarterectomy (CEA), and even bland cerebral infarction. To date, intravenous thrombolysis is still a main therapy of choice for acute ischemic stroke [2]. Ischemic stroke patients will have a much better outcome regarding long-term morbidity if they receive the treatment within 4.5 h after symptom onset. However,

W. Li · H. Dai (✉)
Second Affiliated Hospital, Zhejiang University School of Medicine, Hangzhou, China
e-mail: haibindai@zju.edu.cn

© Springer International Publishing AG, part of Springer Nature 2018
W. Jiang et al. (eds.), *Cerebral Ischemic Reperfusion Injuries (CIRI)*, Springer Series in Translational Stroke Research, https://doi.org/10.1007/978-3-319-90194-7_10

clinical studies have identified a significant increase of mortality when the treatment is administered more than 4.5 h post-stroke. The reason for this temporal delineation is, at least partly, that the perfusion injury counteracts the potential benefit [3].

Consequently, specific treatments that make brain tissue more resistant to ischemia or protect reperfusion-induced brain injury, such as preconditioning and postconditioning of ischemia, remote ischemic conditioning and nitric oxide (NO) [1], have attracted increased attention in clinical safety and efficacy trials.

Disruption in blood-brain barrier (BBB) integrity is one of the most important pathophysiologic events after ischemic stroke. As one of the most well recognized consequences of ischemia-reperfusion injury, hemorrhagic transformation can lead to poorer outcomes of ischemic stroke. In addition, the pathological process affects the clinician treatment decisions for aggressive intervention at initial patient presentation. Although recent animal studies provided much valuable information concerning ischemia-reperfusion injury, involvement of human studies remains at an early stage, especially regarding the dynamic course of reperfusion-induced brain injury and clinical characteristics in acute ischemic stroke. Fortunately, with the development of multimodal imaging technology, which benefits the specific quantification of changes in brain tissue, the study of chronology for serial perfusion continues to improve, which may facilitate the development of novel treatments for ischemia-reperfusion injury in humans [4].

2   Diabetic Ischemia and Reperfusion

Diabetes and cerebral ischemia constitute common disorders that often arise together [5]. Patients with diabetes mellitus (DM) are more susceptible to increased risk of stroke, as well as post-stroke mortality. Cerebral ischemia in diabetics produces severe brain damage (Fig. 10.1), which results in poorer functional recovery [6–8]. Recently, a meta-analysis comprising 102 prospective studies (including 698,782 participants), has found that compared to people without diabetes, the hazard ratio for cerebral ischemia was 2.3 (95% confidence interval (CI) 2.0–2.7) in people with diabetes [9]. In addition, claims data from Taiwan's National Health Insurance reported that compared to non-diabetes patients, DM patients had an increased risk of stroke (adjusted hazard ratio: 1.75; 95% CI: 1.64–1.86). Moreover, this increased risk was significant in males and females, as well as all age groups [10]. Furthermore, both disability and fatal outcome after the stroke event are up-regulated in DM patients [11]. The Multiple Risk Factor Intervention Trial (MRFIT) showed that the risk of stroke-induced mortality increased 2.8-fold (95% CI 2.0–3.7) among DM patients, even after adjusting for cardiovascular risk factors, age, race, and income [12]. It is worth noting that the risk factor between DM patients and stroke recurrence is also obviously increased. A recent published meta-analysis has reported that compared to healthy people, the risk of stroke recurrence in DM patients was significantly higher (hazard ratio, 1.45; 95% CI, 1.32–1.59) [13].

Diabetes is also associated with increased ischemia-reperfusion injury [14, 15]. A retrospective study of the National Institute of Neurological Disorders and Stroke demonstrated that, in acute ischemic stroke, serum glucose level and diabetes predict tissue

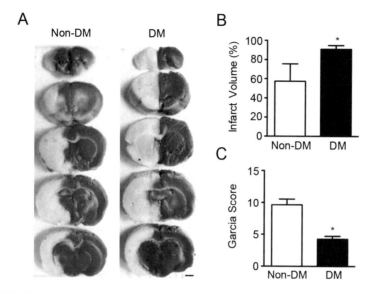

Fig. 10.1 Hyperglycemia enhances cerebral ischemia-induced brain injury. (**a**) Representative images for TTC stained ischemic brain infarctions. Bar = 1 mm. (**b**) Ischemic infarct volumes were quantified at 24 h after stroke. (**c**) Neurological score on day 1 after stroke. Data are mean±SEM. *P < 0.05, DM = Diabetes mellitus

plasminogen activator (tPA) related intracerebral hemorrhage [16]. Hyperglycemia prior to reperfusion potentially decreases the beneficial effect of tPA, given that hyperglycemia may act as an inhibitor of fibrinolysis. A clinical study that comprised 58,265 acute ischemic stroke patients treated with tPA revealed that both acute and chronic hyperglycemia are related to increased mortality and poor clinical outcomes [17]. Hyperglycaemia decreases the effectiveness of thrombolysis, and increases the risk of haemorrhage following thrombolysis [18, 19]. Guidelines recommend against tPA use in the 3- to 4.5-h window in patients with previous stroke or diabetes based on insufficient evidence concerning its effectiveness, as those patients were excluded from the European Cooperative Acute Stroke Study (ESASS) III trial [17, 20].

3   Mechanisms of Cerebral Vascular Injury in Diabetic Ischemia and Reperfusion

During the acute period after cerebral ischemia, uncontrolled inflammation is a major mediator of cerebrovascular injury and brain damage [21]. Compared to healthy controls, inflammation and oxidative stress significantly increased in DM patients. Recent clinical studies revealed that the expression of proinflammatory proteins such as TNF-α, interleukin-1 (IL-1), IL-6 and monocyte chemoattractant protein-1 (MCP-1) are much higher in diabetic patient plasma. A recent experimental study also demonstrated that at 12 h of reperfusion following 45 min transient middle cerebral artery occlusion (MCAO), several inflammatory cytokines expression at the cortex are significantly increased in db/db mice when compared to db/+ mice (IL-1β, 164%; IL-6, 84%; MCP-1, 65%; MIP-1α, 135%) [22]. TNF-α, IL-1 and IL-6 play critical roles in ischemia-induced cerebral vascular injury have been demonstrated by numerous studies showing that reduced expression of IL-6 is accompanied by a better neurological outcome [23]. Moerover, both *in vivo* and *in vitro* experiments have identified the important function of MCP-1 on (MCAO)/reperfusion-induced (BBB) breakdown [24, 25]. A recent study by *El-Sahar AE et al.* showed that, in the diabetic condition, acute oxidative stress is increased in rat brain following cerebral ischemic reperfusion injury, and this alteration is associated with the exacerbation of ischemia-induced infract size [26]. More recently, our group demonstrated that methylglyoxal (MGO), a reactive dicarbonyl accumulated in diabetic patients, significantly increases 4 h oxygen-glucose deprivation (OGD)/20 h reperfusion-induced primary human brain microvascular endothelial cells cytotoxicity by down-regulating glutathione (GSH) production and up-regulating reactive oxygen species (ROS) release [27].

Including the upregulated proinflammatory cytokines, increased infiltration of inflammatory cells, such as leukocytes, and white blood cells, into ischemic brain tissue has also been demonstrated in several diabetic ischemia-reperfusion animal models [28]. Cerebral vascular endothelial cells are activated in the setting of cerebral ischemia, leading to production of adhesion molecules, which enables inflammatory cells to attach to endothelial cells and then migrate to the ischemic tissue to further release proinflammatory cytokines and activate local microglial [29, 30]. Using immunohistochemistry and Western blot methods, *Ding CN et al.* showed that the number of intercellular cell adhesion molecule-1 (ICAM-1) expressed in microvascular of cortex is obviously increased at 3 days of reperfusion in diabetic rats compared to non-diabetic rats [31]. Protein kinase C (PKC) is known to mediate the expression of ICAM-1. Previously published literature also demonstrated that hyperglycemia and diabetes can activate PKC, which promotes cerebral endothelial dysfunction with increased inflammatory-endothelium interactions [32, 33]. In addition to ICAM-1, db/db mice also exhibit significantly higher increases at cortex under reperfusion period after 45 min or 2 h/MCAO of other adhesion molecules, such as P-selectin and E-selectin [22].

The expression of pro-inflammatory cytokines and inflammatory mediators, such as IL-1β, IL-6, TNF-α, MCP-1, CCL-2, iNOS and ICAM-1, in the cerebral vascular is regulated by the induction of transcription factors during ischemic reperfusion inflammation [34, 35]. Thus, the transcription factor NFκB becomes a key regulator in cerebral ischemic reperfusion associated with regulating cell death and inflammation [36]. At 3 h reperfusion after MCAO, translocation of NFκB from cytoplasm to nucleus is detected in the cortex tissue of rats. Interestingly, the translocation of NFκB is significantly enhanced in diabetic rat cortex compared to non-diabetic rat cortex. This trend can be observed even at 24 h reperfusion after MCAO. The nuclear translocation of NFκB will further affect the mRNA expression of COX-2, iNOS, and ICAM-1 [37].

Matrix metalloproteinases (MMPs), as proteolytic enzymes, could degrade all components of the extracellular matrix (ECM) around the blood vessels [38].

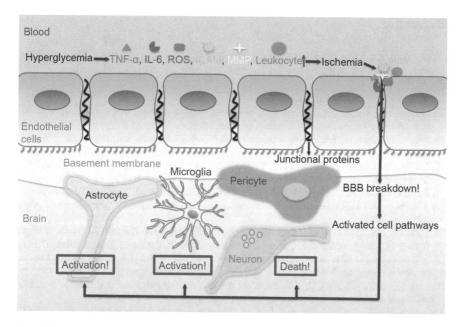

Fig. 10.2 Mechanisms of cerebral vascular injury in diabetic ischemia and reperfusion. Schematic representation of mechanisms involved in aggravating brain damage following cerebral ischemia under diabetic condition

MMPs-induced (BBB) leakage during cerebral ischemia may aggravate ischemic brain tissue to bleeding during reperfusion [39]. Using a model of 1 h MACO/23 h reperfusion, *Kamada H et al.* found that hyperglycemia induced by streptozotocin could enhance the level of MMP-9 activity compared to normal control rats [40]. In another diabetic animal model, the Goto-Kakizaki (GK) rat model, *Elgebaly MM et al.* also demonstrated that hyperglycemic augments 3 h MCAO/21 h reperfusion-induced stimulation of MMP-9 activity in isolated cerebral vessels [41].

Taken together, diabetes and its associated hypoglycemia are linked to cerebral ischemia-induced mortality and poor functional recovery. A number of studies showed that diabetes enhances inflammation, oxidative stress, and MMPs activity, which are associated with exacerbated cerebral vascular damage after ischemia/reperfusion injury (Fig. 10.2). In addition, a few literature published recently reported that autophagy and hyperglycemia-induced advanced glycation end products (AGEs) and its receptor (RAGE) also contribute to diabetes-enhanced cerebral ischemia/reperfusion injury [42–44]. However, the precise mechanisms in diabetic-exacerbated cerebral ischemia injury remain unknown. Thus, a better understanding of cerebral vascular injury in diabetic ischemia may provide novel therapeutic approaches for the treatment and prevention of diabetes-associated stroke damage.

References

1. Eltzschig HK, Eckle T. Ischemia and reperfusion--from mechanism to translation. Nat Med. 2011;17(11):1391–401.
2. Turc G, Isabel C, Calvet D. Intravenous thrombolysis for acute ischemic stroke. Diagn Interv Imaging. 2014;95(12):1129–33.
3. Shi J, Liu Y, Duan Y, et al. A new idea about reducing reperfusion injury in ischemic stroke: gradual reperfusion. Med Hypotheses. 2013;80(2):134–6.
4. Nour M, Scalzo F, Liebeskind DS. Ischemia-reperfusion injury in stroke. Interv Neurol. 2013;1(3-4):185–99.
5. Luitse MJ, Biessels GJ, Rutten GE, Kappelle LJ. Diabetes, hyperglycaemia, and acute ischaemic stroke. Lancet Neurol. 2012;11(3):261–71.
6. Bejot Y, Giroud M. Stroke in diabetic patients. Diabetes Metab. 2010;36(Suppl 3):S84–7.
7. Kissela B, Air E. Diabetes: impact on stroke risk and poststroke recovery. Semin Neurol. 2006;26(1):100–7.
8. Hill MD. Stroke and diabetes mellitus. Handb Clin Neurol. 2014;126:167–74.
9. Sarwar N, Gao P, Seshasai SR, et al. Diabetes mellitus, fasting blood glucose concentration, and risk of vascular disease: a collaborative meta-analysis of 102 prospective studies. Lancet. 2010;375(9733):2215–22.
10. Liao CC, Shih CC, Yeh CC, et al. Impact of diabetes on stroke risk and outcomes: two nationwide retrospective cohort studies. Medicine (Baltimore). 2015;94(52):e2282.
11. Kaarisalo MM, Raiha I, Sivenius J, et al. Diabetes worsens the outcome of acute ischemic stroke. Diabetes Res Clin Pract. 2005;69(3):293–8.
12. Stamler J, Vaccaro O, Neaton JD, Wentworth D. Diabetes, other risk factors, and 12-yr cardiovascular mortality for men screened in the Multiple Risk Factor Intervention Trial. Diabetes Care. 1993;16(2):434–44.
13. Shou J, Zhou L, Zhu S, Zhang X. Diabetes is an independent risk factor for stroke recurrence in stroke patients: a meta-analysis. J Stroke Cerebrovasc Dis. 2015;24(9):1961–8.
14. Kruyt ND, Biessels GJ, Devries JH, Roos YB. Hyperglycemia in acute ischemic stroke: pathophysiology and clinical management. Nat Rev Neurol. 2010;6(3):145–55.
15. Kelly-Cobbs AI, Prakash R, Li W, et al. Targets of vascular protection in acute ischemic stroke differ in type 2 diabetes. Am J Physiol Heart Circ Physiol. 2013;304(6):H806–15.
16. Demchuk AM, Morgenstern LB, Krieger DW, et al. Serum glucose level and diabetes predict tissue plasminogen activator-related intracerebral hemorrhage in acute ischemic stroke. Stroke. 1999;30(1):34–9.
17. Masrur S, Cox M, Bhatt DL, et al. Association of Acute and Chronic Hyperglycemia with acute ischemic stroke outcomes post-thrombolysis: findings from get with the guidelines-stroke. J Am Heart Assoc. 2015;4(10):e002193.
18. Alvarez-Sabin J, Molina CA, Montaner J, et al. Effects of admission hyperglycemia on stroke outcome in reperfused tissue plasminogen activator--treated patients. Stroke. 2003;34(5):1235–41.
19. Ribo M, Molina C, Montaner J, et al. Acute hyperglycemia state is associated with lower tPA-induced recanalization rates in stroke patients. Stroke. 2005;36(8):1705–9.
20. Hacke W, Kaste M, Bluhmki E, et al. Thrombolysis with alteplase 3 to 4.5 hours after acute ischemic stroke. N Engl J Med. 2008;359(13):1317–29.
21. del Zoppo GJ. Inflammation and the neurovascular unit in the setting of focal cerebral ischemia. Neuroscience. 2009;158(3):972–82.
22. Tureyen K, Bowen K, Liang J, Dempsey RJ, Vemuganti R. Exacerbated brain damage, edema and inflammation in type-2 diabetic mice subjected to focal ischemia. J Neurochem. 2011;116(4):499–507.
23. Strecker JK, Minnerup J, Gess B, Ringelstein EB, Schabitz WR, Schilling M. Monocyte chemoattractant protein-1-deficiency impairs the expression of IL-6, IL-1beta and G-CSF after transient focal ischemia in mice. PLoS One. 2011;6(10):e25863.

24. Strecker JK, Minnerup J, Schutte-Nutgen K, Gess B, Schabitz WR, Schilling M. Monocyte chemoattractant protein-1-deficiency results in altered blood-brain barrier breakdown after experimental stroke. Stroke. 2013;44(9):2536–44.
25. Dimitrijevic OB, Stamatovic SM, Keep RF, Andjelkovic AV. Effects of the chemokine CCL2 on blood-brain barrier permeability during ischemia-reperfusion injury. J Cereb Blood Flow Metab. 2006;26(6):797–810.
26. El-Sahar AE, Safar MM, Zaki HF, Attia AS, Ain-Shoka AA. Neuroprotective effects of piogli-tazone against transient cerebral ischemic reperfusion injury in diabetic rats: modulation of anti-oxidant, anti-inflammatory, and anti-apoptotic biomarkers. Pharmacol Rep. 2015;67(5):901–6.
27. Li W, Chen Z, Yan M, He P, Chen Z, Dai H. The protective role of isorhamnetin on human brain microvascular endothelial cells from cytotoxicity induced by methylglyoxal and oxygen-glucose deprivation. J Neurochem. 2016;136(3):651–9.
28. Panes J, Kurose I, Rodriguez-Vaca D, et al. Diabetes exacerbates inflammatory responses to ischemia-reperfusion. Circulation. 1996;93(1):161–7.
29. Enzmann G, Mysiorek C, Gorina R, et al. The neurovascular unit as a selective barrier to polymorphonuclear granulocyte (PMN) infiltration into the brain after ischemic injury. Acta Neuropathol. 2013;125(3):395–412.
30. Jing L, Wang JG, Zhang JZ, et al. Upregulation of ICAM-1 in diabetic rats after transient forebrain ischemia and reperfusion injury. J Inflamm. 2014;11(1):35.
31. Ding C, He Q, Li PA. Diabetes increases expression of ICAM after a brief period of cerebral ischemia. J Neuroimmunol. 2005;161(1-2):61–7.
32. Das Evcimen N, King GL. The role of protein kinase C activation and the vascular complica-tions of diabetes. Pharmacol Res. 2007;55(6):498–510.
33. Booth G, Stalker TJ, Lefer AM, Scalia R. Mechanisms of amelioration of glucose-induced endo-thelial dysfunction following inhibition of protein kinase C in vivo. Diabetes. 2002;51(5):1556–64.
34. Yi JH, Park SW, Kapadia R, Vemuganti R. Role of transcription factors in mediating post-ischemic cerebral inflammation and brain damage. Neurochem Int. 2007;50(7-8):1014–27.
35. Sandireddy R, Yerra VG, Areti A, Komirishetty P, Kumar A. Neuroinflammation and oxida-tive stress in diabetic neuropathy: futuristic strategies based on these targets. Int J Endocrinol. 2014;2014:674987.
36. Russo MA, Sansone L, Carnevale I, et al. One special question to start with: can HIF/NFkB be a target in inflammation? Endocr Metab Immune Disord Drug Targets. 2015;15(3):171–85.
37. Iwata N, Okazaki M, Nakano R, Kasahara C, Kamiuchi S, Suzuki F, Iizuka H, Matsuzaki H, Hibino Y. Diabetes-mediated exacerbation of neuronal damage and inflammation after cerebral ischemia in rat: protective effects of water-soluble extract from culture medium of ganodermalucidum mycelia. In: Advances in the preclinical study of ischemic stroke. Rijeka, Croatia: InTech; 2012.
38. Jian Liu K, Rosenberg GA. Matrix metalloproteinases and free radicals in cerebral ischemia. Free Radic Biol Med. 2005;39(1):71–80.
39. Yang Y, Rosenberg GA. Matrix metalloproteinases as therapeutic targets for stroke. Brain Res. 2015;1623:30–8.
40. Kamada H, Yu F, Nito C, Chan PH. Influence of hyperglycemia on oxidative stress and matrix metalloproteinase-9 activation after focal cerebral ischemia/reperfusion in rats: relation to blood-brain barrier dysfunction. Stroke. 2007;38(3):1044–9.
41. Elgebaly MM, Prakash R, Li W, et al. Vascular protection in diabetic stroke: role of matrix metalloprotease-dependent vascular remodeling. J Cereb Blood Flow Metab. 2010;30(12):1928–38.
42. Li W, Xu H, Hu Y, et al. Edaravone protected human brain microvascular endothelial cells from methylglyoxal-induced injury by inhibiting AGEs/RAGE/oxidative stress. PLoS One. 2013;8(9):e76025.
43. Li W, Liu J, He P, et al. Hydroxysafflor yellow A protects methylglyoxal-induced injury in the cultured human brain microvascular endothelial cells. Neurosci Lett. 2013;549:146–50.
44. Fang L, Li X, Zhong Y, et al. Autophagy protects human brain microvascular endothelial cells against methylglyoxal-induced injuries, reproducible in a cerebral ischemic model in diabetic rats. J Neurochem. 2015;135(2):431–40.

Chapter 11
Ischemia/Reperfusion Damage in Diabetic Stroke

Poornima Venkat, Michael Chopp, and Jieli Chen

Abstract Stroke is a leading cause of death and long-term disability. Patients with diabetes mellitus suffer from an increased risk of cardiovascular and cerebrovascular diseases including ischemic stroke. Diabetic stroke patients sustain worse neurological deficits and battle high mortality rates. Diabetes triggers a detrimental pathophysiological cascade resulting in severe vascular dysfunction and I/R injury which result in poor outcome after stroke in this population. The various aspects of diabetic stroke induced vascular and reperfusion damage and the underlying mechanisms are discussed in this chapter.

Keywords Stroke · Diabetes mellitus · Hyperglycemia · Brain vascular injury · Ischemia reperfusion injury

1 Epidemiology

Diabetes mellitus (DM) is a chronic metabolic disorder, characterized by hyperglycemia, and caused by insulin deficiency (type 1 DM) and/or, insulin resistance (type 2 DM). In 2010, an estimated 285 million adults suffered from DM [1]. The global burden of DM is increasing alarmingly, and an estimated 439 million adults aged 20–79 years

P. Venkat
Neurology, Henry Ford Hospital, Detroit, MI, USA

M. Chopp
Department of Physics, Oakland University, Rochester, MI, USA

Neurology, Henry Ford Hospital, Detroit, MI, USA

J. Chen (✉)
Neurology, Henry Ford Hospital, Detroit, MI, USA

Neurological & Gerontology Institute, Tianjin Medical University General Hospital, Tianjin, China

Neurology Research, Henry Ford Hospital, Detroit, MI, USA
e-mail: jieli@neuro.hfh.edu

© Springer International Publishing AG, part of Springer Nature 2018   171
W. Jiang et al. (eds.), *Cerebral Ischemic Reperfusion Injuries (CIRI)*, Springer Series
in Translational Stroke Research, https://doi.org/10.1007/978-3-319-90194-7_11

are projected to suffer from DM by 2030 [1, 2]. Between 2010 and 2030, developing countries will experience a dramatic 69% increase in adults with DM while developed countries will also face a 20% increase adults in DM [1]. About 90–95% of all DM patients suffer from T2DM [3]. DM patients suffer from an increased risk of cardiovascular and cerebrovascular diseases including ischemic stroke [4]. Stroke is a leading cause of death and long-term disability. Every year, in the United States, an estimated 795,000 people suffer from a new or recurrent stroke [2]. Approximately 87% of stroke patients sustain ischemic stroke while 13% suffer from hemorrhagic stroke [2]. Approximately 30% of acute stroke patients suffer from either pre-existing diabetes or newly diagnosed diabetes [4, 5]. Although DM increases the risk and incidence of ischemic stroke across all ages, races and gender; in particular, DM patients <65 years of age are at a higher risk of stroke [2, 6]. Additionally, diabetic stroke patients sustain worse neurological deficits [7] and battle high mortality rates [8]. Women have a higher lifetime risk of stroke as well as decreased survival after diabetic stroke [2]. While mortality rates are ~29% in stroke patients with normal glucose levels, acute stroke in hyperglycemic patients significantly increases mortality to 45% in patients with pre-existing DM, and to 78% in patients with hyperglycemia but no history of DM [8].

The poor prognosis in diabetic stroke patients may at least in part, be attributed to aggravated stroke pathology and ischemia reperfusion (I/R) injury [9, 10]. The severe microvascular and macrovascular damage induced by DM and aggravated by stroke often leads to the impairment of other end organs such as the kidneys, eyes, and peripheral nervous system [11]. Long term recovery of neurological functional is also hindered by recurrent strokes [12]. A large number of experimental studies provide support to the premise that DM triggers a detrimental pathophysiological cascade resulting in severe vascular dysfunction and I/R injury which result in poor outcome after stroke in this population [13–15]. The various aspects of this cascade and the underlying mechanisms are discussed in the following sections.

2 Ischemia/Reperfusion Damage in Diabetic Stroke

2.1 Reperfusion Strategies After Stroke

Thrombolysis and mechanical recanalization are major reperfusion strategies after stroke. Following ischemic stroke, timely recanalization of the occluded artery is critical to restore regional cerebral blood flow (CBF) and salvage ischemic brain tissue. The two important milestones in interventional stroke care include the advent of thrombolytic therapy with tissue plasminogen activator (tPA) and mechanical thrombectomy [16]. Acute administration of intravenous recombinant tPA (IV rtPA) remains the only FDA approved pharmacological treatment for stroke. In 1995, the landmark clinical trial conducted by the National Institute of Neurological Disorders and Stroke demonstrated the efficacy of IV rtPA in improving neurological functional outcome after stroke when tPA was administered intravenously within 0–3 h of stroke onset [17]. Subsequently, the effectiveness of tPA in extended time frames have been investigated, and benefits are still reported when tPA is administered within 4.5 h of stroke onset in carefully selected patients [18]. The administration of tPA in stroke patients with large intracranial artery occlusions is associated with poor recanalization, neurological deficits and risk of spontaneous intracerebral hemorrhage. Therefore, the other major recanalization strategy after stroke includes endovascular approaches such as mechanical thrombectomy for patients who exhibit large intracranial artery occlusions with salvageable penumbral tissue [19]. The major class of FDA approved mechanical devices includes coil retrievers, aspiration devices, and most recently stent retrievers [19]. Mechanical thrombectomy is often employed as an adjunct therapy with tPA in carefully selected patients [19].

2.2 Challenges in Reperfusion Treatments After Stroke

The major challenge of tPA treatment in stroke care is the narrow time window for treatment initiation. A majority of patients fall outside this time frame and unfortunately, even when hospitalized within this time period, only a fraction of eligible stroke patients receive tPA as per recommended guidelines [20]. Patients for tPA therapy even within the 0–3 h time frame should be carefully selected and the exclusion criteria includes patients who are susceptible to intracerebral hemorrhage, patients who suffered from a stroke or traumatic brain injury in the preceding 3 months, patients consuming anti coagulants, etc. among others [18]. Additionally, the exclusion criteria for use of tPA in the 3–4.5 h time window includes patients older than 80 years of age, patients consuming oral anticoagulants, patients with a history of stroke and DM as well as patients who sustain severe stroke with ischemic injury to more than one third of the middle cerebral artery territory [18]. The major challenges for endovascular mechanical thrombectomy include the lack of routine screening and imaging to identify stroke patients with large arterial occlusion, inappropriate selection of patients for thrombectomy such as patients without large arterial occlusion or with inaccessible clots; and infrequent use of modern mechanical devices resulting in poor recanalization [19]. All reperfusion therapies involve a risk of intracerebral hemorrhage following recanalization [19]. In addition, mechanical thrombectomy can also lead to subarachnoid hemorrhage due to injury to the wall of blood vessel during operation of mechanical device and complications during clot retrieval leading to thrombus fragmentation and subsequent occlusion of a distal vessel [19, 21].

2.3 I/R injury in Diabetic Stroke

Several detrimental effects of reperfusion in diabetic stroke have been documented by clinical and experimental studies [22–25]. Due to the brain's high metabolic and energy requirements and limited energy storage reservoir, any disruption to CBF impairs its functioning and leads to rapid damage and death of brain cells. During ischemia, the deficiency of oxygen and nutrients increases the metabolic need of the ischemic penumbra, and creates a microenvironment in which the restoration of CBF leads to secondary thrombosis, inflammation, and oxidative stress causing secondary tissue damage and expansion of infarct volume beyond the initial ischemic insult [26]. When the infarct volume after stroke is large, I/R injury can lead to hemorrhagic infarct conversion [27]. To protect the microcirculation from ischemic damage, in the early phase after stroke, small vessel resistance increases and large vessel resistance decreases [28]. However, this may result in decreased reperfusion and worse outcome after I/R injury [28]. In DM rats subjected to stroke, prolonged ischemia followed by reperfusion results in poor prognosis due to aggravated vascular damage, blood brain barrier (BBB) disruption, cerebral edema formation and hemorrhagic transformation [22]. Diabetic rodents subjected to stroke and reperfusion exhibit worse neurological deficits, higher mortality rates, large infarct area, frequent and worse hemorrhage, ipsilateral hemispheric swelling and edema [23]. Efficacy of tPA treatment is also dependent on stroke subtype, and in DM stroke, a large number of stroke patients suffer from large vessel atherothrombotic stroke, for which tPA treatment is less efficient compared to other stroke subtypes [29].

3 Pathophysiology of I/R Injury in Diabetic Stroke

Prolonged CBF decrease after stroke irreversibly damages neurons, hence the re-establishment of CBF and increasing functional microvasculature in the ischemic penumbra are essential to maintain neural function, and create a hospitable microenvironment for neuronal plasticity and functional recovery. The major adverse effects of I/R injury in diabetic stroke include exacerbated BBB disruption, extensive vascular damage, aggravated inflammatory responses, increased susceptibility to spontaneous intracerebral hemorrhage and formation of cerebral edema. These pathophysiological events and their underlying mechanisms are discussed in the following sections, and summarized in Fig. 11.1.

3.1 Neurovascular Uncoupling After I/R Injury in Diabetic Stroke

The neurovascular unit is a functional unit encompassing the anatomical and metabolic interactions between the neurons, astrocytes and vascular components of the BBB [30]. The BBB serves as a dynamic semi-permeable barrier separating the peripheral circulation and the central nervous system [31, 32]. While allowing influx of hydrophobic molecules and metabolic products by passive diffusion, the BBB prevents the entry of microscopic substances, hydrophilic molecules, and potential neurotoxins [31, 32].

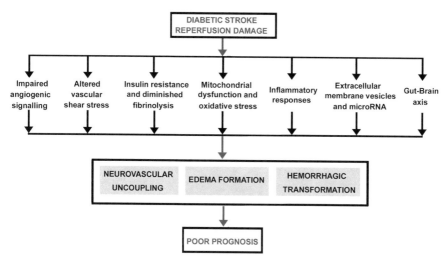

Fig. 11.1 Summary of key pathophysiological changes and mechanisms of reperfusion damage in diabetic stroke

Neurovascular uncoupling and BBB disruption are among the initial steps of the pathophysiological cascade in DM and stroke [33]. In the minutes to hours following ischemia, endothelial swelling and pericyte damage and death lead to irreversible constriction of capillaries and BBB disruption [34, 35]. Stroke in DM patients aggravates BBB disruption; and BBB permeability that usually increases within 7 days post stroke in non-DM conditions has been shown to be extended to 14 days or longer in DM stroke animals [36–38]. A ruptured BBB facilitates the entry of large molecules and the invasion of inflammatory factors, neurotoxins and pathogens into the brain [32], and in a vicious cycle, these inflammatory factors in turn promote BBB disruption and lead to hemorrhagic transformation in diabetic stroke animals [37, 39, 40].

DM induces endothelial dysfunction including impaired blood vessel tone, platelet activation, leukocyte adhesion, thrombogenesis, and inflammation [41]. DM increases vasoconstriction via increasing expression of vasoconstrictor endothelin-1 and decreasing expression of vasodilator Nitric Oxide (NO), leading to vasoconstriction of blood vessels and prolonged CBF decrease [41]. Prolonged attenuation of vasodilation can trigger endothelial dysfunction and aggravate atherosclerosis [42]. DM also creates an environment of high oxidative stress and inflammatory factors, which are conducive to atherosclerosis [43]. DM induced mitochondrial oxidative stress leads to endothelial cell damage, pericyte depletion and BBB leakage [44].

The interactions between the glial and vascular component of the neurovascular unit, i.e. astrocytes and endothelial cell interactions, are needed to regulate brain water content and electrolyte balance both under normal and disease states [30, 45]. In the setting of ischemic brain injury, activation of astrocytes can be both constructive and destructive [46, 47]. In the acute phase of stroke, reactive astrocytes secrete proinflammatory cytokines, inhibit axonal regeneration, and aid in infarct expansion [47]. During the chronic phase after stroke, reactive astrocytes aid in neurite sprouting, synapse formation, rebuilding the BBB and secrete neurotrophic factors that aid in brain repair mechanims [46, 47]. Under the hyperglycemic conditions of DM, post stroke astrocyte activation is suppressed, and there is greater cell death of astrocytes as demonstrated in a rodent model of forebrain ischemia [48, 49]. On the whole, neurovascular uncoupling is a gateway leading to mitochondrial dysfunction and oxidative stress, neuronal death and brain tissue atrophy [50, 51].

3.2   Edema in Diabetic Stroke

The interaction between the astrocytes, the water channel protein Aquaporin-4 and endothelial cells is critical to brain water content regulation as well as post stroke edema resolution [45, 52]. Immediately following ischemic injury, cytotoxic cerebral edema may ensue. In cytotoxic edema, impaired cellular metabolism and dysfunction of sodium and potassium ion pumps lead to accumulation of sodium and increased water uptake, resulting in swelling of brain cells [53]. Cytotoxic edema may give way and/or occur together with vasogenic edema. Vasogenic cerebral edema is a pathological condition in which the intracranial pressure is increased by increasing brain water content in the interstitial space; and can last for several days after stroke [45, 52]. Edema resolution involves cerebral vasculature and cerebrospinal fluid pathways mediated transport of water from the brain parenchyma to the vascular, intra ventricular and subarachnoid compartments via bulk flow [45, 52].

In the evolution of cytotoxic edema, Aquaporin-4 has been implicated in water uptake into the brain tissue, while in vasogenic edema; Aquaporin-4 plays a key role in water reabsorption and clearance [45, 54, 55]. Compared to control wild type mice, Aquaporin-4 deficient mice subjected to stroke exhibit ~30% decrease in cerebral cytotoxic edema; suggesting that during the early phase of ischemia, Aquaporin-4 inhibition could attenuate cytotoxic edema formation [45, 55]. However, Aquaporin-4 deficient mice subjected to a freeze-injury model of vasogenic brain edema, exhibit worse neurological function, increased intracranial pressure and greater brain water content compared to control mice; indicating an essential role of Aquaporin-4 in fluid clearance and reabsorption of vasogenic edema [52]. Hence, altered Aquaporin-4 expression can have adverse effect in edema formation and resolution in diabetic stroke [56]. Middle aged rats induced with DM exhibit a significant decrease of paravascular Aquaporin-4 expression in the hippocampus [57]. Stroke in DM rodents damages the astrocytic end-foot lining around cerebral vessels, and damaged astrocytes exhibit increased withdrawal of the astrocyte end-foot from the cerebral vessel wall [48, 49]. Several studies have reported increased edema formation after diabetic stroke compared to non diabetic stroke, and I/R injury may exacerbate edema after stroke in diabetes [56]. Impaired water channel function and Aquaporin-4 expression may play a key role in edema formation and poor edema resolution in diabetic stroke.

3.3 Hemorrhagic Transformation in Diabetic Stroke

Hemorrhagic transformation (HT) is a commonly encountered critical effect of reperfusion therapy in ischemic stroke [58]. Particularly in DM stroke patients, tPA increases the risk of symptomatic intracerebral hemorrhage [58, 59]. HT may lead to hemorrhagic infarction or hematoma formation. The main predictors of HT after I/R injury include massive infarction volume, edema formation, gray matter injury with greater collateral CBF and hyperglycemia. HT formation is directly related to infarct volume, and DM patients (clinical studies) and DM rats (experimental studies) exhibit greater infarct volume, worse neurological functional outcome, and decreased efficacy of thrombolysis with tPA following stroke compared to non DM stroke subjects [60, 61]. I/R injury in hyperglycemic cats induced a fivefold increase in HT incidence and a 25-fold increase in the extent of hemorrhagic infarction compared to non-DM cats [27]. HT after I/R injury in DM animals has been associated with metabolic alterations and a significant decrease in tissue energy, free radical production, inflammatory responses, and increase in lactate acidosis which damages the cerebral vasculature and facilitates entry of edema fluid and red blood cells extravasation [27, 62, 63]. Compared to non-DM rats, DM rats exhibit a greater CBF reduction and HT when subject to transient stroke with reperfusion injury, but not when subject to permanent ischemic stroke without reperfusion [60, 64]. This indicates that HT formation in DM stroke is largely an adverse effect of reperfusion.

4 Mechanisms of I/R Injury in Diabetic Stroke

4.1 Impaired Angiogenic Signaling

Angiogenesis is the formation and growth of new blood vessels from pre-existing vessels, and although DM induces vigorous angiogenesis, a majority of these vessels are dysfunctional and have poor vessel wall maturity, resulting in a large number of leaky or non perfused blood vessels [13, 37, 65]. Vascular remodeling after stroke via angiogenesis peaks at 7 days after stroke, and has been associated with higher survival rates and improved long-term neurological functional outcome in non-DM stroke patients [66–68]. Angiogenic responses in DM rodents are mediated by vascular endothelial growth factor (VEGF) angiogenic signaling, increased expression of Angiopoietin-2, and increased oxidative stress [65, 69].

VEGF is one of the essential growth factors for angiogenesis, and directly stimulates endothelial cell proliferation and migration [70]. While VEGF administered prior to stroke [71] or at a delayed time point after stroke [72] can be beneficial, during the acute phase of stroke, VEGF increases BBB leakage, cerebral hemorrhage and infarction volume [72]. The detrimental effects of VEGF are pronounced in DM stroke compared non-DM stroke due to the extensive microvascular and macrovascular damage induced by DM, and aggravated by I/R [65, 73].

Angiopoietins are also vascular growth factors that exert key functions in angiogenesis. Angiopoietin-1 is required for the migration, adhesion and survival of endothelial cells [74]. Angiopoietin-1 promotes post stroke angiogenesis, organization and maturation of new blood vessels, and vessel stabilization [75]. Angiopoietin-2 is a natural antagonist of Angiopoietin-1, and together they regulate angiogenesis, vascular stability, vascular permeability and lymphatic integrity [76]. In patients with DM, expression of Angiopoietin-2 and VEGF (but not Angiopoietin-1) was significantly increased compared to non diabetic patients [77]. Interestingly, increased Angiopoietin-2 expression did not correlate with either increased endothelial injury or atherosclerosis while increased VEGF correlated with poorly controlled hyperglycemia (high HbA1C), suggesting a likelihood of a prominent role of VEGF in mediating DM related vascular pathology [77]. Angiopoietin-2 is elevated in stroke patients who also exhibit BBB disruption [76]. Decreased Angiopoietin-1 is related with increased BBB leakage and brain hemorrhagic transformation after stroke in DM mice [78]. Angiopoietin-1 also inhibits pro-inflammatory mediators such as tumor necrosis factor alpha (TNF-α), and interleukins (IL-6, IL-8) [79] that exacerbate vascular damage after stroke in diabetic populations [15, 37].

4.2   Altered Vascular Shear Stress

The walls of the vasculature are continuously subject to shear stress, which is the tangential force exerted by blood flow on the endothelial cell surfaces [80]. Shear stress helps maintain BBB integrity as well as helps create an endothelial transport barrier between blood and underlying tissues [80, 81]. In T2DM patients, acute, short-term hyperglycemia increases shear stress induces platelet activation, adhesion and aggregation on the subendothelium, which accelerates atherosclerosis and increases the risk of arterial thrombosis [82]. High shear stress induced platelet activation is strictly dependent on plasma von Willebrand factor (vWF) which is also significantly increased in the DM population [82]. While laminar blood flow in vessels is atheroprotective; in DM, non-laminar blood flow in vessels induces changes to endothelial gene expression, disturbs endothelial cytoskeletal alignment and wound repair, alters leukocyte adhesion to activated endothelium, as well as alters the vasoreactive, oxidative and inflammatory states of the vessel wall [81]. In addition to being a metabolic disease, DM is also a vascular disease inducing extensive vascular damage. In DM rats, the inability of the naïve and damaged blood vessels to withstand the changes in shear stress during I/R injury can result in vessel wall rupture and bleeding, which can aggravate damage to ischemic tissue and lead to HT [22].

4.3 High Insulin Resistance and Diminished Fibrinolysis

Patients with hypertension, obesity, vascular disorders or who are aged are prone to developing T2DM in which insulin resistance usually precedes disease manisfestation [83]. When DM patients with high levels of insulin resistance are treated with tPA following stroke, reperfusion induces greater infarct size, worse neurological deficits, worse disease prognosis and greater susceptibility to intracerebral hemorrhage [24, 25]. The development of insulin resistance and endothelial dysfunction may be initiated by the release of pro-inflammatory adipokines such as IL-6, TNF-α, PAI-1 (plasminogen activator inhibitor-1) by visceral adipose tissue [84]. Accumulated fat and enhanced adipose tissue-derived PAI-1, induce metabolic, vascular and fibrinolysis abnormalities which enhance macrophage infiltration, chronic inflammation, and free fatty acid release in obese states [85]. Insulin resistance resulting from impaired glycogen synthesis can induce detrimental changes to the vascular beds [86]. In DM rodents subject to cerebral ischemic via hind limb I/R injury, insulin resistance was found to initiate an inflammatory cascade involving microglial activation resulting in greater cortical and hippocampal ischemic damage and neuronal apoptosis [87].

Impaired fibrinolysis is yet another common complication among diabetic subjects, and is related to insulin resistance. During fibrinolysis, plasminogen is released by the liver which upon activation to plasmin by endothelial cell derived tPA, degrades the blood clot by dissecting the fibrin mesh at several points [88]. PAI-1 is an inhibitor of tPA and DM is characterized by impaired fibrinolysis which leads to increased coagulability and increased concentration of PAI-1 in blood as well as tissues [89]. In DM, increased expression of PAI-1 induces insulin resistance and metabolic abnormalities during proinflammatory processes involving several cytokines and chemokines [85]. Decreased fibrinolytic capacity is a major cause for the development and progression of atherosclerotic plaque [90]. As mentioned earlier, a large fraction of DM stroke patients suffer from large vessel athero-thrombotic stroke, and have increased PAI-1 levels that can persist even up to 6 months after stroke [91]. The increase in circulating thrombin-activatable fibrinolysis inhibitor (TAFI) and PAI-1 in DM patients are major contributing factors for failure of thrombolytic therapy [92]. Therefore, targeting PAI-1 and TAFI can promote neuroprotective effects by decreasing fibrin deposition and improving reperfusion after stroke in mice [92].

4.4 Mitochondrial Dysfunction and Oxidative Stress

In DM patients and in animals induced with DM, insulin resistance associated mitochondrial dysfunction and excessive metabolism of substrates such as glucose and fatty acids, induces a state of chronic oxidative stress [93]. Under conditions of high extracellular glucose, the cellular response to oxidative stress is altered and even a level of oxidative stress normally anabolic may have pathological consequences [94]. I/R induces mitochondrial dysfunction resulting in mismatch of adenosine triphosphate (ATP) production and uptake, decreased ATP content, decreased cell proliferation and impaired delivery of glucose and oxygen to brain tissue [95–97]. In addition, DM stroke induces morphological changes such as fragmentation, vacuolation, and cristae disruption of mitochondria [97]. As a result of these functional and structural alterations to the mitochondria, I/R injury in DM increases apoptosis, necrosis, and HT [95–97].

In DM stroke, mitochondrial oxidative stress and an abundant increase in reactive oxygen species (ROS) are key mediators of pathological changes [44]. I/R is accompanied by an increase in oxidative and nitrative stress and production of free radicals that increase susceptibility to HT, vasogenic cerebral edema and neutrophil infiltration into the brain [96]. Oxidative stress triggers recruitment and migration of neutrophils and other leukocytes to the cerebral blood vessels which degrade the basal lamina and disrupt vessel wall and BBB integrity [96]. In DM stroke, hyperglycemia induced oxidative stress and free radical production increases inflammatory factor expression such as matrix metalloproteinase-9 (MMP9) [15, 98]. MMP9 induces extracellular matrix degradation and has been implicated in post stroke BBB disruption in DM rodents [15].

Oxidative stress mechanisms such as increased superoxide production by NADPH (Nicotinamide adenine dinucleotide phosphate) oxidase have also been implicated in reperfusion injury after stroke in DM rats [99]. While ROS and reactive nitrogen species (RNS) are signaling molecules of growth factors that promote angiogenesis; the 'redox window' concept suggests that a mild but not severe increase of ROS and/or RNS stimulates functional angiogenesis [69]. Hence, the surge in ROS generated by mitochondria in DM stroke mediate neurodegeneration and apoptosis [44].

4.5 Inflammatory Responses

In the chronic phase after stroke, mild inflammation can be favorable for brain repair [100], but uncontrolled inflammation in the acute phase after stroke stimulates the release of pro-inflammatory factors from the activated microglia, astrocytes and macrophages which can exacerbate damage to the injured brain tissue in addition to creating an inhospitable environment for endogenous brain repair mechanisms [101]. The inflammatory cascade in the acute phase after injury, induced by

DM, DM induced vasoconstriction, and I/R injury involves an increase of inflammatory mediators and pro-inflammatory cytokines [93, 102]. Pro-inflammatory cytokines inhibit brain repair due to an increased production and activity of free radicals and ROS, creating oxidative stress [103]. Inflammatory responses activate both the innate and adaptive immune systems; and DM, stroke and DM-stroke can impact immune response [23, 104]. In an experimental stroke model in T2DM rodents, I/R induces aggravated neutrophil adhesion in the cerebral microcirculation early after reperfusion, along with aggravated inflammatory responses and poor neurological functional outcome [23]. Adhesion of activated neutrophils to the injured endothelium may induce additional damage to the microvasculature and surrounding brain tissue upon I/R [23].

In DM stroke animals, inflammatory factors such as inflammatory cytokines that activate NF-κB, MMP's, Toll-like receptors (TLR's), High mobility group box-1 (HMGB-1) and receptor for advanced glycation endproducts (RAGE) are significantly increased compared to non DM stroke animals [14, 15]. HMGB1 is a pro-inflammatory mediator secreted early after ischemic injury by immune cells or injured neurons [105, 106]. HMGB1 release can induce microglial activation and trigger an inflammatory cascade after brain injury [105]. In DM mice subject to I/R stroke model, HMGB1 promotes inflammation by increasing expression of cytokines such as interleukins (IL-1β, IL-6), and inflammation-related enzyme inducible nitric oxide synthase (iNOS), which can then lead to secondary brain injury [102]. HMGB1 binds to its receptors TLR2, TLR4 and RAGE [106]. TLRs play a primary role in regulating innate immunity and impact endothelial cell survival and angiogenic responses [107]. Increased TLR4 expression in infiltrating macrophages has been linked to the development of ischemic brain damage, and I/R injury may be mediated in part by HMBG1 via TLR4 signaling [107]. In stroke, HMGB1 triggers MMP9 increase in neurons and astrocytes primarily via TLR4 signalling [108]. In DM stroke animals, RAGE expression is significantly increased, and increased RAGE expression is associated with the pathogenesis of diabetic complications, inflammatory disorders and neurodegenerative diseases [14, 109]. Following I/R injury in DM rodents, DM induced increase in MMP-9 activity and cerebrovascular remodeling have been attributed to greater HT [63]. Tissue response to I/R injury includes regulating the expression of several protein kinases such as calcium/calmodulin-dependent protein kinase II, mitogen-activated protein kinase, family members c-Jun N-terminal kinase, extracellular signal-regulated kinase, protein kinase B, and protein kinase C [110].

4.6 Extracellular Membrane Vesicles

Extracellular Membrane Vesicles (EMVs) such as exosomes and microparticles (MPs), are small membrane vesicles of endosomal and plasma membrane origin, respectively, secreted by many cells into the extracellular space [111]. EMVs contain high levels of RNA and proteins; their protein composition reflects the composition of their parent cell membrane [112]. Exosomes mediate intercellular communication by transporting regulatory nucleotides and proteins between different cell types in the body, thereby affecting the physiology of recipient cells under normal and pathological conditions [113]. EMVs easily pass through the BBB and have been implicated in endothelial dysfunction, inflammation and thrombosis [114]. MPs released by endothelial cells, platelets and leukocytes can aggravate endothelial dysfunction [115]. For instance, endothelial cell derived MPs promote inflammation of the arterial wall and thrombogenicity through cellular cross-talk [116]. Leukocyte-derived MPs can interact with specific adhesion molecules on endothelial cells, stimulate cytokine release and initiate a proinflammatory and procoagulant cascade [116]. MPs can also contribute to inflammation and endothelial dysfunction via modulating nitric oxide and prostacyclin production in endothelial cells, and stimulating cytokine release and tissue factor induction in endothelial cells, as well as monocyte chemotaxis and adherence to the endothelium [112]. Circulating MPs are significantly elevated in diabetic patients, and an analysis of MP expression in diabetic patients has revealed a positive correlation between elevated MP expression level in plasma and vascular complications [116].

4.7   MicroRNAs (MiRs)

MiRs are small non-coding RNA sequences that are emerging as key players in the pathogenesis of diabetes and hyperglycemia-induced vascular damage [117, 118]. They have the capacity to regulate many genes, pathways, and complex biological networks within cells, acting either alone or in concert with one another [119]. MiRs can mediate key pathophysiological processes as modulators of cellular activities during cell growth, differentiation, apoptosis, adhesion, and cell death.

Endothelial cell specific miR-126 has a key role in maintaining vascular integrity and regulating angiogenesis [120]. MiR-126 is among the miRs most consistently associated with DM [117, 121]. MiR-126 level is significantly decreased in the circulating vesicles in plasma of DM patients [117], and in ischemic stroke patients, decrease in miR-126 expression can last up to 24 weeks after stroke [122]. In addition, miR-126 regulates the response of endothelial cells to VEGF [123] and as discussed earlier, VEGF expression during the acute phase following ischemic stroke increases cerebral microvascular perfusion, exacerbates BBB disruption, increases infarct volume and increases the risk of HT [72].

MiR-145 has been associated with DM stroke and reperfusion injury [124, 125]. MiR-143/145 cluster regulates vascular smooth muscle cell differentiation [126]. The MiR-143/145 cluster is up regulated in liver of T2DM mice and may contribute to insulin resistance [127].

MiR-143 and miR-145 are also up regulated in endothelial cells in response to shear stress and can contribute to I/R injury after stroke [126]. MiR-145 expression significantly increases in the circulation and ischemic brain within 24 h of cerebral ischemia, and increased miR-145 level in the circulation has been positively correlated with elevated serum inflammatory factor IL-6 [128, 129]. Inhibition of miR-145 using anti-miR was found to yield moderate neuroprotective effects after transient focal ischemia [130]. MiR-145 acts as a communication molecule between smooth muscle cells and endothelial cells [131]; and reduces endothelial cell proliferation and capillary tube formation in response to growth factors [132]. Hence, miR-145 can modulate endothelial cell function in angiogenesis and vessel stabilization which are important for post stroke recovery [131].

MiR-15a/16-1 cluster can alter BBB function by binding to the 3'-UTRs of major BBB tight junctions proteins such as Claudins [133]. A significant increase of miR-15a/16-1 cluster expression was found in the cerebral vasculature and cultured brain microvascular endothelial cells after in vivo and in vitro ischemia, as well as DM [133]. Over expression of miR-15a/16-1 cluster in transgenic mice decreased Claudin expression and exacerbated ischemic infarct volume and BBB leakage [133]. MiR-15a/16-1 is increased in the ischemic penumbra which inhibits neurovascular remodeling and neurological recovery after stroke in mice [134].

DM also regulates expression of miRs that have been associated with platelet activity such as miR-24, miR-197, miR-191, and miR-223; and angiogenesis such as miR-320a and miR-27b [135]. Up-regulation of miR-150 decreases endothelial cell survival by targeting Tie-2, and decreases the expression of Claudin-5 in the ischemic penumbra resulting in worse BBB function after stroke in rats [136]. Decrease of miR17–92 cluster expression after stroke causes BBB dysfunction due to disruption of vascular coverage by pericytes and astrocyte end-foot processes [137]. Increase of miR-34 after stroke [138] affects mitochondrial function and induces a discontinuous expression of tight junction protein Zona Occludin-1 and aggravates BBB disruption [139]. The profile of circulating miR's is altered in subjects with obesity, pre-diabetes, and diabetes, such as miR-21, miR-24.1, miR-27a, miR-28-3p, miR-29b, miR-30d, miR-34a, miR-93, miR-126, miR-146a, miR-148, miR-150, miR-155, and miR-223, miR-661, miR-571, miR-770-5p, miR-892b and miR-1303 [140, 141]. The exact role of miRs in DM stroke pathophysiology and the use of miRs as biomarkers of disease [142] are of current research interest.

4.8   Gut-Brain Axis

Emerging experimental evidence from animal studies indicate an interaction between the intestinal microbiota, the gut, and the central nervous system, which can affect brain function as well as pathological cascade in neurological diseases such as stroke, traumatic brain injury, etc. [143, 144]. The gut-brain axis consists of neural and humoral pathways with signaling molecules such as cytokines, hormones, and neuropeptides affecting immune response and lymphocyte population [144, 145]. The gut-brain axis is a bidirectional communication pathway in which acute brain lesions induces dysbiosis (disturbance of balance of normal microbiota) of the gut microbiome, which in turn can affect immune and neuroinflammatory responses in the brain and impact functional outcome [146]. Dysbiosis of the gut microbiota has been reported in patients with several diseases including DM and stroke [147, 148]. In stroke patients, the three major beneficial microbes that were depleted include Bacteroides, Prevotella, and Faecalibacterium; while the major enriched microbes include the opportunistic pathogens Enterobacter, Megasphaera, and Desulfovibrio [147]. Increase in the abundance of Lactobacillus ruminis subgroup in the gut, may contribute to inflammation in stroke patients [149]. In mice subject to I/R injury, intestinal T cells traffic from the gut to the meninges of the brain; promote post-ischemic neuroinflammation by increasing pro-inflammatory cytokine IL-17; which can stimulate the production of several other cytokines and chemokines facilitating the infiltration of cytotoxic immune cells and neutrophils into the injured brain [145].

The gut microbial composition differs between DM and non DM patients [150]. It was found that, patients with low bacterial richness (low gene count) were more prone to obesity, insulin resistance, dyslipidemia, and inflammation compared to patients with a high gene count [150]. In a population of Chinese T2DM patients, moderate gut microbial dysbiosis was characterized by a decrease in beneficial bacteria such as butyrate-producing bacteria that has protective role against several types of diseases; and an increase in diverse opportunistic pathogens [151]. In a group of Japanese T2DM patients, compared to control patients, an abundance of gut bacteria was found in the blood stream of T2DM patients, indicating bacterial translocation from the gut to the blood [152]. In T2DM patient, gut dysbiosis increases some microbial functions which leads to increased oxidative stress response, which may mediate the proinflammatory state in DM [151]. In DM mice subject to I/R injury, altering the gut microbiome via probiotic treatment restored the gut microbiome profile, significantly decreased behavioral deficits, decreased blood glucose levels, induced neuroprotection, and increased cell survival [143]. Extensive depletion of the gut microbiota significantly increases post stroke mortality in mice, suggesting that conventional intestinal microbiota provides protection against ischemic damage, and that alterations to the gut microbiota can influence stroke outcome [153]. Future studies are required to understand the mechanisms of gut-brain axis and its impact on stroke in DM patients.

5 Summary

DM aggravates microvascular and macrovascular damage, BBB disruption, and ischemia/reperfusion damage after stroke, thereby creating an inhospitable environment for brain repair. A comprehensive understanding of the pathophysiological changes following stroke in diabetes is required for developing successful treatments strategies. In addition, understanding the acute events, I/R damage and role of insulin resistance following stroke in diabetic populations can enable development of secondary stroke preventative strategies.

Acknowledgements None

Sources of funding This work was supported by National Institute of Neurological Disorders and Stroke R01 NS083078-01A1 (JC) and RO1 NS099030-01 (JC) and R01NS097747 (QJ/JC).

Disclosures None

References

1. Shaw JE, Sicree RA, Zimmet PZ. Global estimates of the prevalence of diabetes for 2010 and 2030. Diabetes Res Clin Pract. 2010;87:4–14.
2. Mozaffarian D, Benjamin EJ, Go AS, Arnett DK, Blaha MJ, Cushman M, et al. Heart disease and stroke statistics-2016 update: a report from the American heart association. Circulation. 2016;133:e38–360.
3. Geiss LS, Wang J, Cheng YJ, et al. Prevalence and incidence trends for diagnosed diabetes among adults aged 20 to 79 years, united states, 1980-2012. JAMA. 2014;312:1218–26.
4. Mast H, Thompson JL, Lee SH, Mohr JP, Sacco RL. Hypertension and diabetes mellitus as determinants of multiple lacunar infarcts. Stroke. 1995;26:30–3.
5. Lindsberg PJ, Roine RO. Hyperglycemia in acute stroke. Stroke. 2004;35:363–4.
6. Kissela BM, Khoury J, Kleindorfer D, Woo D, Schneider A, Alwell K, et al. Epidemiology of ischemic stroke in patients with diabetes: the greater Cincinnati/northern Kentucky stroke study. Diabetes Care. 2005;28:355–9.
7. Yong M, Kaste M. Dynamic of hyperglycemia as a predictor of stroke outcome in the ECASS-II trial. Stroke. 2008;39:2749–55.
8. Candelise L, Landi G, Orazio EN, Boccardi E. Prognostic significance of hyperglycemia in acute stroke. Arch Neurol. 1985;42:661–3.
9. Megherbi SE, Milan C, Minier D, Couvreur G, Osseby GV, Tilling K, et al. Association between diabetes and stroke subtype on survival and functional outcome 3 months after stroke: data from the European biomed stroke project. Stroke. 2003;34:688–94.
10. Ergul A, Hafez S, Fouda A, Fagan SC. Impact of comorbidities on acute injury and recovery in preclinical stroke research: focus on hypertension and diabetes. Transl Stroke Res. 2016;7:248–60.
11. MEMBERS WG, Lloyd-Jones D, Adams RJ, Brown TM, Carnethon M, Dai S, et al. Heart disease and stroke statistics—2010 update: a report from the American heart association. Circulation. 2010;121:e46–e215.
12. Callahan A, Amarenco P, Goldstein LB, Sillesen H, Messig M, Samsa GP, et al. Risk of stroke and cardiovascular events after ischemic stroke or transient ischemic attack in patients with type 2 diabetes or metabolic syndrome: secondary analysis of the stroke prevention by aggressive reduction in cholesterol levels (SPARCL) trial. Arch Neurol. 2011;68:1245–51.
13. Li PA, Gisselsson L, Keuker J, Vogel J, Smith ML, Kuschinsky W, et al. Hyperglycemia-exaggerated ischemic brain damage following 30 min of middle cerebral artery occlusion is not due to capillary obstruction. Brain Res. 1998;804:36–44.
14. Ye X, Chopp M, Liu X, Zacharek A, Cui X, Yan T, et al. Niaspan reduces high-mobility group box 1/receptor for advanced glycation endproducts after stroke in type-1 diabetic rats. Neuroscience. 2011;190:339–45.
15. Chen J, Cui X, Zacharek A, Cui Y, Roberts C, Chopp M. White matter damage and the effect of matrix metalloproteinases in type 2 diabetic mice after stroke. Stroke. 2011;42:445–52.
16. Linfante I, Cipolla MJ. Improving reperfusion therapies in the era of mechanical thrombectomy. Transl Stroke Res. 2016;7:294–302.

17. The national institute of neurological disorders and stroke rt-PA stroke study group. Tissue plasminogen activator for acute ischemic stroke. N Engl J Med. 1995;333:1581–7.
18. Jauch EC, Saver JL, Adams HP Jr, Bruno A, Connors JJ, Demaerschalk BM, et al. Guidelines for the early management of patients with acute ischemic stroke: a guideline for health-care professionals from the American heart association/American stroke association. Stroke. 2013;44:870–947.
19. Ding D. Endovascular mechanical thrombectomy for acute ischemic stroke: a new standard of care. J Stroke. 2015;17:123–6.
20. Reeves MJ, Arora S, Broderick JP, Frankel M, Heinrich JP, Hickenbottom S, et al. Acute stroke care in the us: results from 4 pilot prototypes of the Paul Coverdell national acute stroke registry. Stroke. 2005;36:1232–40.
21. Raychev R, Saver JL. Mechanical thrombectomy devices for treatment of stroke. Neurol Clin Pract. 2012;2:231–5.
22. Ergul A, Elgebaly MM, Middlemore M-L, Li W, Elewa H, Switzer JA, et al. Increased hemorrhagic transformation and altered infarct size and localization after experimental stroke in a rat model type 2 diabetes. BMC Neurol. 2007;7:33.
23. Ritter L, Davidson L, Henry M, Davis-Gorman G, Morrison H, Frye JB, et al. Exaggerated neutrophil-mediated reperfusion injury after ischemic stroke in a rodent model of type 2 diabetes. Microcirculation. 2011;18:552–61.
24. Bas DF, Ozdemir AO, Colak E, Kebapci N. Higher insulin resistance level is associated with worse clinical response in acute ischemic stroke patients treated with intravenous thrombolysis. Transl Stroke Res. 2016;7:167–71.
25. Poppe AY, Majumdar SR, Jeerakathil T, Ghali W, Buchan AM, Hill MD. Admission hyperglycemia predicts a worse outcome in stroke patients treated with intravenous thrombolysis. Diabetes Care. 2009;32:617–22.
26. Pan J, Konstas A-A, Bateman B, Ortolano GA, Pile-Spellman J. Reperfusion injury following cerebral ischemia: pathophysiology, MR imaging, and potential therapies. Neuroradiology. 2007;49:93–102.
27. de Courten-Myers GM, Kleinholz M, Holm P, DeVoe G, Schmitt G, Wagner KR, et al. Hemorrhagic infarct conversion in experimental stroke. Ann Emerg Med. 1992;21:120–6.
28. Ahnstedt H, Sweet J, Cruden P, Bishop N, Cipolla MJ. Effects of early post-ischemic reperfusion and tPA on cerebrovascular function and nitrosative stress in female rats. Transl Stroke Res. 2016;7:228–38.
29. Caso V, Paciaroni M, Venti M, Palmerini F, Silvestrelli G, Milia P, et al. Determinants of outcome in patients eligible for thrombolysis for ischemic stroke. Vasc Health Risk Manag. 2007;3:749–54.
30. Venkat P, Chopp M, Chen J. New insights into coupling and uncoupling of cerebral blood flow and metabolism in the brain. Croat Med J. 2016;57:223–8.
31. Abbott NJ, Patabendige AA, Dolman DE, Yusof SR, Begley DJ. Structure and function of the blood-brain barrier. Neurobiol Dis. 2010;37:13–25.
32. Sandoval KE, Witt KA. Blood-brain barrier tight junction permeability and ischemic stroke. Neurobiol Dis. 2008;32:200–19.
33. Shimizu F, Kanda T. disruption of the blood-brain barrier in inflammatory neurological diseases. Brain Nerve. 2013;65:165–76.
34. Hall CN, Reynell C, Gesslein B, Hamilton NB, Mishra A, Sutherland BA, et al. Capillary pericytes regulate cerebral blood flow in health and disease. Nature. 2014;508:55–60.
35. Fernandez-Klett F, Potas JR, Hilpert D, Blazej K, Radke J, Huck J, et al. Early loss of pericytes and perivascular stromal cell-induced scar formation after stroke. J Cereb Blood Flow Metab. 2013;33:428–39.
36. Belayev L, Busto R, Zhao W, Ginsberg MD. Quantitative evaluation of blood-brain barrier permeability following middle cerebral artery occlusion in rats. Brain Res. 1996;739:88–96.
37. Chen J, Ye X, Yan T, Zhang C, Yang XP, Cui X, et al. Adverse effects of bone marrow stromal cell treatment of stroke in diabetic rats. Stroke. 2011;42:3551–8.

38. Reeson P, Tennant KA, Gerrow K, Wang J, Weiser Novak S, Thompson K, et al. Delayed inhibition of VEGF signaling after stroke attenuates blood-brain barrier breakdown and improves functional recovery in a comorbidity-dependent manner. J Neurosci. 2015;35:5128–43.
39. Borlongan CV, Glover LE, Sanberg PR, Hess DC. Permeating the blood brain barrier and abrogating the inflammation in stroke: implications for stroke therapy. Curr Pharm Des. 2012;18:3670–6.
40. Denes A, Ferenczi S, Kovacs KJ. Systemic inflammatory challenges compromise survival after experimental stroke via augmenting brain inflammation, blood- brain barrier damage and brain oedema independently of infarct size. J Neuroinflammation. 2011;8:164.
41. Hadi HAR, Suwaidi JA. Endothelial dysfunction in diabetes mellitus. Vasc Health Risk Manag. 2007;3:853–76.
42. Chen R, Ovbiagele B, Feng W. Diabetes and stroke: epidemiology, pathophysiology, pharmaceuticals and outcomes. Am J Med Sci. 2016;351:380–6.
43. Aronson D, Rayfield EJ. How hyperglycemia promotes atherosclerosis: molecular mechanisms. Cardiovasc Diabetol. 2002;1:1.
44. Price TO, Eranki V, Banks WA, Ercal N, Shah GN. Topiramate treatment protects blood-brain barrier pericytes from hyperglycemia-induced oxidative damage in diabetic mice. Endocrinology. 2012;153:362–72.
45. Haj-Yasein NN, Vindedal GF, Eilert-Olsen M, Gundersen GA, Skare O, Laake P, et al. Glial-conditional deletion of aquaporin-4 (aqp4) reduces blood-brain water uptake and confers barrier function on perivascular astrocyte endfeet. Proc Natl Acad Sci U S A. 2011;108:17815–20.
46. Li L, Lundkvist A, Andersson D, Wilhelmsson U, Nagai N, Pardo AC, et al. Protective role of reactive astrocytes in brain ischemia. J Cereb Blood Flow Metab. 2008;28:468–81.
47. Li Y, Liu Z, Xin H, Chopp M. The role of astrocytes in mediating exogenous cell-based restorative therapy for stroke. Glia. 2014;62:1–16.
48. Jing L, Mai L, Zhang J-Z, Wang J-G, Chang Y, Dong J-D, et al. Diabetes inhibits cerebral ischemia-induced astrocyte activation - an observation in the cingulate cortex. Int J Biol Sci. 2013;9:980–8.
49. Jing L, He Q, Zhang JZ, Li PA. Temporal profile of astrocytes and changes of oligodendrocyte-based myelin following middle cerebral artery occlusion in diabetic and non-diabetic rats. Int J Biol Sci. 2013;9:190–9.
50. Marlatt MW, Lucassen PJ, Perry G, Smith MA, Zhu X. Alzheimer's disease: cerebrovascular dysfunction, oxidative stress, and advanced clinical therapies. J Alzheimers Dis. 2008;15:199–210.
51. Chen B, Friedman B, Cheng Q, Tsai P, Schim E, Kleinfeld D, et al. Severe blood–brain barrier disruption and surrounding tissue injury. Stroke. 2009;40:e666-e674.
52. Papadopoulos MC, Manley GT, Krishna S, Verkman AS. Aquaporin-4 facilitates reabsorption of excess fluid in vasogenic brain edema. FASEB J. 2004;18:1291–3.
53. Rosenberg GA. Ischemic brain edema. Prog Cardiovasc Dis. 1999;42:209–16.
54. Stokum JA, Gerzanich V, Simard JM. Molecular pathophysiology of cerebral edema. J Cereb Blood Flow Metab. 2016;36:513–38.
55. Manley GT, Fujimura M, Ma T, Noshita N, Filiz F, Bollen AW, et al. Aquaporin-4 deletion in mice reduces brain edema after acute water intoxication and ischemic stroke. Nat Med. 2000;6:159–63.
56. Ergul A, Kelly-Cobbs A, Abdalla M, Fagan SC. Cerebrovascular complications of diabetes: focus on stroke. Endocr Metab Immune Disord Drug Targets. 2012;12:148–58.
57. Zhang L, Chopp M, Zhang Y, Xiong Y, Li C, Sadry N, et al. Diabetes mellitus impairs cognitive function in middle-aged rats and neurological recovery in middle-aged rats after stroke. Stroke. 2016;47:2112.
58. Alvarez-Sabin J, Molina CA, Montaner J, Arenillas JF, Huertas R, Ribo M, et al. Effects of admission hyperglycemia on stroke outcome in reperfused tissue plasminogen activator-treated patients. Stroke. 2003;34:1235–41.

59. Demchuk AM, Morgenstern LB, Krieger DW, Linda Chi T, Hu W, Wein TH, et al. Serum glucose level and diabetes predict tissue plasminogen activator–related intracerebral hemorrhage in acute ischemic stroke. Stroke. 1999;30:34–9.
60. Fan X, Qiu J, Yu Z, Dai H, Singhal AB, Lo EH, et al. A rat model of studying tissue-type plasminogen activator thrombolysis in ischemic stroke with diabetes. Stroke. 2012;43:567–70.
61. Ning R, Chopp M, Yan T, Zacharek A, Zhang C, Roberts C, et al. Tissue plasminogen activator treatment of stroke in type-1 diabetes rats. Neuroscience. 2012;222:326–32.
62. Chen CH, Anatol M, Zhan Y, Liu WW, Ostrowki RP, Tang J, et al. Hydrogen gas reduced acute hyperglycemia-enhanced hemorrhagic transformation in a focal ischemia rat model. Neuroscience. 2010;169:402–14.
63. Elgebaly MM, Prakash R, Li W, Ogbi S, Johnson MH, Mezzetti EM, et al. Vascular protection in diabetic stroke: role of matrix metalloprotease-dependent vascular remodeling. J Cereb Blood Flow Metab. 2010;30:1928–38.
64. Quast MJ, Wei J, Huang NC, Brunder DG, Sell SL, Gonzalez JM, et al. Perfusion deficit parallels exacerbation of cerebral ischemia/reperfusion injury in hyperglycemic rats. J Cereb Blood Flow Metab. 1997;17:553–9.
65. Prakash R, Li W, Qu Z, Johnson MA, Fagan SC, Ergul A. Vascularization pattern after ischemic stroke is different in control versus diabetic rats: relevance to stroke recovery. Stroke. 2013;44:2875–82.
66. Arenillas JF, Sobrino T, Castillo J, Davalos A. The role of angiogenesis in damage and recovery from ischemic stroke. Curr Treat Options Cardiovasc Med. 2007;9:205–12.
67. Navarro-Sobrino M, Rosell A, Hernandez-Guillamon M, Penalba A, Boada C, Domingues-Montanari S, et al. A large screening of angiogenesis biomarkers and their association with neurological outcome after ischemic stroke. Atherosclerosis. 2011;216:205–11.
68. Wei L, Erinjeri JP, Rovainen CM, Woolsey TA. Collateral growth and angiogenesis around cortical stroke. Stroke. 2001;32:2179–84.
69. Abdelsaid M, Prakash R, Li W, Coucha M, Hafez S, Johnson MH, et al. Metformin treatment in the period after stroke prevents nitrative stress and restores angiogenic signaling in the brain in diabetes. Diabetes. 2015;64:1804–17.
70. Hoeben A, Landuyt B, Highley MS, Wildiers H, Van Oosterom AT, De Bruijn EA. Vascular endothelial growth factor and angiogenesis. Pharmacol Rev. 2004;56:549–80.
71. Zechariah A, ElAli A, Doeppner TR, Jin FY, Hasan MR, Helfrich I, et al. Vascular endothelial growth factor promotes pericyte coverage of brain capillaries, improves cerebral blood flow during subsequent focal cerebral ischemia, and preserves the metabolic penumbra. Stroke. 2013;44:1690.
72. Zhang ZG, Zhang L, Jiang Q, Zhang R, Davies K, Powers C, et al. VEGF enhances angiogenesis and promotes blood-brain barrier leakage in the ischemic brain. J Clin Investig. 2000;106:829–38.
73. Kolluru GK, Bir SC, Kevil CG. Endothelial dysfunction and diabetes: effects on angiogenesis, vascular remodeling, and wound healing. Int J Vasc Med. 2012;2012:918267.
74. Suri C, Jones PF, Patan S, Bartunkova S, Maisonpierre PC, Davis S, et al. Requisite role of angiopoietin-1, a ligand for the tie2 receptor, during embryonic angiogenesis. Cell. 1996;87:1171–80.
75. Zacharek A, Chen J, Cui X, Li A, Li Y, Roberts C, et al. Angiopoietin1/tie2 and VEGF/Flk1 induced by MSC treatment amplifies angiogenesis and vascular stabilization after stroke. J Cereb Blood Flow Metab. 2007;27:1684–91.
76. Gurnik S, Devraj K, Macas J, Yamaji M, Starke J, Scholz A, et al. Angiopoietin-2-induced blood-brain barrier compromise and increased stroke size are rescued by VE-PTP-dependent restoration of tie2 signaling. Acta Neuropathol. 2016;131:753–73.
77. Lim HS, Lip GY, Blann AD. Angiopoietin-1 and angiopoietin-2 in diabetes mellitus: relationship to VEGF, glycaemic control, endothelial damage/dysfunction and atherosclerosis. Atherosclerosis. 2005;180:113–8.

78. Cui X, Chopp M, Zacharek A, Ye X, Roberts C, Chen J. Angiopoietin/tie2 pathway mediates type 2 diabetes induced vascular damage after cerebral stroke. Neurobiol Dis. 2011;43:285–92.
79. Wang YQ, Song JJ, Han X, Liu YY, Wang XH, Li ZM, et al. Effects of angiopoietin-1 on inflammatory injury in endothelial progenitor cells and blood vessels. Curr Gene Ther. 2014;14:128.
80. Tarbell JM. Shear stress and the endothelial transport barrier. Cardiovasc Res. 2010;87:320–30.
81. Cunningham KS, Gotlieb AI. The role of shear stress in the pathogenesis of atherosclerosis. Lab Invest. 2004;85:9–23.
82. Gresele P, Guglielmini G, De Angelis M, Ciferri S, Ciofetta M, Falcinelli E, et al. Acute, short-term hyperglycemia enhances shear stress-induced platelet activation in patients with type ii diabetes mellitus. J Am Coll Cardiol. 2003;41:1013–20.
83. Kernan WN, Inzucchi SE, Viscoli CM, Brass LM, Bravata DM, Horwitz RI. Insulin resistance and risk for stroke. Neurology. 2002;59:809–15.
84. Cheng C, Daskalakis C. Association of adipokines with insulin resistance, microvascular dysfunction, and endothelial dysfunction in healthy young adults. Mediators Inflamm. 2015;2015:594039.
85. Kaji H. Adipose tissue-derived plasminogen activator inhibitor-1 function and regulation. Compr Physiol. 2016;6:1873–96.
86. Rizk NN, Rafols JA, Dunbar JC. Cerebral ischemia-induced apoptosis and necrosis in normal and diabetic rats: effects of insulin and c-peptide. Brain Res. 2006;1096:204–12.
87. Liu H, Ou S, Xiao X, Zhu Y, Zhou S. Diabetes worsens ischemia-reperfusion brain injury in rats through gsk-3β. Am J Med Sci. 2015;350:204–11.
88. Chapin JC, Hajjar KA. Fibrinolysis and the control of blood coagulation. Blood Rev. 2015;29:17–24.
89. Trost S, Pratley RE, Sobel BE. Impaired fibrinolysis and risk for cardiovascular disease in the metabolic syndrome and type 2 diabetes. Curr Diab Rep. 2006;6:47–54.
90. Lijnen HR, Ds C. Impaired fibrinolysis and the risk for coronary heart disease. Circulation. 1996;94:2052–4.
91. Jotic A, Milicic T, Covickovic Sternic N, Kostic VS, Lalic K, Jeremic V, et al. Decreased insulin sensitivity and impaired fibrinolytic activity in type 2 diabetes patients and nondiabetics with ischemic stroke. Int J Endocrinol. 2015;2015:934791.
92. Wyseure T, Rubio M, Denorme F, Martinez de Lizarrondo S, Peeters M, Gils A, et al. Innovative thrombolytic strategy using a heterodimer diabody against TAFI and PAI-1 in mouse models of thrombosis and stroke. Blood. 2015;125:1325–32.
93. Dokken BB. The pathophysiology of cardiovascular disease and diabetes: beyond blood pressure and lipids. Diabetes Spectr. 2008;21:160–5.
94. Poulsen RC, Knowles HJ, Carr AJ, Hulley PA. Cell differentiation versus cell death: extracellular glucose is a key determinant of cell fate following oxidative stress exposure. Cell Death Dis. 2014;5:e1074.
95. Sims NR, Muyderman H. Mitochondria, oxidative metabolism and cell death in stroke. Biochim Biophys Acta. 2010;1802:80–91.
96. Doyle KP, Simon RP, Stenzel-Poore MP. Mechanisms of ischemic brain damage. Neuropharmacology. 2008;55:310–8.
97. Mishiro K, Imai T, Sugitani S, Kitashoji A, Suzuki Y, Takagi T, et al. Diabetes mellitus aggravates hemorrhagic transformation after ischemic stroke via mitochondrial defects leading to endothelial apoptosis. PLoS One. 2014;9:e103818.
98. Tang J, Li YJ, Li Q, Mu J, Yang DY, Xie P. Endogenous tissue plasminogen activator increases hemorrhagic transformation induced by heparin after ischemia reperfusion in rat brains. Neurol Res. 2010;32:541–6.
99. Won SJ, Tang XN, Suh SW, Yenari MA, Swanson RA. Hyperglycemia promotes tissue plasminogen activator-induced hemorrhage by increasing superoxide production. Ann Neurol. 2011;70:583–90.

100. Kim JY, Kawabori M, Yenari MA. Innate inflammatory responses in stroke: mechanisms and potential therapeutic targets. Curr Med Chem. 2014;21:2076–97.
101. Whitney NP, Eidem TM, Peng H, Huang Y, Zheng JC. Inflammation mediates varying effects in neurogenesis: relevance to the pathogenesis of brain injury and neurodegenerative disorders. J Neurochem. 2009;108:1343–59.
102. Wang C, Jiang J, Zhang X, Song L, Sun K, Xu R. Inhibiting hmgb1 reduces cerebral ischemia reperfusion injury in diabetic mice. Inflammation. 2016;39:1862–70.
103. di Penta A, Moreno B, Reix S, Fernandez-Diez B, Villanueva M, Errea O, et al. Oxidative stress and proinflammatory cytokines contribute to demyelination and axonal damage in a cerebellar culture model of neuroinflammation. PLoS One. 2013;8:e54722.
104. Pickup JC. Inflammation and activated innate immunity in the pathogenesis of type 2 diabetes. Diabetes Care. 2004;27:813–23.
105. Kim JB, Sig Choi J, Yu YM, Nam K, Piao CS, Kim SW, et al. Hmgb1, a novel cytokine-like mediator linking acute neuronal death and delayed neuroinflammation in the postischemic brain. J Neurosci. 2006;26:6413–21.
106. Qiu J, Nishimura M, Wang Y, Sims JR, Qiu S, Savitz SI, et al. Early release of hmgb-1 from neurons after the onset of brain ischemia. J Cereb Blood Flow Metab. 2008;28:927–38.
107. Yang Q-W, Lu F-L, Zhou Y, Wang L, Zhong Q, Lin S, et al. Hmbg1 mediates ischemia–reperfusion injury by TRIF-adaptor independent toll-like receptor 4 signaling. J Cereb Blood Flow Metab. 2011;31:593–605.
108. Qiu J, Xu J, Zheng Y, Wei Y, Zhu X, Lo EH, et al. High-mobility group box 1 promotes metalloproteinase-9 upregulation through toll-like receptor 4 after cerebral ischemia. Stroke. 2010;41:2077–82.
109. Maillard-Lefebvre H, Boulanger E, Daroux M, Gaxatte C, Hudson BI, Lambert M. Soluble receptor for advanced glycation end products: a new biomarker in diagnosis and prognosis of chronic inflammatory diseases. Rheumatology. 2009;48:1190–6.
110. Bright R, Mochly-Rosen D. The role of protein kinase c in cerebral ischemic and reperfusion injury. Stroke. 2005;36:2781–90.
111. Chen J, Venkat P, Zacharek A, Chopp M. Neurorestorative therapy for stroke. Front Hum Neurosci. 2014;8:382.
112. Puddu P, Puddu GM, Cravero E, Muscari S, Muscari A. The involvement of circulating microparticles in inflammation, coagulation and cardiovascular diseases. Can J Cardiol. 2010;26:e140–5.
113. Colombo M, Raposo G, Thery C. Biogenesis, secretion, and intercellular interactions of exosomes and other extracellular vesicles. Annu Rev Cell Dev Biol. 2014;30:255–89.
114. Kalani A, Tyagi A, Tyagi N. Exosomes: mediators of neurodegeneration, neuroprotection and therapeutics. Mol Neurobiol. 2014;49:590–600.
115. Mesri M, Altieri DC. Leukocyte microparticles stimulate endothelial cell cytokine release and tissue factor induction in a jnk1 signaling pathway. J Biol Chem. 1999;274:23111–8.
116. Tramontano AF, Lyubarova R, Tsiakos J, Palaia T, DeLeon JR, Ragolia L. Circulating endothelial microparticles in diabetes mellitus. Mediators Inflamm. 2010;2010:250476.
117. Zampetaki A, Kiechl S, Drozdov I, Willeit P, Mayr U, Prokopi M, et al. Plasma microrna profiling reveals loss of endothelial mir-126 and other micrornas in type 2 diabetes. Circ Res. 2010;107:810–7.
118. Shantikumar S, Caporali A, Emanueli C. Role of micrornas in diabetes and its cardiovascular complications. Cardiovasc Res. 2012;93:583–93.
119. Lim LP, Lau NC, Garrett-Engele P, Grimson A, Schelter JM, Castle J, et al. Microarray analysis shows that some micrornas downregulate large numbers of target mRNAs. Nature. 2005;433:769–73.
120. Wang S, Aurora AB, Johnson BA, Qi X, McAnally J, Hill JA, et al. The endothelial-specific microrna mir-126 governs vascular integrity and angiogenesis. Dev Cell. 2008;15:261–71.
121. Zampetaki A, Willeit P, Drozdov I, Kiechl S, Mayr M. Profiling of circulating micrornas: from single biomarkers to re-wired networks. Cardiovasc Res. 2012;93:555–62.

122. Long G, Wang F, Li H, Yin Z, Sandip C, Lou Y, et al. Circulating mir-30a, mir-126 and let-7b as biomarker for ischemic stroke in humans. BMC Neurol. 2013;13:1–10.
123. Fish JE, Santoro MM, Morton SU, Yu S, Yeh RF, Wythe JD, et al. Mir-126 regulates angiogenic signaling and vascular integrity. Dev Cell. 2008;15:272–84.
124. Cui C, Ye X, Chopp M, Venkat P, Zacharek A, Yan T, et al. Mir-145 regulates diabetes-bone marrow stromal cell-induced neurorestorative effects in diabetes stroke rats. Stem Cells Transl Med. 2016;26:2015–0349.
125. Altintas O, Ozgen Altintas M, Kumas M, Asil T. Neuroprotective effect of ischemic preconditioning via modulating the expression of cerebral miRNAs against transient cerebral ischemia in diabetic rats. Neurol Res. 2016;38:1003–11.
126. Zhao W, Zhao S-P, Zhao Y-H. Microrna-143/-145 in cardiovascular diseases. Biomed Res Int. 2015;2015:9.
127. Jordan SD, Kruger M, Willmes DM, Redemann N, Wunderlich FT, Bronneke HS, et al. Obesity-induced overexpression of mirna-143 inhibits insulin-stimulated Akt activation and impairs glucose metabolism. Nat Cell Biol. 2011;13:434–46.
128. Jia L, Hao F, Wang W, Qu Y. Circulating mir-145 is associated with plasma high-sensitivity c-reactive protein in acute ischemic stroke patients. Cell Biochem Funct. 2015;33:314–9.
129. Weiss JBW, Eisenhardt SU, Stark GB, Bode C, Moser M, Grundmann S. Micrornas in ischemia-reperfusion injury. Am J Cardiovasc Dis. 2012;2:237–47.
130. Dharap A, Bowen K, Place R, Li LC, Vemuganti R. Transient focal ischemia induces extensive temporal changes in rat cerebral micrornaome. J Cereb Blood Flow Metab. 2009;29:675–87.
131. Climent M, Quintavalle M, Miragoli M, Chen J, Condorelli G, Elia L. Tgfbeta triggers mir-143/145 transfer from smooth muscle cells to endothelial cells, thereby modulating vessel stabilization. Circ Res. 2015;116:1753–64.
132. Krupinski J, Kaluza J, Kumar P, Kumar S, Wang JM. Role of angiogenesis in patients with cerebral ischemic stroke. Stroke. 1994;25:1794–8.
133. Zhu T, Song J, Hamblin MH, Chen YE, Yin K-J. Abstract 145: mir-15a/16-1 mediates blood-brain barrier dysfunction in ischemic stroke. Stroke. 2016;47:A145.
134. Yin K-J, Hamblin M, Zhang J, Zhu T, Chen YE. Abstract tp117: mir-15a/16-1 cluster inhibits angiogenesis in mouse after ischemic stroke. Stroke. 2013;44:ATP117.
135. Willeit P, Zampetaki A, Dudek K, Kaudewitz D, King A, Kirkby NS, et al. Circulating micrornas as novel biomarkers for platelet activation. Circ Res. 2013;112:595–600.
136. Fang Z, He Q-W, Li Q, Chen X-L, Baral S, Jin H-J, et al. Microrna-150 regulates blood–brain barrier permeability via tie-2 after permanent middle cerebral artery occlusion in rats. FASEB J. 2016;30:2097.
137. Zhang R, Chopp M, Roberts C, Teng H, Wei M, Zhang L, et al. Abstract 49: deletion of miRNA 17-92 in cerebral endothelial cells induces disruption of bbb. Stroke. 2014;45:A49.
138. T-y L, J-y L. Increased expression of mir-34a-5p and clinical association in acute ischemic stroke patients and in a rat model. Med Sci Monit. 2016;22:2950–5.
139. Bukeirat M, Sarkar SN, Hu H, Quintana DD, Simpkins JW, Ren X. Mir-34a regulates blood–brain barrier permeability and mitochondrial function by targeting cytochrome c. J Cereb Blood Flow Metab. 2016;36:387–92.
140. Nunez Lopez YO, Garufi G, Seyhan AA. Altered levels of circulating cytokines and micrornas in lean and obese individuals with prediabetes and type 2 diabetes. Mol Biosyst. 2016;13:106.
141. Wang C, Wan S, Yang T, Niu D, Zhang A, Yang C, et al. Increased serum micrornas are closely associated with the presence of microvascular complications in type 2 diabetes mellitus. Sci Rep. 2016;6:20032.
142. Peng G, Yuan Y, Wu S, He F, Hu Y, Luo B. Microrna let-7e is a potential circulating biomarker of acute stage ischemic stroke. Transl Stroke Res. 2015;6:437–45.
143. Sun J, Wang F, Ling Z, Yu X, Chen W, Li H, et al. Clostridium butyricum attenuates cerebral ischemia/reperfusion injury in diabetic mice via modulation of gut microbiota. Brain Res. 2016;1642:180–8.

144. Bercik P, Collins SM, Verdu EF. Microbes and the gut-brain axis. Neurogastroenterol Motil. 2012;24:405–13.
145. Benakis C, Brea D, Caballero S, Faraco G, Moore J, Murphy M, et al. Commensal microbiota affects ischemic stroke outcome by regulating intestinal γδt cells. Nat Med. 2016;22:516–23.
146. Singh V, Roth S, Llovera G, Sadler R, Garzetti D, Stecher B, et al. Microbiota dysbiosis controls the neuroinflammatory response after stroke. J Neurosci. 2016;36:7428–40.
147. Yin J, Liao SX, He Y, Wang S, Xia GH, Liu FT, et al. Dysbiosis of gut microbiota with reduced trimethylamine-n-oxide level in patients with large-artery atherosclerotic stroke or transient ischemic attack. J Am Heart Assoc. 2015;4:pii: e002699.
148. Karlsson FH, Tremaroli V, Nookaew I, Bergstrom G, Behre CJ, Fagerberg B, et al. Gut metagenome in European women with normal, impaired and diabetic glucose control. Nature. 2013;498:99–103.
149. Yamashiro K, Tanaka R, Urabe T, Ueno Y, Yamashiro Y, Nomoto K, et al. Gut dysbiosis is associated with metabolism and systemic inflammation in patients with ischemic stroke. PLoS One. 2017;12:e0171521.
150. Le Chatelier E, Nielsen T, Qin J, Prifti E, Hildebrand F, Falony G, et al. Richness of human gut microbiome correlates with metabolic markers. Nature. 2013;500:541–6.
151. Qin J, Li Y, Cai Z, Li S, Zhu J, Zhang F, et al. A metagenome-wide association study of gut microbiota in type 2 diabetes. Nature. 2012;490:55–60.
152. Sato J, Kanazawa A, Ikeda F, Yoshihara T, Goto H, Abe H, et al. Gut dysbiosis and detection of "live gut bacteria" in blood of Japanese patients with type 2 diabetes. Diabetes Care. 2014;37:2343–50.
153. Winek K, Engel O, Koduah P, Heimesaat MM, Fischer A, Bereswill S, et al. Depletion of cultivatable gut microbiota by broad-spectrum antibiotic pretreatment worsens outcome after murine stroke. Stroke. 2016;47:1354–63.

Chapter 12
Current Understanding of Pathology and Therapeutic Status for CADASIL

Suning Ping and Li-Ru Zhao

Abstract Recently, it has been drawn an increased attention on a hereditary form of stroke and vascular dementia named cerebral autosomal dominant arteriopathy with subcortical infarcts and leukoencephalopathy (CADASIL). Because of poor understanding of the pathogenesis of this disease, the treatment that specifically delay or stop the pathological progression of CADASIL has not yet been developed. This chapter provides an update on CADASIL research. The pathological features and possible molecular mechanisms of CADASIL are outlined, the involvement of vascular endothelial cells in pathological progression of CADASIL is introduced, and emerging studies in treatment research for CADASIL is reviewed and discussed.

Keywords CADASIL · Notch3 mutation · Vascular smooth muscle cells · Endothelial cells · Capillary thrombosis · Pathology · Treatment

Abbreviations

CADASIL	Cerebral autosomal dominant arteriopathy with subcortical infarcts and leukoencephalopathy
EGFrs	Epidermal growth factor-like repeats
G-CSF	Granulocyte colony stimulating factor
GOM	Granular osmiophilic materials
MA	Migraine with aura
N3ECD	NOTCH3 N-terminal extracellular domain

S. Ping · L.-R. Zhao (✉)
Department of Neurosurgery, State University of New York, Upstate Medical University, Syracuse, NY, USA
e-mail: ZHAOL@upstate.edu

© Springer International Publishing AG, part of Springer Nature 2018
W. Jiang et al. (eds.), *Cerebral Ischemic Reperfusion Injuries (CIRI)*, Springer Series in Translational Stroke Research, https://doi.org/10.1007/978-3-319-90194-7_12

N3TMIC   NOTCH3 transmembrane intracellular domain
SCF   Stem cell factor
VSMCs   Vascular smooth muscle cells

1 Introduction

Cerebral autosomal dominant arteriopathy with subcortical infarcts and leuko-encephalopathy (CADASIL) is a hereditary cerebral small vascular disease. CADASIL has been recognized as the leading cause of vascular cognitive impairment and a major contributor to inherited stroke in adults [1]. CADASIL primarily affects the pial arteries, smaller penetrating arteries and arterioles, and it is characterized by progressive damage and loss of vascular smooth muscle cells (VSMCs) from the media [2]. Mutations in the NOTCH3 gene encoding the Notch3 receptor have been identified as the cause of this disease [3]. Since the diagnostic approach for genetic identification of NOTCH3 gene mutation was established in 2000 [4], the number of reported and diagnosed cases has been increased rapidly within recent years. It remains poorly understood, however, how the mutant NOTCH3 gene leads to structural and functional impairments in the cerebral VSMCs. There is currently lack of specific treatment that can stop or delay the progression of CADASIL. Recently, numerous studies have been conducted to explore possible pathological processes and putative treatment for restricting the progression of CADASIL. Here, we provide an overview of CADASIL, including the discovery of CADASIL and current understanding of the genetics, biochemistry, clinical features, and pathological changes for this inherited small vascular disease. Finally, the status of clinical managements and preclinical research for developing therapeutic interventions are summarized and discussed.

2 History

The first CADASIL case can be traced back to 1955. It was reported as "hereditary Binswanger's disease" by Bogaert [5]. This case was later verified as CADASIL by genetic analysis of old tissue samples. Until 1990s, the linkage of CADASIL to chromosome 19 was identified in 1993 by Tournier and coworkers [6], and later in 1996 the causative gene defects of NOTCH3 on 19p13 was discovered by Joutel and colleagues [3]. This discovery was a breakthrough in the research of CADASIL and offered solid foundations for the diagnosis and understanding of this inherited small vascular disease.

3   Molecular Genetics

CADASIL is an autosomal dominantly inherited disease with strong-stereotyped mutations in the NOTCH3 gene. Most individuals with CADASIL have a parent or first-degree relatives with similar symptoms. However, there are still a few patients with de novo mutations [4, 7, 8]. NOTCH3 gene has 33 exons, which encode Notch3 receptor comprising 2321 amino acids. Notch3 is initially synthesized as a full length precursor, which comprises a large extracellular domain containing 34 epidermal growth factor-like repeats (EGFrs), a single transmembrane domain and an intracellular domain containing seven ankyrin repeats flanked by two nuclear localization. After being proteolytically cleaved, the mature heterodimeric transmembrane receptor forms. The mature receptor consists of an N-terminal extracellular domain (N3ECD, 210 kDa) and a transmembrane intracellular domain (N3TMIC, 97 kDa), which is non-covalently attached to the N3ECD [9, 10].

The pathological mutations are all located in exons 2–24 in the N3ECD, which are prominently clustered in exons3 and 4 encoding EGFrs2-5. Most of the mutations (>95%) are missense mutations that cause substitution in the cysteine sequence, while there are still splice site mutation or small in-frame deletions [5]. Remarkably, all these mutations lead to the odd number of cysteine residues in the EGFrs, which appears to be crucial in the pathogenesis of CADASIL [1]. However, recently, increasing clinical reports have shown that patients with CADASIL clinical features carry a non-cysteine mutation in NOTCH3 [11–13]. The relationship between CADASIL syndromes with non-cysteine mutation, however, still remains unclear.

According to the characteristics of the Notch3 mutation, genetic sequence of 23 exons of NOTCH3 has become the gold standard for diagnosis of CADASIL. Screening of the 2–24 exons that encode the 34 EGFrs shows a 100% specificity and sensitivity to detect the odd number of cysteine residues. This genetic test provides a diagnostic tool for CADASIL. It has been revealed that this test is useful to identify CADASIL for the patients having a characteristic clinical syndrome and neuroimaging features, particularly when the patients have no hypertension or there is a positive family history [14].

4   Clinical Syndromes and Neuroimaging Features

The major clinical syndromes of CADASIL include migraine with aura, recurrent ischemic strokes, mood disturbance and apathy, and cognitive decline and dementia. Migraine with aura (MA) is presented in almost half of patients. It is generally the first clinical manifestation and not associated with the severity of this disease. The peak age of MA onset is between 16 and 30 years in women and between 31 and 40 years in men [15]. Recurrent ischemic strokes affect more than 85% CADASIL patients. It is the most common manifestation, while the age of the first stroke varies between individuals. The youngest ever reported was only 11 years old [16], and the oldest patient was almost 70 years old [17]. Mood disturbance and apathy are found in nearly 50% patients. Patients with CADASIL generally present apathy in elder age. Clinical studies have shown that there is a significant association between apathy and a lower score of overall quality of life or a higher load of white matter and lacunar lesions [18]. A reduction of cortical surface rather than thickness has been found to be related to the apathy in CADASIL patients [19]. Cognitive decline and dementia are the second most frequent clinical features of CADASIL. The earliest sign in most cases is the impairments in executive function and processing speed. Nearly all patients after 50 years have cognitive decline and dementia. The manifestation can appear in isolation, while they mostly occur in succession. Overall, CADASIL is a severe medical condition, which especially affects young and middle-aged adults [20, 21].

Magnetic resonance imaging is one of the best approaches to detect subcortical infarcts and leukoencephalopathy. The earliest and most frequent abnormalities are the areas of increased signal on T2 weighted imaging or fluid attenuated inversion recovery. With the progression of this disease, the abnormalities become more diffused and mostly occur in the specific external capsule and the anterior part of the temporal lobes. These abnormalities are highly suggestive of CADASIL. Lacunar infarcts appeared on T1 weighted images occur later in life. The lacunar infarcts also have diagnostic value for CADASIL [22].

5 Pathology

In accordance with the image findings, macroscopic examinations of the brain show the typical features of chronic small artery disease. There are diffused lesions in hemispheric white matter. Multiple small lacunar infarcts are found predominantly in the white matter and deep grey matter as well as in the brain stem. In some cases, micro-bleeds are reported in occipital lobe, showing an atypical image in CADASIL [23]. In the cortex, which was thought to be unaffected, there is also widespread neuronal apoptosis [24].

Microscopic and ultrastructural investigations show a specific arteriopathy affecting mainly the small penetrating cerebral and leptomeningeal arteries. The arteriopathy is characterized by a thickening of arterial wall, especially in small arteries from cerebral white matter. Fibrous connective tissues around the arteries accumulate in tunica adventitia, resulting in lumen stenosis or complete obliteration [25]. The VSMCs have predominant morphological alterations and these cells eventually degenerate from the vessel wall [1]. The non-amyloid granular osmiophilic materials (GOM), the specific ultrastructural pathological feature of CADASIL, are located extracellularly and close to the cell surface of smooth muscle cells and pericytes [26]. Besides, GOM deposits can be detected in arteries of many other organs. In clinics, the presence of GOM found by electron microscopy in skin biopsy samples has become one of the diagnostic tests for CADASIL [27]. However, because the availability for electron microscopy examination is limited, immunohistochemical demonstration of

N3ECD accumulations is proffered as an alternative. In the brain, N3ECD deposits have been observed in the tunica media of cortical arteries. This pathological feature may indicate the progression of CADASIL at presympomatic stage. At advanced stages, N3ECD are also found on veins and capillaries in the white matter and grey matter together with leptomeningeal arteries [28].

6   Pathogenesis

Numerous studies have shown that significant decreases in the cerebral blood flow in the white matter [29, 30] with impaired glucose metabolism [30] occur in the brains of CADASIL patients, suggesting chronic subcortical ischemia in CADASIL. The impaired cerebral hemodynamics has been proposed to be the results of both structural and functional changes in brain arteries. In the white matter, the unequivocal arteriolar stenosis and the formation of vascular occlusion or thrombosis are most likely the dominant causes of the ischemic lesions. Besides, as the arterioles are thickened and covered by the increased fibrotic walls, even if they are not stenosis, there is still a risk of occurring infarction. Due to the vascular fibrosis and degeneration of the VSMCs, these arterioles are gradually losing their compliance and autoregulation. Emerging functional studies have indicated that delayed increases of blood flow response to vasodilation stimulus happen in CADASIL patients [31–34].

Moreover, in CADASIL, the blood brain barrier is multifocal impaired because of the common appearance of multiple microbleeds. Generally, CADASIL has been considered as a smooth muscle cell-related disease, the endothelial cells should be spared from this disease. However, recent studies in both CADASIL patients and transgenic mouse models have demonstrated that the abnormal changes of endothelial cells are involved in the pathogenesis of CADASIL. To track bone marrow-derived cells, the bone marrow cells isolated from UBC-GFP mice are transplanted to CADASIL transgenic mice. After the reconstitution of the hematopoietic system, the CADASIL mice are sacrificed for determining bone marrow-derived cells in the brain. Through this experiment, we have noted that the transplanted bone marrow-derived cells (GFP positive cells) coexpress endothelial marker (CD31) in the brain capillaries, suggesting that they are bone marrow-derived endothelial cells. In addition, some capillaries are infilled by the bone marrow-derived cells (GFP positive cells). Through confocal imaging, these infilled bone marrow-derived cells are identified as the thrombosis occluding the capillaries (Fig. 12.1). These findings suggest that the thrombosis may be a result of the impairment of the endothelium. Many other researchers have also observed similar findings of endothelium impairment in CADASIL. It has been reported the endothelial dysfunction with impaired vasoreactivity in CADASIL patients [31, 33]. Using mouse models of CADASIL, emerging evidence has shown that the pericyte pathology leads to the endothelial dysfunction and impairments of blood brain barrier [35, 36]. Together, both clinical and preclinical studies have demonstrated that degeneration and dysfunction not only occur in the VSMCs but it also happens in the endothelial cells in the condition

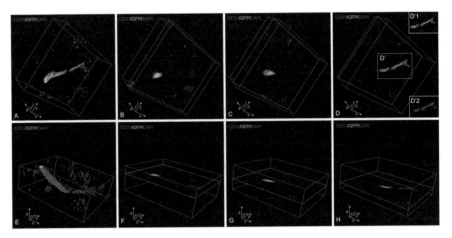

Fig. 12.1 Confocal images of the cerebral capillaries in CADASIL mice. The bone marrow of CADASIL mice is replaced with the bone marrow of UBC-GPF transgenic mice for tracking bone marrow-derived cells. The brains of CADASIL mice (~22 months old) are collected and then cut into 30 μm thick sections. The brain sections are processed for immunofluorescence staining. CD31 (red): endothelial marker for visualizing the distribution of capillaries; GFP (green): bone marrow-derived cells; DAPI: nuclear counterstain. (**a, e**) the general view of one of the capillaries in 3 dimensional (3D) images. (**b, c, f, g**) Z-stack scanning images display the occluded capillaries by GFP positive bone marrow-derived cells. Note that the cross section of the vessel is filled with GFP positive cells, suggesting that the bone marrow-derived blood cells occlude the capillaries. (**d, h**) the 3D images showing the cross sections of capillaries. Note that this part of the capillary is co-expressing of endothelial cells marker (CD31) and GFP, suggesting that the endothelial cells of the capillary are derived from the transplanted bone marrow cells

of CADASIL. The mechanisms as to how both VSMCs and endothelial cells are involved in the pathogenesis of CADASIL, however, remain unclear.

Since Joutel and coworkers reported the NOTCH3 mutation as the causative gene defects of CADASIL syndrome in 1996, a great effort has been made to understand the pathological progression of CADASIL. Over the past two decades, many studies have been done to characterize the pathological changes in both CADASIL patients and animal models. However, it still remains to be elucidated as to how mutations in NOTCH3 lead to the disorder of the cerebral vascular system. NOTCH3 is one of the four Notch receptors, which are present in most multicellular organisms. Notch signaling pathway plays a central role in the development. Expression studies have shown that Notch3 is predominantly expressed in the VSMCs, especially in small arteries [37]. The full length Notch3 is synthesized as a single polypeptide chain precursor, which undergoes constitutive proteolytic processing and forms a mature heterodimeric receptor. The Notch3 receptor consists of an N-terminal extracellular subunit and a C-terminal transmembrane subunit. Once the Notch3 receptor is activated, the Notch3 intracellular domain is released and translocated into the nucleus where interacting with the transcription factor RBP-Jk and co-activators [38]. CADASIL-related Notch3 mutations cause N3ECD accumulation around VSMCs and pericytes of brain arteries and capillaries. However, no N3TMIC aggregation has

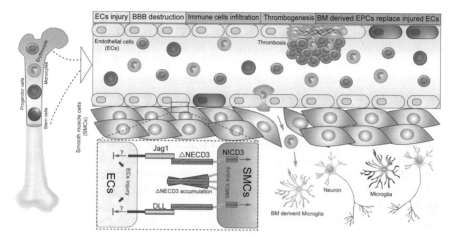

Fig. 12.2 Possible pathogenesis in the progression of CADASIL disease. CADASIL is characterized by the degeneration of vascular smooth muscle cells (VSMCs) in small-to-middle-sized cerebral arteries. The main cause for this arterial pathology is the mutation of Notch3 extracellular
domain (ΔNECD3). Both clinical and preclinical studies have demonstrated that dysfunction and
degeneration occur not only in VSMCs but also in endothelial cells in the condition of CADASIL;
however, the mechanisms as to how both VSMCs and endothelial cells are involved in the pathogenesis of CADASIL still remain unclear. We hypothesize that the mutant NECD3 and the toxic
NECD3 accumulation not only lead to the VSMCs degeneration, but they also trigger pathological
cascades of Jag1/ΔNECD3 and DLL/ΔNECD3 signaling, which finally cause the injuries of endothelial cells (ECs) in the cerebral arteries. The dysfunctions of ECs may result in the destructions
and opening of the blood–brain barrier (BBB), leading to the thrombosis formation and infiltration
of immune cells to the brain parenchymal. During the injuries, the repair processes are also occurring in the cerebral arteries. The injured ECs are replaced by bone marrow-derived endothelial
progenitor cells (EPCs), making the repair of injured-vessels possible. More studies need to be
applied to further examine the hypothesis and search for the therapeutic strategies to restrict the
pathological progression of CADSIL

been found in CADASIL. Studies using reporter gene assays in cultured cells have
demonstrated that CADASIL-associated Notch3 mutant alleles can activate RBP-Jk
transcription at wild type levels [39]. Thus, an abnormal function of the aggregated
mutant N3ECD protein in protein-protein interaction is a likely mechanism for
pathogenesis of CADASIL. As a result of accumulation of GOM surrounding the
VSMCs, the pathological interaction between endothelial cells and VSMCs may also
be involved in the pathogenesis of CADSIL (see schematic summary in Fig. 12.2).

7   Treatment

At present, there is no proven effective treatment for CADASIL; only symptomatic
therapy is available. No smoking is recommended for all CADASIL patients [5]. If
migraine with aura once happens, the treatment is to give the classical migraine

drugs, while ergotamines and triptans are not recommended due to their vasocon-strictory effects [40]. Prevention of ischemic attacks is similar to usual prevention approach for non-cardioembolic ischemic stroke. However, anticoagulants and anti-aggregants have been demonstrated to have no positive effects on CADASIL, because of the increased risk of intracerebral haemorrhage [41]. In hypercholester-olaemia patients, statins are suggested because of their well-established preventive effects in vascular diseases [5].

There are a few studies to explore therapeutic strategies for CADASIL. Using a transgenic mouse model of CADASIL, our research group has demonstrated the efficacy of two hematopoietic growth factors, stem cell factor (SCF) and granulo-cyte colony stimulating factor (G-CSF), in restricting pathological progression of CADASIL. Our findings have revealed that SCF in combination with G-CSF (SCF+G-CSF) improves spatial learning and memory in the mice carrying human Notch3 mutant genes (CADASIL mice). In addition, SCF+G-CSF also increases angiogenesis and neurogenesis, restricts CADASIL-associated degeneration of VSMCs in small arteries and pathological changes in cerebral endothelial cells, and inhibits neural progenitor cell loss and cerebral apoptosis in CADASIL mice [42]. These findings suggest that SCF+G-CSF treatment restricts the pathological pro-gression of CADASIL. Using the approach of gene therapy, other researchers have explored the possibility of interfering CADASIL pathological progression by pre-venting the expression of mutated pathogenic allele of Notch3. Rutten and col-leagues [43] have proposed a model for a selection of exon skip in hoping to develop a new approach to treat CADASIL. In their in vitro study using CADASIL patient-derived cerebral VSMCs, mutant Notch3 exons were specifically skipped without mRNA transcription and protein expression, while this approach appeared to have no destruction in the structure and function of Notch3. They hypothesized that the exclusion of the mutant EGFr domain from Notch3 could abolish the detrimental effect of the unpaired cysteine and prevent toxic Notch3 accumulation. However, this hypothesis has not yet been examined in vivo. Clearly, before performing clini-cal trials for these putative therapeutic approaches, much work still needs to be done at the preclinical level for clarifying the mechanisms underlying the therapeutic effectiveness of these treatments and to address safety concerns for gene therapy in animal models of CADASIL.

8   Concluding Remarks

During the past two decades, the major breakthrough in CADASIL research is the discovery of causative gene and the method for diagnosis of CADASIL. A great effort has been made in understanding the pathology of CADASIL. However, the pathogenesis of CADASIL still remains unclear. There are many open questions that need to be addressed, such as how Notch3 mutation leads to VSMC degenera-tion, how the Notch3 mutation in the VSMCs causes cerebral endothelial cell degeneration, and how these degenerations affect neuron loss and cognitive decline.

Developing therapeutic strategies for this devastating disease has just began. More efforts should be made in future to investigate the therapeutic interventions for slowing the pathological progression of CADASIL.

Acknowledgements This work was supported by American Heart Association (15GRNT25700284).

Funding information This work was supported by American Heart Association (15GRNT25700284).

References

1. Joutel A. Pathogenesis of CADASIL. Bioessays. 2011;33(1):73–80.
2. Pantoni L. Cerebral small vessel disease: from pathogenesis and clinical characteristics to therapeutic challenges. Lancet Neurol. 2010;9(7):689–701.
3. Joutel A, Corpechot C, Ducros A, Vahedi K, Chabriat H, Mouton P, Alamowitch S, Domenga V, Cecillion M, Marechal E, Maciazek J, Vayssiere C, Cruaud C, Cabanis EA, Ruchoux MM, Weissenbach J, Bach JF, Bousser MG, Tournier-Lasserve E. Notch3 mutations in CADASIL, a hereditary adult-onset condition causing stroke and dementia. Nature. 1996;383(6602):707–10. https://doi.org/10.1038/383707a0.
4. Joutel A, Dodick DD, Parisi JE, Cecillon M, Tournier-Lasserve E, Bousser MG. De novo mutation in the Notch3 gene causing CADASIL. Ann Neurol. 2000b;47(3):388–91.
5. Tikka S, Baumann M, Siitonen M, Pasanen P, Poyhonen M, Myllykangas L, Viitanen M, Fukutake T, Cognat E, Joutel A, Kalimo H. CADASIL and CARASIL. Brain Pathol. 2014;24(5):525–44. https://doi.org/10.1111/bpa.12181.
6. Tournier-Lasserve E, Joutel A, Melki J, Weissenbach J, Lathrop GM, Chabriat H, Mas JL, Cabanis EA, Baudrimont M, Maciazek J, et al. Cerebral autosomal dominant arteriopathy with subcortical infarcts and leukoencephalopathy maps to chromosome 19q12. Nat Genet. 1993;3(3):256–9. https://doi.org/10.1038/ng0393-256.
7. Coto E, Menendez M, Navarro R, Garcia-Castro M, Alvarez V. A new de novo Notch3 mutation causing CADASIL. Eur J Neurol. 2006;13(6):628–31. https://doi.org/10.1111/j.1468-1331.2006.01337.x.
8. Stojanov D, Grozdanovic D, Petrovic S, Benedeto-Stojanov D, Stefanovic I, Stojanovic N, Ilic DN. De novo mutation in the NOTCH3 gene causing CADASIL. Bosn J Basic Med Sci. 2014;14(1):48–50.
9. Joutel A, Andreux F, Gaulis S, Domenga V, Cecillon M, Battail N, Piga N, Chapon F, Godfrain C, Tournier-Lasserve E. The ectodomain of the Notch3 receptor accumulates within the cerebrovasculature of CADASIL patients. J Clin Invest. 2000a;105(5):597–605. https://doi.org/10.1172/jci8047.
10. Rutten JW, Haan J, Terwindt GM, van Duinen SG, Boon EM, Lesnik Oberstein SA. Interpretation of NOTCH3 mutations in the diagnosis of CADASIL. Expert Rev Mol Diagn. 2014;14(5):593–603. https://doi.org/10.1586/14737159.2014.922880.
11. Brass SD, Smith EE, Arboleda-Velasquez JF, Copen WA, Frosch MP. Case records of the Massachusetts General Hospital. Case 12-2009. A 46-year-old man with migraine, aphasia, and hemiparesis and similarly affected family members. N Engl J Med. 2009;360(16):1656–65. https://doi.org/10.1056/NEJMcpc0810839.
12. Cognat E, Baron-Menguy C, Domenga-Denier V, Cleophax S, Fouillade C, Monet-Lepretre M, Dewerchin M, Joutel A. Archetypal Arg169Cys mutation in NOTCH3 does not drive the pathogenesis in cerebral autosomal dominant arteriopathy with sub-

cortical infarcts and leucoencephalopathy via a loss-of-function mechanism. Stroke. 2014a;45(3):842–9. https://doi.org/10.1161/strokeaha.113.003339.

13. Cognat E, Herve D, Joutel A. Response to letter regarding article, "Archetypal Arg169Cys mutation in NOTCH3 does not drive the pathogenesis in cerebral autosomal dominant arteriopathy with subcortical infarcts and leucoencephalopathy via a loss-of-function mechanism". Stroke. 2014b;45(7):e129. https://doi.org/10.1161/strokeaha.114.005616.

14. Peters N, Opherk C, Bergmann T, Castro M, Herzog J, Dichgans M. Spectrum of mutations in biopsy-proven CADASIL: implications for diagnostic strategies. Arch Neurol. 2005;62(7):1091–4. https://doi.org/10.1001/archneur.62.7.1091.

15. Guey S, Mawet J, Herve D, Duering M, Godin O, Jouvent E, Opherk C, Alili N, Dichgans M, Chabriat H. Prevalence and characteristics of migraine in CADASIL. Cephalalgia. 2016;36:1038. https://doi.org/10.1177/0333102415620909.

16. Granild-Jensen J, Jensen UB, Schwartz M, Hansen US. Cerebral autosomal dominant arteriopathy with subcortical infarcts and leukoencephalopathy resulting in stroke in an 11-year-old male. Dev Med Child Neurol. 2009;51(9):754–7. https://doi.org/10.1111/j.1469-8749.2008.03241.x.

17. Watanabe M, Adachi Y, Jackson M, Yamamoto-Watanabe Y, Wakasaya Y, Shirahama I, Takamura A, Matsubara E, Kawarabayashi T, Shoji M. An unusual case of elderly-onset cerebral autosomal dominant arteriopathy with subcortical infarcts and leukoencephalopathy (CADASIL) with multiple cerebrovascular risk factors. J Stroke Cerebrovasc Dis. 2012;21(2):143–5. https://doi.org/10.1016/j.jstrokecerebrovasdis.2010.05.008.

18. Reyes S, Viswanathan A, Godin O, Dufouil C, Benisty S, Hernandez K, Kurtz A, Jouvent E, O'Sullivan M, Czernecki V, Bousser MG, Dichgans M, Chabriat H. Apathy: a major symptom in CADASIL. Neurology. 2009;72(10):905–10. https://doi.org/10.1212/01.wnl.0000344166.03470.f8.

19. Jouvent E, Reyes S, Mangin JF, Roca P, Perrot M, Thyreau B, Herve D, Dichgans M, Chabriat H. Apathy is related to cortex morphology in CADASIL. A sulcal-based morphometry study. Neurology. 2011;76(17):1472–7. https://doi.org/10.1212/WNL.0b013e31821810a4.

20. Chabriat H, Joutel A, Dichgans M, Tournier-Lasserve E, Bousser MG. Cadasil. Lancet Neurol. 2009;8(7):643–53. https://doi.org/10.1016/s1474-4422(09)70127-9.

21. Herve D, Chabriat H. CADASIL. J Geriatr Psychiatry Neurol. 2010;23(4):269–76. https://doi.org/10.1177/0891988710383570.

22. Herve D, Godin O, Dufouil C, Viswanathan A, Jouvent E, Pachai C, Guichard JP, Bousser MG, Dichgans M, Chabriat H. Three-dimensional MRI analysis of individual volume of Lacunes in CADASIL. Stroke. 2009;40(1):124–8. https://doi.org/10.1161/strokeaha.108.520825.

23. Trikamji B, Thomas M, Hathout G, Mishra S. An unusual case of cerebral autosomal-dominant arteriopathy with subcortical infarcts and leukoencephalopathy with occipital lobe involvement. Annals Indian Acad Neurol. 2016;19(2):272–4. https://doi.org/10.4103/0972-2327.173403.

24. Viswanathan A, Gray F, Bousser MG, Baudrimont M, Chabriat H. Cortical neuronal apoptosis in CADASIL. Stroke. 2006;37(11):2690–5. https://doi.org/10.1161/01.STR.0000245091.28429.6a.

25. Kalimo H, Ruchoux MM, Viitanen M, Kalaria RN. CADASIL: a common form of hereditary arteriopathy causing brain infarcts and dementia. Brain Pathol. 2002;12(3):371–84.

26. Lewandowska E, Dziewulska D, Parys M, Pasennik E. Ultrastructure of granular osmiophilic material deposits (GOM) in arterioles of CADASIL patients. Folia Neuropathol. 2011;49(3):174–80.

27. Cotrutz CE, Indrei A, Badescu L, Dacalu C, Neamtu M, Dumitrescu GF, Stefanache F, Petreus T. Electron microscopy analysis of skin biopsies in CADASIL disease. Rom J Morphol Embryol. 2010;51(3):455–7.

28. Joutel A, Favrole P, Labauge P, Chabriat H, Lescoat C, Andreux F, Domenga V, Cecillon M, Vahedi K, Ducros A, Cave-Riant F, Bousser MG, Tournier-Lasserve E. Skin biopsy immunostaining with a Notch3 monoclonal antibody for CADASIL diagnosis. Lancet. 2001;358(9298):2049–51. https://doi.org/10.1016/s0140-6736(01)07142-2.

29. Joutel A, Monet-Lepretre M, Gosele C, Baron-Menguy C, Hammes A, Schmidt S, Lemaire-Carrette B, Domenga V, Schedl A, Lacombe P, Hubner N. Cerebrovascular dysfunction and micro-

circulation rarefaction precede white matter lesions in a mouse genetic model of cerebral ischemic small vessel disease. J Clin Invest. 2010;120(2):433–45. https://doi.org/10.1172/jci39733.

30. Tuominen S, Miao Q, Kurki T, Tuisku S, Pöyhönen M, Kalimo H, Viitanen M, Sipilä HT, Bergman J, Rinne JO. Positron emission tomography examination of cerebral blood flow and glucose metabolism in young CADASIL patients. Stroke. 2004;35(5):1063–7.

31. Campolo J, De Maria R, Frontali M, Taroni F, Inzitari D, Federico A, Romano S, Puca E, Mariotti C, Tomasello C, Pantoni L, Pescini F, Dotti MT, Stromillo ML, De Stefano N, Tavani A, Parodi O. Impaired vasoreactivity in mildly disabled CADASIL patients. J Neurol Neurosurg Psychiatry. 2012;83(3):268–74. https://doi.org/10.1136/jnnp-2011-300080.

32. De Maria R, Campolo J, Frontali M, Taroni F, Federico A, Inzitari D, Tavani A, Romano S, Puca E, Orzi F, Francia A, Mariotti C, Tomasello C, Dotti MT, Stromillo ML, Pantoni L, Pescini F, Valenti R, Pelucchi C, Parolini M, Parodi O. Effects of sapropterin on endothelium-dependent vasodilation in patients with CADASIL: a randomized controlled trial. Stroke. 2014;45(10):2959–66. https://doi.org/10.1161/strokeaha.114.005937.

33. Peters N, Freilinger T, Opherk C, Pfefferkorn T, Dichgans M. Enhanced L-arginine-induced vasoreactivity suggests endothelial dysfunction in CADASIL. J Neurol. 2008;255(8):1203–8. https://doi.org/10.1007/s00415-008-0876-9.

34. Pfefferkorn T, von Stuckrad-Barre S, Herzog J, Gasser T, Hamann GF, Dichgans M. Reduced cerebrovascular CO(2) reactivity in CADASIL: a transcranial Doppler sonography study. Stroke. 2001;32(1):17–21.

35. Ghosh M, Balbi M, Hellal F, Dichgans M, Lindauer U, Plesnila N. Pericytes are involved in the pathogenesis of cerebral autosomal dominant arteriopathy with subcortical infarcts and leukoencephalopathy. Ann Neurol. 2015;78(6):887–900. https://doi.org/10.1002/ana.24512.

36. Kofler NM, Cuervo H, Uh MK, Murtomaki A, Kitajewski J. Combined deficiency of Notch1 and Notch3 causes pericyte dysfunction, models CADASIL, and results in arteriovenous malformations. Sci Rep. 2015;5:16449. https://doi.org/10.1038/srep16449.

37. Wang T, Baron M, Trump D. An overview of Notch3 function in vascular smooth muscle cells. Prog Biophys Mol Biol. 2008;96(1-3):499–509. https://doi.org/10.1016/j.pbiomolbio.2007.07.006.

38. Monet M, Domenga V, Lemaire B, Souilhol C, Langa F, Babinet C, Gridley T, Tournier-Lasserve E, Cohen-Tannoudji M, Joutel A. The archetypal R90C CADASIL-NOTCH3 mutation retains NOTCH3 function in vivo. Hum Mol Genet. 2007;16(8):982–92. https://doi.org/10.1093/hmg/ddm042.

39. Low WC, Santa Y, Takahashi K, Tabira T, Kalaria RN. CADASIL-causing mutations do not alter Notch3 receptor processing and activation. Neuroreport. 2006;17(10):945–9. https://doi.org/10.1097/01.wnr.0000223394.66951.48.

40. Chabriat H, Bousser MG. Cerebral autosomal dominant arteriopathy with subcortical infarcts and leukoencephalopathy. Handb Clin Neurol. 2008;89:671–86. https://doi.org/10.1016/s0072-9752(07)01261-4.

41. Choi JC, Kang SY, Kang JH, Park JK. Intracerebral hemorrhages in CADASIL. Neurology. 2006;67(11):2042–4. https://doi.org/10.1212/01.wnl.0000246601.70918.06.

42. Liu XY, Gonzalez-Toledo ME, Fagan A, Duan WM, Liu Y, Zhang S, Li B, Piao CS, Nelson L, Zhao LR. Stem cell factor and granulocyte colony-stimulating factor exhibit therapeutic effects in a mouse model of CADASIL. Neurobiol Dis. 2015;73:189–203. https://doi.org/10.1016/j.nbd.2014.09.006.

43. Rutten JW, Dauwerse HG, Peters DJ, Goldfarb A, Venselaar H, Haffner C, van Ommen GJ, Aartsma-Rus AM, Lesnik Oberstein SA. Therapeutic NOTCH3 cysteine correction in CADASIL using exon skipping: in vitro proof of concept. Brain. 2016;139(Pt 4):1123–35. https://doi.org/10.1093/brain/aww011.

Chapter 13
Blood Pressure and Cerebral Ischemic Reperfusion Injury

Weijian Jiang, Chen Li, Mohamad Orabi, and Wuwei Feng

Abstract Recanalization and reperfusion injury in ischemic stroke patients casts shadows on current stroke management. The potential mechanisms of reperfusion injury remain obscure and need further investigation. This article summarized several clinically important mechanisms and especially discussed a potential role of blood pressure control in reperfusion injury.

Keywords Stroke · Blood · Pressure · Reperfusion

1 Introduction

Acute treatment of an ischemic stroke depends on the restoration of blood flow in a small treatment window, to reverse the damage to the penumbral tissue using intravenous alteplase and/or mechanical thrombectomy. Ischemic penumbra has been defined as severely hypoperfused, nonfunctional, but still viable tissue surrounding the irreversibly infarct core. The efficiency of reperfusion depends on severity and duration of ischemia. While reperfusion can improve the functional outcome for some patients, however; others have acute deterioration of ischemic brain tissue that parallels and antagonizes the benefit of restoring perfusion known as "Cerebral Ischemic Reperfusion Injury (CIRI)" or "Cerebral Hyperperfusion Syndrome (CHS)." The term CHS has often been used interchangeably with "Cerebral Reperfusion Injury". Sometimes the latter term is more appropriate to describe the pathogenesis of patients who have acute deterioration of ischemic brain tissue after restoring perfusion. Patients with perfusion increase of >100%

W. Jiang (✉) · C. Li
Vascular Neurosurgery Department of New Era Stroke Care and Research Institute,
The PLA Rocket Force General Hospital, Beijing, China
e-mail: cjr.jiangweijian@vip.163.com

M. Orabi · W. Feng
MUSC Stroke Center, Department of Neurology, Medical University of South Carolina,
Charleston, SC, USA

© Springer International Publishing AG, part of Springer Nature 2018   205
W. Jiang et al. (eds.), *Cerebral Ischemic Reperfusion Injuries (CIRI)*, Springer Series
in Translational Stroke Research, https://doi.org/10.1007/978-3-319-90194-7_13

compared with the baseline after recanalization therapy are more likely to have this situation [1]. Clinically, patients presented with throbbing severe ipsilateral headache (typically frontotemporal or periorbital, but sometimes diffuse headache, eye and face pain), nausea and vomit, visual disturbance and even altered mental status. Seizures (focal or generalized type), focal neurological deficits, and intracerebral or subarachnoid hemorrhage can also occur.

The pathophysiology of CIRI is a result of a complex series of events at molecular and cellular levels due to activation of immune system, complement, platelets and coagulation cascade [2]. Additional consequences of reperfusion injury include disruption of blood brain barrier and the loss of cerebral autoregulation as result of further damage from ischemia and reperfusion.

This chapter focuses on the epidemiology of CIRI, the relationship between blood pressure, cerebral autoregulation and reperfusion injury. This is followed by a discussion on prevention and treatment of CIRI. In the end, we will highlight the research direction on such topic.

2 Epidemiology of Cerebral Ischemic Reperfusion Injury (CIRI) and Cerebral Hyperperfusion Syndrome (CHS)

Van Mook et al. reviewed the incidence of CHS and found it varied widely from 0.2% to 18.9% depending on the definition of CHS, sample size, and inclusion criteria of the cohort [1]. Nevertheless, most studies reported CHS occurs no more than 3.0%. It can occur anytime in the first 28 days after carotid endarterectomy or stenting; several studies reported the onset of CHS could be within several hours to a couple of days after the procedure. In most studies, patients with CHS had evidence of hyperperfusion increasing 100% from the baseline to the postprocedure, such as CEA [1].

Ogasawara et al. investigated and compared the rate of intracranial hemorrhage (ICH) associated with CHS between patients undergoing CEA and those whom having CAS [3]. In the subgroup of patients undergoing CEA, the incidence of CHS was 1.9%, with 0.4% experiencing ICH. In those patients undergoing CAS, the incidence of CHS was 1.1%, with 0.7% suffering ICH. Similar findings were seen by Moulakakis et al. where the incidence of CHS after CEA was estimated to at 1.9%, with ICH of 0.37% [3].

CIRI refers to the injury that occurs to the ischemic brain tissue that abolishes the benefit of restoring perfusion, regardless how the reperfusion happened—spontaneous recanalization, alteplase or mechanical thrombectomy.

Recanalization following vessel occlusion can result in three outcomes. In a subset of patients, both recanalization and reperfusion successfully lead to tissue salvage and functional recovery. On the other side, recanalization can result in no-reflow phenomenon which refers to a state of "successful mechanical recanalization but with no flow restoration to the ischemic tissue" which has been reported post-recanalization of myocardial ischemic tissue. In addition, recanalization may lead to reperfusion injury.

3 Cerebral Autoregulation

Cerebral perfusion pressure values are determined via the equation (CPP = MAP − ICP), where MAP is the mean arterial pressure and is equal to 1/3 of pulse pressure plus the diastolic pressure. Cerebral blood flow (CBF) depends on cerebral perfusion pressure and vascular resistance. It equals to cerebral perfusion pressure divided by cerebrovascular resistance (CBF = CPP/CVR). CVR is influenced by the diameter of intracranial arteries and blood viscosity. In normal circumstances, CBF stays constant across a wide range of CPP values by the mechanism called "cerebrovascular autoregulation". In another word, CVR decrease as CPP decrease, and vice versa. It is an energy-dependent process that occurs at the arteriolar level requiring adenosine triphosphate (ATP) within the vessel wall to either constrict or dilate. In the setting of acute brain injury, such as, ischemic stroke or hemorrhagic stroke, cerebral autoregulation can be dysfunctional—CBF changes linearly passive to CPP values [4]. This linear relationship between CPP and CBF poses a challenge to stroke patients as CBF may drop below the critical ischemic threshold of 20 mL/100 g/min. In 1968, Waltz et al. first described the impact of changes in systemic blood pressure on CBF in the ischemic and non-ischemic cortex in a cat MCAO stroke model, they hypothesized that cerebral autoregulation cerebral was impaired in the setting of induced ischemia [5].

Because the arteries in the brain cannot sustain prolonged vasoconstriction after an injury, the upper limit of the autoregulation will be lost eventually. When this occurs, an appearance of "sausage stringing"—an alternating pattern of dilated arterial segments with focal regions of constriction, can be observed on the DSA. The dilated segments are regions of passive dilatation while the constricted segments are regions of sustained autoregulation. Further increase in CPP leads to an expansion along the full length of the arterioles with subsequent increase in CBF. Unfortunately, this situation is always accompanied by cerebrovascular endothelium damages as well as blood–brain barrier disruptions, which can cause extravasation of plasma proteins through the vessel wall with resultant tissue edema [6].

The upper and lower limits of autoregulation can be changeable. They depend on both physiological stimuli and disease states [7]. Activation of the sympathetic nerves can make an upward shift of both the lower and upper limits. It is a potentially protective response as a result of sympathetic activation. The upward shift of both the lower and upper limits (right-shifted autoregulatory curve) in patients with chronic hypertension allows them to better tolerate higher blood pressures before the upward breakout happens [8]. However, this protective response could have a detrimental effect if ischemia occurs at a relatively lower blood pressure state. The downward shift of both the lower and upper limits (normally left-shifted autoregulatory curve) typically occurs in children. It means that children can be easily at risk of the upward breakout of the autoregulation at lower MAP than adults. As a result, children are more vulnerable to hypertensive encephalopathy [9, 10].

4 Pathophysiology of Hyperperfusion

The pathophysiology of CIRI is not entirely clear. Several mechanisms may contribute to the pathophysiology of hyperperfusion and CIRI, such as molecular mechanisms of leukocyte, platelet-mediated reperfusion injury and complement-mediated reperfusion injury [11]. However, Muhammad U. Farooq et al. consider that dysregulation of the cerebral vascular system and hypertension play a significant role in increasing CBF [2].

First, endothelial dysfunction mediated by free oxygen radicals can result in impaired autoregulation in traumatic brain injury, ischemic stroke, or severe stenosis or occlusion of cerebral arteries [12]. It is reported that patients with impaired cerebral autoregulation are sensitive to fluctuations of CPP [13]. Jorgensen et al. found that mean flow velocities of the ipsilateral middle cerebral artery (MCA) are pressure dependent. Cerebral blow velocity returns to normal if arterial pressure after CEA is reduced patient symptoms can resolve accordingly [14]. Similarly, a breakthrough in cerebral autoregulation can cause rapid increase in CPP which leads to hypertensive encephalopathy. This suggests CIRI, CHS, hypertensive urgency and hypertension emergency may share the common pathway [15]. Bernstein and colleagues performed an autopsy on a succumbed patient from CIRI, they presumed that the chronic cerebral ischemia distal to the high-grade stenotic carotid artery led to chronic vasodilatation, loss of autoregulation, and an absence of arterial vasoconstriction to protect the capillary bed [16]. Nitric oxide is a possible mediator of impaired autoregulation in CIRI as it can cause vasodilatation and increase the permeability of cerebral vessels [17]. Free radicals can induce damage to the cerebrovascular endothelium and result in postoperative hyperperfusion. Pretreatment a free-radical scavenger (edaravone) was shown to reduce the occurrence of CHS after CEA (16.0% vs. 2.0%; p = 0.031) in a proof-of-concept study led by Ogasawara et al. [18].

Secondly, baroreceptor-reflex breakdown may also be associated with the development of CIRI. The baroreceptor-reflex buffers acute fluctuations in arterial blood pressure. Baroreceptor-reflex can disappear after receptor denervation, an infrequent adverse event from CEA [19]. Uncontrolled hypertension in the setting of baroreceptor-reflex breakdown can impair cerebral autoregulation and increase cerebral perfusion.

Finally, an axon reflex-like trigeminovascular reflex has also been implicated in the pathophysiology of CHS [20]. The trigeminovascular system has a cerebroprotective mechanism which bring the vascular tone back to the baseline level after exposure to multiple vasoconstrictors. The possible mechanism likely involves with the release and increase of vasoactive neuropeptides in cerebral blood system. Its response can be attenuated by trigeminal ganglionectomy [20, 21].

5   Hypertension and CIRI

Several conditions can predispose stroke patients for CIRI or CHS [3, 22–27]. Cerebral vascular reserve reduction, postoperative hypertension, and hyperperfusion lasting a few hours to several days after recanalization therapy seem to be the critical risk factors [3, 22–27]. In animal experiments of the effects of hypertension on cerebral and mesenteric vessels [28], Byrom et al. first proposed that acutely increased perfusion pressures produced excessive vasoconstriction of the cerebral vasculature that resulted in ischemia and cytotoxic edema in the borderzone between arterial territories [28].

Hyperperfusion encephalopathy is an acute to the subacute phenomenon which result from several days of exposure to increased systemic pressures or cerebral perfusion pressures. However, several researchers found that vascular remodeling in chronic hypertension maybe protective against cerebral hyperperfusion [29]. Chronic hypertension-related vascular pathology typically manifests as an increased arterial wall thickness and a decreased intraluminal diameter as a result of lipohyalinosis. Additionally, chronic hypertension can result in fibrinoid necrosis and microaneurysms in small terminal arteries [30]. A sudden, acute elevation of blood pressure in these patients could result in a lacunar infarction or hematoma in the vascular territories of these end arteries (i.e. basal ganglia, thalamus, centrum semi-ovale and the pons). It has recently been reported that chronic essential hypertension is accompanied with a reduction of NO synthesis in the vascular endothelium [31].

The underlying mechanisms for CBF autoregulation in the human are not entirely understood. The traditional view is a combination of neurogenic, myogenic, and metabolic factors [6]. Recently, "NO" has been perceived as another critical component. Several animal studies showed that autoregulation is still preserved after sympathetic and parasympathetic denervation. This phenomenon suggests that neurogenic factors are likely not of primary importance [32]. Different hypotheses provide divergent interpretations. The myogenic hypothesis states that smooth muscle in the resistance arteries directly responds to alterations in perfusion pressure by contracting muscles during "the pressure increase" and relaxing muscles during "the pressure decrease". The metabolic hypothesis declares that reductions in CBF induce the release of vasoactive substances including carbon dioxide, hydrogen ions, oxygen, adenosine, potassium and calcium, from the brain which subsequently stimulates the dilatation of cerebral resistance arteries.

If the intracranial vessels fail to adequately compensate for the rapid increases in blood pressure, the cerebral autoregulation will be impaired eventually. Cerebral autoregulation in response to the changes in systemic blood pressure is believed to occur predominantly at the level of the small capillary and arteriole which has both myogenic and neurogenic components [33]. However, it appears that purely myogenic autoregulation, the mechanism whereby increased intravascular pressure results in depolarization of vascular smooth muscle and vasoconstriction, plays a relatively minor role in controlling CBF at a higher systemic blood pressures [34, 35]. Myogenic autoregulation occurs in the smaller arterioles, and it appears to be primarily responsible for cerebral autoregulation at systemic blood pressures at the upper level of normal physiologic ranges [35]. Autoregulation at systemic blood pressures exceeding the limits of myogenic autoregulation is likely to be mediated at the level of the large arteriole by perivascular sympathetic autonomic nerves in the adventitia of the vascular wall [33]. Furthermore, immunohistochemical study has shown a regional variability in the innervation of cerebral arteries. The vertebrobasilar system, particularly the vessels supplying the occipital lobe, possess sparse sympathetic innervation and are therefore less well protected than the remainder of the brain [36].

6   Hypertension, Cerebral Hyperperfusion, and CIRI

Neal et al. found that if intravascular pressures raise to the point just below what required to cause the rupture of the capillary wall, the permeability of the endothelium increased strikingly, [37] Fluid leaks through the capillary wall actively to relieve the intravascular pressures preventing hematoma from occurring. This fluid accumulation is not cytotoxic or vasogenic edema but better termed "hydrostatic edema," because the etiology of this condition is fluid dynamics.

Another mechanism involved in the pathogenesis of CIRI is the pressure exerted on the blood-brain barrier (BBB) with hypertension [38]. The BBB is susceptible to hypertensive urgency. During the hypertension urgency, elevated intraluminal pressure induces spontaneous vascular stress to BBB, and as a result, the BBB permeability increase. Endothelial structure and function modifies after long of exposure to chronic hypertension, several vasoactive substances (cytokines, endothelin, nitric oxide and reactive oxygen species) were released to compromise vascular tone and BBB permeability that leads to "vasogenic edema" [38]. Wardlaw et al. investigated the integrity of BBB in patients with recent lacunar infarct [39]. They found that more leakage of intravenous gadolinium in the white matter of 51 patients with lacunar infarcts than the 46 patients with cortical infarcts [39]. This study showed that BBB hyperpermeability coupled with dysfunctional endothelium did exist in patients with small vessel disease.

7   Blood Pressure Control and Prevention of CIRI

Early recognition of CHS/CIRI is important as it may be reversed at an super early stage. Given that the implication of hypertension in CIRI/CHS, restricted blood pressure control has been proposed. As a matter of fact, it is widely adopted in the clinical practice, especially in the postoperative phase to prevent CIRI. The accompanying therapies include treatment of cerebral edema with osmotic agents and anticonvulsant agent for seizure. However, there are no definitive guidelines for the target blood pressure in CIRI patients and how long blood pressure needs to be controlled [1, 16, 40–42]. The target BP level and the extent of BP reduction have no consensus, but the lowering of BP by ~10/5 mm Hg can bring demonstrable benefits [43]. Such evidence still need to be formally established from carefully designed clinical trial.

8   Choice of Antihypertensive Agents

Again, there has been no guideline for antihypertensive therapy to prevent CIRI or CHS. Theoretically, drugs without direct effects on cerebral blood flow could be advantageous [1].

The β1-adrenergic antagonists (metoprolol) frequently are used to control blood pressure in patients with brain injury because it can reduce arterial blood pressure with little effect on the intracranial pressure within the autoregulatory range [44]. After CEA, the mixed-adrenergic antagonist and α-adrenergic antagonist (labetalol) are widely used to prevent CHS [14]. In general, they can decreases the CPP and MAP about 30% lower than the baseline without any effects on CBF [45]. Clonidine, a central-acting sympatholytic agent, is also used to control blood pressure after recanalization therapy for acute ischemic stroke [46]. It is an α2-adrenergic agonist which results in vasorelaxation with subsequent decreases in arterial blood pressure, heart rate and cardiac output. Direct vasodilators (nitroprusside or glycerol trini-trate) and calcium antagonists (amlodipine) are contraindicated [23, 44].

9   Summary

Cerebral Ischemic Reperfusion Injury (CIRI), similar to Cerebral Hyperperfusion Syndrome (CHS) has been observed in acute ischemic stroke patients after recana-lization therapy and is associated with poor outcomes. Hypertension and impaired cerebral autoregulation have been implicated in the pathophysiological process although the exact mechanisms remain unclear. Restricted blood pressure control has been proposed and widely used in the clinical practice to prevent CIRI despite the fact that it has not been systematically investigated and is lacking evidence from well-design clinical trials. This highlights the tremendous needs for further investi-gations in such topic.

References

1. van Mook WN, Rennenberg RJ, Schurink GW, et al. Cerebral hyperperfusion syndrome. Lancet Neurol. 2005;4:877–88.
2. Farooq MU, Goshgarian C, Min J, Gorelick PB. Pathophysiology and management of reper-fusion injury and hyperperfusion syndrome after carotid endarterectomy and carotid artery stenting. Exp Transl Stroke Med. 2016;8:7.

3. Ogasawara K, Yukawa H, Kobayashi M, et al. Prediction and monitoring of cerebral hyperperfusion after carotid endarterectomy by using single-photon emission computerized tomography scanning. J Neurosurg. 2003;99:504–10.
4. Rose JC, Mayer SA. Optimizing blood pressure in neurological emergencies. Neurocrit Care. 2006;4:98.
5. Waltz AG. Effect of blood pressure on blood flow in ischemic and in nonischemic cerebral cortex. The phenomena of autoregulation and luxury perfusion. Neurology. 1968;18:613–21.
6. Markus HS. Cerebral perfusion and stroke. J Neurol Neurosurg Psychiatry. 2004;75:353–61.
7. Strandgaard S, Olesen J, Skinhoj E, Lassen NA. Autoregulation of brain circulation in severe arterial hypertension. Br Med J. 1973;1:507–10.
8. Dawson SL, Panerai RB, Potter JF. Serial changes in static and dynamic cerebral autoregulation after acute ischaemic stroke. Cerebrovasc Dis. 2003;16:69–75.
9. Pryds O, Edwards AD. Cerebral blood flow in the newborn infant. Arch Dis Child Fetal Neonatal Ed. 1996;74:F63–9.
10. Tyszczuk L, Meek J, Elwell C, Wyatt JS. Cerebral blood flow is independent of mean arterial blood pressure in preterm infants undergoing intensive care. Pediatrics. 1998;102:337–41.
11. Pan J, Konstas AA, Bateman B, Ortolano GA, Pile-Spellman J. Reperfusion injury following cerebral ischemia: pathophysiology, MR imaging, and potential therapies. Neuroradiology. 2007;49:93–102.
12. Symon L, Branston NM, Strong AJ. Autoregulation in acute focal ischemia. An experimental study. Stroke. 1976;7:547–54.
13. Jiang WJ. Cerebral perfusion imaging. The carotid and supra-aortic trunks: diagnosis, angioplasty and stenting. 2nd ed. Chichester: John Wiley & Sons, Ltd; 2011. p. 76–85.
14. Jorgensen LG, Schroeder TV. Defective cerebrovascular autoregulation after carotid endarterectomy. Eur J Vasc Surg. 1993;7:370–9.
15. Vaughan CJ, Delanty N. Hypertensive emergencies. Lancet. 2000;356:411–7.
16. Bernstein M, Fleming JF, Deck JH. Cerebral hyperperfusion after carotid endarterectomy: a cause of cerebral hemorrhage. Neurosurgery. 1984;15:50-6.
17. Janigro D, West GA, Nguyen TS, Winn HR. Regulation of blood-brain barrier endothelial cells by nitric oxide. Circ Res. 1994;75:528–38.
18. Ogasawara K, Inoue T, Kobayashi M, Endo H, Fukuda T, Ogawa A. Pretreatment with the free radical scavenger edaravone prevents cerebral hyperperfusion after carotid endarterectomy. Neurosurgery. 2004;55:1060–7.
19. Timmers HJ, Wieling W, Karemaker JM, Lenders JW. Baroreflex failure: a neglected type of secondary hypertension. Neth J Med. 2004;62:151–5.
20. Macfarlane R, Moskowitz MA, Sakas DE, Tasdemiroglu E, Wei EP, Kontos HA. The role of neuroeffector mechanisms in cerebral hyperperfusion syndromes. J Neurosurg. 1991;75:845–55.
21. Edvinsson L, McCulloch J, Kingman T, Uddman R. On the functional role of the trigemino-cerebrovascular system in the regulation of cerebral circulation. In: Neural regulation of the brain cerebral circulation. New York, NY: Elsevier; 1986. p. 407–18.
22. Sundt TM Jr, Sharbrough FW, Piepgras DG, Kearns TP, Messick JM Jr, O'Fallon WM. Correlation of cerebral blood flow and electroencephalographic changes during carotid endarterectomy: with results of surgery and hemodynamics of cerebral ischemia. Mayo Clin Proc. 1981;56:533–43.
23. Sbarigia E, Speziale F, Giannoni MF, Colonna M, Panico MA, Fiorani P. Post-carotid endarterectomy hyperperfusion syndrome: preliminary observations for identifying at risk patients by transcranial Doppler sonography and the acetazolamide test. Eur J Vasc Surg. 1993;7:252–6.
24. Hosoda K, Kawaguchi T, Shibata Y, et al. Cerebral vasoreactivity and internal carotid artery flow help to identify patients at risk for hyperperfusion after carotid endarterectomy. Stroke. 2001;32:1567–73.
25. Nielsen MY, Sillesen HH, Jorgensen LG, Schroeder TV. The haemodynamic effect of carotid endarterectomy. Eur J Vasc Endovasc Surg. 2002;24:53–8.

26. Ascher E, Markevich N, Schutzer RW, Kallakuri S, Jacob T, Hingorani AP. Cerebral hyperperfusion syndrome after carotid endarterectomy: predictive factors and hemodynamic changes. J Vasc Surg. 2003;37:769–77.
27. Yoshimoto T, Houkin K, Kuroda S, Abe H, Kashiwaba T. Low cerebral blood flow and perfusion reserve induce hyperperfusion after surgical revascularization: case reports and analysis of cerebral hemodynamics. Surg Neurol. 1997;48:132–8. discussion 8–9.
28. Byrom FB. The pathogenesis of hypertensive encephalopathy and its relation to the malignant phase of hypertension; experimental evidence from the hypertensive rat. Lancet. 1954;267:201–11.
29. Schwartz RB. Hyperperfusion encephalopathies: hypertensive encephalopathy and related conditions. Neurologist. 2002;8:22–34.
30. Okazaki H. Fundamentals of neuropathology. Morphologic basics of neurologic disorders. New York, NY: Igaku-Shoin; 1989.
31. Moncada S, Higgs A. The L-arginine-nitric oxide pathway. N Engl J Med. 1993;329:2002–12.
32. Busija DW, Heistad DD. Factors involved in the physiological regulation of the cerebral circulation. Rev Physiol Biochem Pharmacol. 1984;101:161–211.
33. Beausang-Linder M, Bill A. Cerebral circulation in acute arterial hypertension--protective effects of sympathetic nervous activity. Acta Physiol Scand. 1981;111:193–9.
34. Dacey RG Jr, Duling BR. A study of rat intracerebral arterioles: methods, morphology, and reactivity. Am J Phys. 1982;243:H598–606.
35. Ngai AC, Winn HR. Modulation of cerebral arteriolar diameter by intraluminal flow and pressure. Circ Res. 1995;77:832–40.
36. Edvinsson L, Owman C, Sjoberg NO. Autonomic nerves, mast cells, and amine receptors in human brain vessels. A histochemical and pharmacological study. Brain Res. 1976;115:377–93.
37. Neal CR, Michel CC. Openings in frog microvascular endothelium induced by high intravascular pressures. J Physiol. 1996;492(Pt 1):39–52.
38. Johansson BB. Hypertension mechanisms causing stroke. Clin Exp Pharmacol Physiol. 1999;26:563–5.
39. Wardlaw JM, Doubal F, Armitage P, et al. Lacunar stroke is associated with diffuse blood-brain barrier dysfunction. Ann Neurol. 2009;65:194–202.
40. Yoshimoto T, Shirasaka T, Yoshizumi T, Fujimoto S, Kaneko S, Kashiwaba T. Evaluation of carotid distal pressure for prevention of hyperperfusion after carotid endarterectomy. Surg Neurol. 2005;63:554–7. discussion 7–8.
41. Coutts SB, Hill MD, Hu WY. Hyperperfusion syndrome: toward a stricter definition. Neurosurgery. 2003;53:1053–8. discussion 8–60.
42. Dalman JE, Beenakkers IC, Moll FL, Leusink JA, Ackerstaff RG. Transcranial Doppler monitoring during carotid endarterectomy helps to identify patients at risk of postoperative hyperperfusion. Eur J Vasc Endovasc Surg. 1999;18:222–7.
43. Sacco RL, Adams R, Albers G, et al. Guidelines for prevention of stroke in patients with ischemic stroke or transient ischemic attack: a statement for healthcare professionals from the American Heart Association/American Stroke Association Council on Stroke: co-sponsored by the Council on Cardiovascular Radiology and Intervention: the American Academy of Neurology affirms the value of this guideline. Circulation. 2006;113:e409–49.
44. Tietjen CS, Hurn PD, Ulatowski JA, Kirsch JR. Treatment modalities for hypertensive patients with intracranial pathology: options and risks. Crit Care Med. 1996;24:311–22.
45. Muzzi DA, Black S, Losasso TJ, Cucchiara RF. Labetalol and esmolol in the control of hypertension after intracranial surgery. Anesth Analg. 1990;70:68-71.
46. Ahn SS, Marcus DR, Moore WS. Post-carotid endarterectomy hypertension: association with elevated cranial norepinephrine. J Vasc Surg. 1989;9:351–60.

Chapter 14
Collateral Circulation and Cerebral Reperfusion After Ischemic Stroke

Qinghai Huang, Wanling Wen, Myles McCrary, and Ling Wei

Abstract As treatment of acute ischemic stroke embraces its age of reperfusion therapy, it has become increasingly evident that cerebral collateral circulation in tissue injury development at early stage of brain ischemia is of great importance. This article presents some clinical, anatomic and physiological findings on cerebral collaterals and unsolved problems as well, aiming to provide an overview of cerebral collaterals as a promising diagnostic and therapeutic target of reperfusion injury for both clinicians and translational medicine researchers.

Keywords Ischemic stroke · Collateral circulation · Reperfusion

1   The Collateral Circulation and Cerebral Ischemia/ Reperfusion in Acute Ischemic Stroke (AIS)

The progression of brain tissue damage after artery occlusion is inversely correlated to the amount of residual blood flow. Cerebral collateral circulation refers to the subsidiary network of vascular channels that stabilize cerebral blood flow when a principal conduit fails [1]. The cerebral vessel network is

---

A comprehensive review from clinical evidence to bench side studies and therapeutic perspectives for acute ischemic stroke.

Q. Huang (✉)
Department of Neurosurgery, Second Military Medical University Changhai Hospital, Shanghai, China
e-mail: ocinhqh@163.com

W. Wen
Department of neurology, Hospital of People's Liberation Army, Beijing, China

M. McCrary · L. Wei (✉)
Department of Neurology, Emory University School of Medicine, Atlanta, GA, USA

Department of Anesthesiology, Emory University School of Medicine, Atlanta, GA, USA
e-mail: lwei7@emory.edu

© Springer International Publishing AG, part of Springer Nature 2018
W. Jiang et al. (eds.), *Cerebral Ischemic Reperfusion Injuries (CIRI)*, Springer Series in Translational Stroke Research, https://doi.org/10.1007/978-3-319-90194-7_14

complex and consists of plexuses of cortical arteries, arterioles, penetrating arterioles, capillaries, venules, and veins. The connections at the Circle of Willis and anastomoses between terminals of major artery trees (for example, the leptomeningeal and pial anastomoses) ensure the penetrating arterioles can be sourced rapidly by blood flow from other regions when occlusions take place. While the hemodynamics of collaterals and their relationship to tissue perfusion has not been fully explored, the compensatory effect is evident for their significant role in stroke prognosis, infarct progression, and tissue fate after reperfusion in acute ischemic stroke (AIS) ascribed to large artery occlusion [2]. The cerebral ischemic reperfusion injury (CIRI) is associated with extent and duration of primary ischemic insult [3–6].

1.1   *Collateral Circulation and Stroke Severity*

The amount of collateral circulation is related to functional outcomes in the natural history of anterior circulation large artery occlusion. In a prospective study of 126 untreated patients with anterior circulation large artery occlusion, Lima et al. [7] found that the pattern of leptomeningeal collaterals assessed by conventional CT angiograms (CTA) were independent predictors of good clinical outcome on 6 months follow up (odds ratio, 2.37 [95% CI, 1.08–5.20]). In another study, Seyman et al. [8] found that the compensatory ability of collaterals in AIS patients (<12 h) is inversely correlated with final cortical infarct volume and long term modified Rankin Scale (mRS) score. A post-hoc analysis of data from the Warfarin-Aspirin Symptomatic Intracranial Disease (WASID) study also revealed increased risk of ischemic stroke in patients with poorer collaterals (HR = 4.78, CI 1.55–14.7, p = 0.002). However, the absolute relationship between collateral grade, baseline severity, and infarction volume remains controversial. For instance, in one study involving AIS patients who underwent emergent endovascular treatment, there was no significant association between collateral grade and pretreatment infarct volume or National Institutes of Health Stroke Scale (NIHSS) score [9–11]. In another study which enrolled 60 patients with anterior circulation AIS (<12 h), there was no difference in the rate of good functional outcomes in patients with poor or good collaterals after reperfusion [12]. These conflicting studies suggest the significance of collateral grade on stroke severity may be time-dependent, or might rely on whether reperfusion is achieved.

1.2   *Collateral Circulation and Infarct Progression*

How does collateral grade affect infarction expansion? Cheripelli et al. [13] examined the collateral state and penumbra volume of 144 AIS patients within 6 h of stroke onset. The patients were grouped according to duration from stroke onset to brain imaging time. While little significant association between penumbra volume over time in patients with good collaterals was found, there was a trend towards reduction in penumbra proportion among those with poor collaterals [13]. This suggests that targeting the salvageable penumbra region may provide greater benefits for patients with poor collaterals. Collateral quality is strongly correlated with the rate of penumbra tissue loss during the early stages of stroke progression. Jung et al. [11] collected data from 44 AIS patients with M1 or M2 occlusion and calculated the rate of penumbra tissue loss using sequent MRI images. In patients who were successfully reperfused (without complication), there was a strong association between collateral quality and total penumbral tissue loss. Patients with poor collaterals (grade 0) lost 27% of their penumbral tissue, while patients with grade 1 lost only 11%, and patients with good collaterals actually saw an increase in their penumbra volume (−2% loss). Likewise, the rates of penumbral tissue loss were also associated with collateral quality. They concluded that rich collaterals can remarkably slow penumbra tissue loss and potentially extend the therapeutic window for reperfusion treatment [11]. This is supported by evidence linking residual blood flow and oxygen metabolism/extraction to the ability to salvage hypo-perfused brain tissue after infarction [14]. Using positron emission tomography studies in a swine model, Sakoh et al. showed that animals subjected to only mild hypo-perfusion showed no oxygen-metabolism deficiencies within 6 h, while moderately and severely ischemic tissues lost viability after only 3 and 1 h, respectively [14].

1.3 Collateral Circulation and Tissue Fate Following Reperfusion

Our understanding of the significance of cerebral collateral circulation has grown due to the increasing prevalence of reperfusion therapy following acute ischemic stroke. The benefits of reperfusion are magnified in patients with good collaterals. A meta-analysis of 2652 subjects who received endovascular reperfusion showed that patients with good collaterals had higher rates of favorable outcomes, reduced risks of peri-procedural symptomatic intracranial hemorrhage, and reduced 3 month mortality compared to patients with poor collaterals [15]. Another study showed that poor collateral circulation is a predictor of malignant infarction after reperfusion therapy [16]. In a study involving a cohort of 207 AIS patients with M1 occlusion that received endovascular treatment within 8 h, a shorter onset-to-reperfusion time was significantly associated with higher rate of favorable outcomes in those with poor collaterals, but the difference was not significant in the group with good collaterals [17]. Taken together, these studies suggest that patients with good collaterals appear to have a longer time window for reperfusion therapy, and those with poor collaterals were more likely to have dramatic reversion of functional outcome after prompt reperfusion [12]. Importantly, the beneficial effects of collateral blood flow are only prominent in patients who achieve reperfusion [18].

1.4   Collateral Blood Flow and Recovery

Collateralization and neovascularization are also very important to recovery after ischemic stroke. Post-stroke angiogenesis is a recognized phenomenon that occurs in effort to maintain long-term tissue remodeling and homeostasis [19]. In a rodent model of ischemic stroke, Wei et al. showed that establishment of collaterals after stroke was integral to functional recovery [20]. Arteriogenesis and angiogenesis in the penumbra region can facilitate later neurogenesis and likely plays a role in neuroplasticity [19, 21]. Methods and treatments to promote vascularization after stroke are being explored.

2   The Anatomic and Functional Variability of Collateral Circulation

2.1   Anatomic Features and Regulatory Mechanisms of Collaterals

The anatomic variation of the Circle of Willis and the number and diameter of the leptomeningeal anastomoses (LMAs) can vary substantially between individuals [1]. These differences are primarily determined by genetics, however, vessel remodeling and extent of collateralization can be influenced by shear stress, chronic ischemia, metabolic disorders, aging, and other factors [22–25]. Genetics may play a significant role in determining variability in collateral circulation. Variations in native LMAs exist between different mouse strains and are correlated to extent of ischemic injury induced by permanent focal ischemia [26, 27]. Recently, the Faber lab identified the gene at the locus determinant of collateral extent 1 (Dce1/Candq1, chromosome 7) is related to the number and diameter of collaterals in mice, and also to the severity of ischemic injury after experimental infarction [23, 28–30]. Kao et al. confirmed the role of the Dce1 locus in the evolution of brain infarction in mice using multimodal MRI [31]. They found that mice with collateral-rich genetic backgrounds had reduced perfusion-deficits volumes, smaller infarct volumes at 5 h after occlusion, and smaller lesion volumes at 24 h compared to genetically collateral-poor mice [31]. In a follow up study, they found that the variations in the gene *Rabep2* (Rab GTPase-effector binding protein 2) are responsible for these phenotypes. Rabep2 is involved in VEGF-A/VEGFR2 signaling and plays a significant role in collaterogenesis during embryonic development, and subsequently, is a key player in collateral extent in adult mice [29]. In a registry study of 206 M1 ± ICA occluded patients, metabolic syndrome (OR 3.22 95% CI 1.69–6.15, p < 0.001), hyperuricemia (per 1 mg/dl OR 1.35 95% CI 1.12–1.62, p < 0.01), and older age (per 10 years, OR 1.34 95% CI 1.02–1.77, p = 0.03) were independent predictors of poor leptomeningeal collateral status at baseline [24]. Another report showed that the completeness of the ipsilateral posterior Circle of Willis and glucose levels at admission are predictive of LA quality, which implies that both congenital and acquired factors are involved in the quality of LMAs [32]. Chan et al. explored the reactivity of LMAs in rats. They showed that LMAs from rats with spontaneous hypertension were vaso-constrictive and had impaired vasodilatory responses [33]. This provides evidence suggesting targeting LMAs may be a viable strategy to improve collateral circulation in AIS, at least in patients with chronic hypertension (Table 14.1).

Table 14.1 Useful mouse strains for studying collateral arteries

Mouse strain	Features	Representative images
Tie2-GFP	Tie2-GFP labels endothelial cells in nearly all stages of development [34, 35]	Tie2-GFP mouse embryo illustrating arteries throughout the body. Image modified from Motoike et al. [34]
α-SMA-GFP	α-SMA-GFP labels α-smooth muscle actin in vascular smooth muscle for visualization of arteries and arterioles	Increased diameter of ipsilateral collaterals was shown 21 days after stroke in mice
BALB/c	Greatly reduced number of collaterals, reduced collateral diameters, fewer anastomoses, lower perfusion after hindlimb ligation [26]. Causal gene identified to be Rabep2 [29]	Fluorescent arteriogram of the pial circulation in C57BL6 (left) versus BALB/c mice (right) demonstrates marked reduction in collaterals noted by red arrows [26]. Image modified from Chalothorn et al. [26]

2.2   The Dynamic Pattern of Collateral Circulation during AIS

The leptomeningeal collateral channels are believed to be recruited immediately after AIS due to the formation of a pressure gradient caused by arterial occlusion. However, collateral flow can be highly variable in the hours, days, and weeks after stroke. A number of studies have highlighted changes in collateral circulation patterns and their relevance to outcomes in acute stroke patients. In a study including patients who did not undergo reperfusion 3–5 days after AIS, Campbell et all showed that deterioration in collateral quality is associated with infarct growth [36]. Using laser speckle imaging, Wang et al. described three kinds of dynamic changes in leptomeningeal arteriole flow after middle cerebral artery occlusion: persistent LMAs, impermanent LMAs, and transient LMAs [37]. Persistent LMAs were patent observable over 3 h, while impermanent LOs diminished after 90–150 min, and transient LMAs that were no longer observable after 90 min. Importantly, these dynamic changes in collateral circulation were correlated to changes in regional blood flow. An early study in monkeys by Meyer et al. found that after MCA occlusion, collateral flow can fluctuate to meet the metabolic needs of the ischemic tissue; however the collateral flow is variable, and eventually stabilizes after ~2 weeks [38]. Some proposed mechanisms of collateral collapse include blood pressure reduction, dysfunction of auto-regulation, collateral vessel thrombosis, and venous steal. While collateral failure is known to be associated with infarct progression, it is unsure whether the effect is causative. These studies suggest that targeting collateral flow during the initial stages of AIS may be a viable therapeutic strategy.

3   Assessment of Cerebral Collaterals

Multiple imaging modalities are available for the evaluation of cerebral collateral circulation for acute ischemic stroke. The advantages and disadvantages of each modality are reviewed elsewhere [39–42]. Briefly, digital subtraction angiography (DSA) stands out for its high spatial and temporal resolution and is considered by many to be the gold standard for collateral assessment in acute ischemic stroke [39]. However, DSA has many major drawbacks. First, it is invasive. Second, not all sources of collateral flow may be visualized by a single injection, which can lead to collateral underestimation. Furthermore, exposure time may not be long enough to present the entire course of contrast in the region affected by ischemia. Finally, the total area perfused and the corresponding calculated hemodynamic parameters can be affected by the pressure and rate of contrast injection.

Computational tomography (CT) angiography is one of the most frequently applied imaging modalities for evaluating vessels in both clinical trials and in practice. The advantages of using CT angiography for acute ischemic stroke include wide availability, short acquisition time, and a linear correlation between contrast concentration and regional blood flow to estimate perfusion. Single-phase

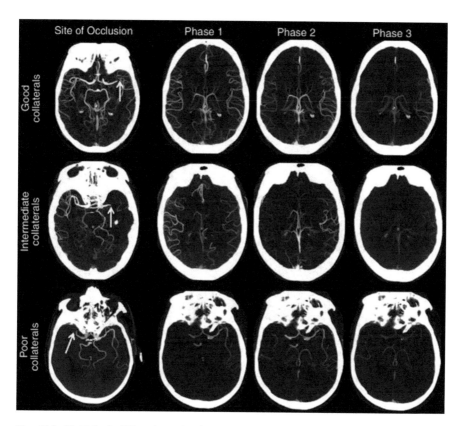

Fig. 14.1 Multiphasic CT angiography demonstrating the extent of collateral blood flow. Flow from good collaterals can be seen as early as non-occluded blood flow on the contralateral side. Intermediate collateral flow peaks later, but can be as abundant. Poor collaterals show weak vessel filling in all phases. (Source: Menon et al. Radiology, 2015 [43])

CT angiography has limited capabilities for the assessment of collaterals since the delayed flow from the leptomeningeal anastomoses is underestimated [43]. Multiphase CT angiography is more capable of providing the temporal resolution needed to assess collateral status (Fig. 14.1). Four dimensional (time-invariant) CT angiography provides even more resolution than multi-phase CT angiography and may be a useful alternative to digital subtraction angiography for clinical assessment of collateral flow (Fig. 14.2).

Traditional magnetic resonance imaging (MRI) techniques are of limited use for evaluating collateral flow due to their low spatial resolution and their reduced sensitivity to impeded blood flow. Fluid-attenuated inversion recovery (FLAIR) imaging may be used to visualize distal hyperintense vessels in acute ischemic stroke patients, however, their significance is not fully clarified. The presence of distal hyperintense vessels may be associated with smaller final infarct volumes and better clinical outcomes [45]. Other advanced imaging techniques such as contrast enhanced MR angiography and arterial spin labeling may also have a space in

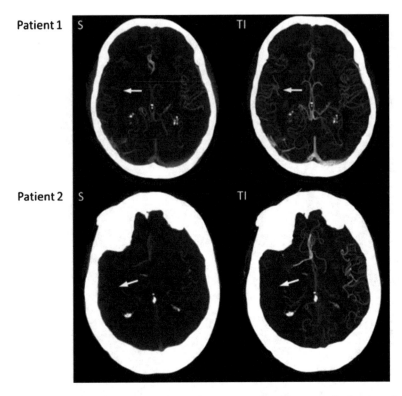

Fig. 14.2 Single phase (S) and time-invariant CT angiography with temporally fused maximum intensity projection (TI) images demonstrating the leptomeningeal collateral flow (white arrow) of two patients with right middle cerebral artery occlusion. (Source: d'Esterre et al. Stroke, 2017 [44])

evaluating collateral flow in the setting of acute ischemic stroke. The minutia of these methods are reviewed by Raymond et al. [42].

3.1   Using Collateral Parameters to Guide the Treatment of AIS Patients

The American Society of Interventional and Therapeutic Neuroradiology (ASITN/ ISR) collateral grading system has related collateral features to clinical outcomes, and can be used as a guide for clinical decision making for endovascular treatment in AIS patients [46]. Parameters of primary importance to grading collateral flow include the abundance of collaterals, their rate of filling, and total area they cover [47, 48]. One issue in standardizing the assessment of collateral status is that most scales are qualitative. This may create difficulties in comparing vessel assessments. Wen et al. [49] and Kawano et al. [48] have used time-intensity curve data from digital subtraction angiography and CT angiography to empirically quantify collateral status. Specifically, they utilized peak contrast density, filling velocity, and contrast time delay to predict clinical outcomes. Future quantitative studies on leptomeningeal hemodynamics are expected. Collateral status has become a potential parameter for patient selection of reperfusion therapy [50, 51]. Ribo et al. showed that good pial collateral circulation predicts a better clinical response to intra-arterial treatment beyond a 5-h treatment window [50]. A survey of clinical trials by Ginsberg et al. suggested that collateral circulation is a critical determinant of both stroke severity and of clinical improvement, and that the expanded neuroprotection conferred by good collaterals should be considered in guiding reperfusion therapy [51].

3.2   The Relation of Vein Hemodynamics to Collateral Status in AIS

Collateral status after arterial collusion can also be reflected in the pattern of delayed cortical or deep vein contrast filling. The asymmetric appearance of veins on dynamic CT angiography can predict poor collateral flow and negative clinical outcomes in AIS patients [52, 53]. Delayed late-venous phase cortical vein filling was associated with collateral quality by assessment of dynamic CT angiography [54]. Despite these findings, the importance of vein hemodynamics and their relevance to collateral circulation in AIS patients is not fully understood.

4   Collaterals as Therapeutic Targets—Evidence and Concerns

The viability of cerebral collateral flow after acute proximal artery occlusion illustrates the possibility of enhancing collateral enhancement as a therapeutic strategy to preserve tissue before reperfusion. Such methods would effectively convert what would be ischemic core tissue into salvageable penumbral tissue. Are there safe and effective ways to augment collateral flow in such a short window? How fast can leptomeningeal anastamoses remodel to accommodate sufficient collateral flow? Is it possible to develop and enhance collateral channels in individuals at high risk for ischemic stroke?

4.1   Blood Pressure (BP)

Blood pressure is an important regulator of collateral flow. Early work by Meyer et al. found that the major cortical collateral circulation requires a minimum of 80–85 mmHg in an experimental MCAO model [55]. Below this pressure, the major collateral branches from the anterior cerebral artery fail to dilate after MCAO. It is generally accepted that very low blood pressure is harmful in acute ischemic stroke, and that low BP may even warrant careful inotropic support [56]. However, the possibility of drug-induced hypertension for alleviating ischemic injury via enhancing collateral circulation is still under debate. This is due to the association of hypertension with infarct progression and hemorrhagic transformation, although this is mainly in patients with chronic hypertension [57, 58]. Some animal studies show potential benefits of inducing hypertension and subsequently improving collateral flow. For example, Shin et al. found that hypertensive therapy can drastically reduce infarct volume (48%) if given during the duration of ischemia [59]. They also showed that hypertensive therapy led to increased blood flow, oxyhemoglobin concentrations, and cerebral metabolic rate in both the core and penumbra regions. Future research is needed to find strategies to balance the potential benefits of hypertension-enhanced collateral flow in acute ischemic stroke with the serious risks.

4.2   Head Position

A meta-analysis of upright (head position at 30°) versus flat positioning for ischemic stroke patients revealed that blood flow velocity can be increased in side of the brain affected by ischemic stroke in patients lying flat [60]. However, the clinical relevance of this finding is unknown. A cluster randomized multicenter clinical trial has been initiated to investigate this question [61].

4.3   Partial Aortic Occlusion

Partial occlusion of the descending aorta can increase flow through the common carotids and improve brain perfusion [62]. Interestingly, this treatment can lead to a persistent increase in cerebral blood flow. This hemodynamic phenomenon is not fully understood. Some possible explanations include shunting of blood from the splanchnic reserve, and increased pLMAsma epinephrine/norepinephrine [62–64]. A pilot study has been implemented to test the feasibility of partial aortic occlusion in the clinic [65].

4.4   High-Dose Albumin

High-dose albumin treatment may be a useful method to increase collateral circulation after stroke. In rodent models, intravenous administration of 25% albumin after focal transient ischemia was found to increase regional blood flow, reverse stagnation and thrombosis, and decrease corpuscular adherence within cortical venules during the reperfusion phase [66]. DeFazio et al. further revealed that high-dose albumin can double collateral perfusion in mice with sparse collateralization [67]. Contrarily, there are no benefits associated with both isovolemic or hypervolemic hemodilution therapy [68]. Unfortunately, a pilot clinical trial testing high-dose albumin treatment for AIS was stopped prematurely due to lack of efficacy [69]. For this phase 3 clinical trial, 25% albumin was administered within 5 h after stroke for patients with a baseline NIHSS score of 6 or greater. There appeared to be an enrichment in adverse effects such as pulmonary edema and intracranial hemorrhage in the albumin treatment arm. Future studies are needed to clarify the potential role of high-dose albumin therapy for acute ischemic stroke.

4.5 Sphenopalatine Ganglion Stimulation

The nervous system plays a key role in regulating cerebral blood flow. Parasympathetic innervation arises from the superior salivatory nucleus via the sphenopalatine and otic ganglia to provide vasodilatory responses to the anterior cerebral circulation. In animal models, stimulation of the sphenopalatine ganglion can result in an immediate vasodilatory response in the ipsilateral cerebral vessels to increase cortical blood flow by ~40% [70, 71]. Parasympathetic stimulation and the subsequent vasodilation may be useful for the treatment of stroke. To test this possibility, a clinical trial has been initiated, but the results have not yet been reported [72].

4.6 Vasoactive Agents

Myogenic tone at the leptomeningeal anastomoses (LMAs) and penetrating arterioles significantly contributes to vessel resistance and blood flow to the cortex. The LMAs are presumed to be the bottleneck for collateral cerebral blood flow, however, most studies have focused on the vasoactive responses of the parenchymal and penetrating arterioles near or at the area of ischemia. Previous work by Meyer et al. showed that collateral vessels were responsive to intrinsic vasoactive agents such as oxygen, carbon dioxide, and acidic metabolites in the setting of focal cerebral ischemia. Little is known about the mechanisms behind LA regulation and their responsiveness to vasoactive agents. Chan et al. investigated the vasoactive properties of leptomeningeal arteries ex vivo. The LMAs in hypertensive rodents were less responsive to pressure changes and vasodilatory stimuli after focal ischemic stroke than normal, which leads to a greater perfusion deficit [33]. Additionally, Chan's work suggested that the perivascular innervation may play a role in dictating the LMA dilatory response, and may be a potential therapeutic target.

Nitric oxide (NO) is also a promising candidate for the treatment of acute ischemic stroke. Numerous preclinical experiments in both permanent and transient ischemic models have illustrated that treatment with NO donors can improve cerebral

Table 14.2 Published clinical trials on collateral enhancement therapy for the treatment of acute ischemic stroke

Trial title	Intervention	Year published	Inclusion criteria	Effects on collateral flow	Primary result
ENOS [75]	Glyceryl Trinitrate	2016	Randomized ischemic or hemorrhagic stroke patients within 48 h after ictus	Perfusion not affected	Safe but not effective in improving functional outcomes
ALIAS [69]	25% albumin intravenous administration	2013	NIHSS > 6; able to treat within 5 h; i.v. tPA or other intra-arterial treatments were permitted	Not evaluated	No difference between groups; increased risk of intracranial hemorrhage and pulmonary edema
SENTIS [65]	Partial aortic occlusion	2011	Symptomatic AIS patients; device could be implanted within 14 h; patients who received i.v. tPA or other intra-arterial treatment were excluded	Not evaluated	Safe but not effective in improving functional outcomes
HeadPoST [61, 76]	Head position	2017	Clinical diagnosis of acute stroke, including intracerebral hemorrhage; excluded if head position could not be maintained	Not evaluated	Safe but not effective in improving functional outcomes

perfusion [73]. Terpolilli et al. showed that inhaled NO can selectively dilate both arterioles and veins in the ischemic brain without significantly affecting the normal tissue to promote penumbral perfusion and reduce infarct volume in sheep [74].

As with other neuroprotective strategies that are struggling to move from bench to bedside, it is paramount to address the issues that prevent collateral enhancement therapy for AIS from translating to the clinic. One such effort was performed by Beretta et al. [77]. In a preclinical rodent model, they investigated four different 'collateral therapeutics', including phenylephrine to induce hypertension, polygeline to increase intravascular volume load, acetazolamide to promote cerebral arteriolar vasodilation, and head down tilt (HDT) to encourage cerebral blood flow. They attributed the highest efficacy and safety profile to the HDT group. Despite these positive findings in rodent models, a recent clinical trial found no significant difference between treatment groups for AIS patients (Table 14.2) [76]. Despite these results, there is still ample room for the discovery and translation of therapies which enhance collateral circulation. As a whole, preclinical studies suggest that augmenting collateral blood flow after acute ischemic stroke remains a promising therapeutic target to preserve the penumbra and alleviate ischemic reperfusion injury, particularly in the early stages of AIS management. More effort is needed clarify how these findings can be applied successfully to stroke patients.

References

1. Liebeskind DS. Collateral circulation. Stroke. 2003;34(9):2279–84.
2. Shuaib A, et al. Collateral blood vessels in acute ischaemic stroke: a potential therapeutic target. Lancet Neurol. 2011;10(10):909–21.
3. Neumann-Haefelin T, et al. Serial MRI after transient focal cerebral ischemia in rats: dynamics of tissue injury, blood-brain barrier damage, and edema formation. Stroke. 2000;31(8):1965–72. discussion 1972–3
4. Jones TH, et al. Thresholds of focal cerebral ischemia in awake monkeys. J Neurosurg. 1981;54(6):773–82.
5. Marchal G, Young AR, Baron JC. Early postischemic hyperperfusion: pathophysiologic insights from positron emission tomography. J Cereb Blood Flow Metab. 1999;19(5):467–82.
6. Heiss WD, et al. Repeat positron emission tomographic studies in transient middle cerebral artery occlusion in cats: residual perfusion and efficacy of postischemic reperfusion. J Cereb Blood Flow Metab. 1997;17(4):388–400.
7. Lima FO, et al. Prognosis of untreated strokes due to anterior circulation proximal intracranial arterial occlusions detected by use of computed tomography angiography. JAMA Neurol. 2014;71(2):151–7.
8. Seyman E, et al. The collateral circulation determines cortical infarct volume in anterior circulation ischemic stroke. BMC Neurol. 2016;16(1):206.
9. Al-Ali F, et al. Capillary index score in the interventional management of stroke trials I and II. Stroke. 2014;45(7):1999–2003.
10. Liebeskind DS, et al. Collaterals at angiography and outcomes in the Interventional Management of Stroke (IMS) III trial. Stroke. 2014;45(3):759–64.
11. Jung S, et al. Factors that determine penumbral tissue loss in acute ischaemic stroke. Brain. 2013;136:3554–60.
12. Marks MP, et al. Effect of collateral blood flow on patients undergoing endovascular therapy for acute ischemic stroke. Stroke. 2014;45(4):1035–9.
13. Cheripelli BK, et al. What is the relationship among penumbra volume, collaterals, and time since onset in the first 6 h after acute ischemic stroke? Int J Stroke. 2016;11(3):338–46.
14. Sakoh M, et al. Relationship between residual cerebral blood flow and oxygen metabolism as predictive of ischemic tissue viability: sequential multitracer positron emission tomography scanning of middle cerebral artery occlusion during the critical first 6 hours after stroke in pigs. J Neurosurg. 2000;93(4):647–57.
15. Leng X, et al. Impact of collaterals on the efficacy and safety of endovascular treatment in acute ischaemic stroke: a systematic review and meta-analysis. J Neurol Neurosurg Psychiatry. 2016;87(5):537–44.
16. Flores, A., et al., Poor collateral circulation assessed by multiphase computed tomographic angiography predicts malignant middle cerebral artery evolution after reperfusion therapies. Stroke, 2015. 46(11):3149-3153
17. Hwang YH, et al. Impact of time-to-reperfusion on outcome in patients with poor collaterals. AJNR Am J Neuroradiol. 2015;36(3):495–500. https://doi.org/10.3174/ajnr.A4151. Epub 2014 Nov 6
18. Bouts MJ, et al. Lesion development and reperfusion benefit in relation to vascular occlusion patterns after embolic stroke in rats. J Cereb Blood Flow Metab. 2014;34(2):332–8.

19. Liu J, et al. Vascular remodeling after ischemic stroke: mechanisms and therapeutic potentials. Prog Neurobiol. 2014;115:138–56.
20. Wei L, et al. Collateral growth and angiogenesis around cortical stroke. Stroke. 2001;32(9):2179–84.
21. Arai K, et al. Brain angiogenesis in developmental and pathological processes: neurovascular injury and angiogenic recovery after stroke. FEBS J. 2009;276(17):4644–52.
22. van Raamt AF, et al. The fetal variant of the circle of Willis and its influence on the cerebral collateral circulation. Cerebrovasc Dis. 2006;22(4):217–24.
23. Zhang H, et al. Wide genetic variation in the native pial collateral circulation is a major determinant of variation in severity of stroke. J Cereb Blood Flow Metab. 2010;30(5):923–34.
24. Menon BK, et al. Leptomeningeal collaterals are associated with modifiable metabolic risk factors. Ann Neurol. 2013;74(2):241–8.
25. Faber JE, et al. Aging causes collateral rarefaction and increased severity of ischemic injury in multiple tissues. Arterioscler Thromb Vasc Biol. 2011;31(8):1748–56.
26. Chalothorn D, et al. Collateral density, remodeling, and VEGF-A expression differ widely between mouse strains. Physiol Genomics. 2007;30(2):179–91.
27. Keum S, Marchuk DA. A locus mapping to mouse chromosome 7 determines infarct volume in a mouse model of ischemic stroke. Circ Cardiovasc Genet. 2009;2(6):591–8.
28. Sealock R, et al. Congenic fine-mapping identifies a major causal locus for variation in the native collateral circulation and ischemic injury in brain and lower extremity. Circ Res. 2014;114(4):660–71.
29. Lucitti JL, et al. Variants of Rab GTPase-effector binding protein-2 cause variation in the collateral circulation and severity of stroke. Stroke. 2016;47(12):3022–31.
30. Wang S, et al. Genetic architecture underlying variation in extent and remodeling of the collateral circulation. Circ Res. 2010;107(4):558–68.
31. Kao YJ, et al. Role of genetic variation in collateral circulation in the evolution of acute stroke: a multimodal magnetic resonance imaging study. Stroke. 2017;48(3):754–61.
32. van Seeters T, et al. Determinants of leptomeningeal collateral flow in stroke patients with a middle cerebral artery occlusion. Neuroradiology. 2016;58(10):969–77.
33. Chan SL, et al. Pial collateral reactivity during hypertension and aging: understanding the function of collaterals for stroke therapy. Stroke. 2016;47(6):1618–25. https://doi.org/10.1161/STROKEAHA.116.013392. Epub 2016 Apr 21
34. Motoike T, et al. Universal GFP reporter for the study of vascular development. Genesis. 2000;28(2):75–81.
35. Coles JA, et al. Where are we? The anatomy of the murine cortical meninges revisited for intravital imaging, immunology, and clearance of waste from the brain. Prog Neurobiol. 2017;156:107–48.
36. Campbell BC, et al. Failure of collateral blood flow is associated with infarct growth in ischemic stroke. J Cereb Blood Flow Metab. 2013;33(8):1168–72.
37. Wang Z, et al. Dynamic change of collateral flow varying with distribution of regional blood flow in acute ischemic rat cortex. J Biomed Opt. 2012;17(12):125001.
38. Meyer JS. Circulatory changes following occlusion of the middle cerebral artery and their relation to function. J Neurosurg. 1958;15(6):653–73.
39. Bang OY, Goyal M, Liebeskind DS. Collateral circulation in ischemic stroke: assessment tools and therapeutic strategies. Stroke. 2015;46(11):3302–9.
40. McVerry F, Liebeskind DS, Muir KW. Systematic review of methods for assessing leptomeningeal collateral flow. AJNR Am J Neuroradiol. 2012;33(3):576–82.
41. Martinon E, et al. Collateral circulation in acute stroke: assessing methods and impact: a literature review. J Neuroradiol. 2014;41(2):97–107.
42. Raymond SB, Schaefer PW. Imaging brain collaterals: quantification, scoring, and potential significance. Top Magn Reson Imaging. 2017;26(2):67–75.
43. Menon BK, et al. Multiphase CT angiography: a new tool for the imaging triage of patients with acute ischemic stroke. Radiology. 2015;275(2):510–20.

44. d'Esterre CD, et al. Regional comparison of multiphase computed tomographic angiography and computed tomographic perfusion for prediction of tissue fate in ischemic stroke. Stroke. 2017;48(4):939–45.
45. Lee KY, et al. Distal hyperintense vessels on FLAIR: an MRI marker for collateral circulation in acute stroke? Neurology. 2009;72(13):1134–9.
46. Zaidat OO, et al. Recommendations on angiographic revascularization grading standards for acute ischemic stroke: a consensus statement. Stroke. 2013;44(9):2650–63.
47. Beyer SE, et al. Predictive value of the velocity of collateral filling in patients with acute ischemic stroke. J Cereb Blood Flow Metab. 2015;35(2):206–12.
48. Kawano H, et al. Relationship between collateral status, contrast transit, and contrast density in acute ischemic stroke. Stroke. 2016;47(3):742–9.
49. Wen WL, et al. Parametric digital subtraction angiography imaging for the objective grading of collateral flow in acute middle cerebral artery occlusion. World Neurosurg. 2016;88:119–25.
50. Ribo M, et al. Extending the time window for endovascular procedures according to collateral pial circulation. Stroke. 2011;42(12):3465–9.
51. Ginsberg MD. Expanding the concept of neuroprotection for acute ischemic stroke: the pivotal roles of reperfusion and the collateral circulation. Prog Neurobiol. 2016;145–146:46–77.
52. Parthasarathy R, et al. Prognostic evaluation based on cortical vein score difference in stroke. Stroke. 2013;44(10):2748–54.
53. van den Wijngaard IR, et al. Cortical venous filling on dynamic computed tomographic angiography: a novel predictor of clinical outcome in patients with acute middle cerebral artery stroke. Stroke. 2016;47(3):762–7.
54. Bhaskar S, et al. Delay of late-venous phase cortical vein filling in acute ischemic stroke patients: associations with collateral status. J Cereb Blood Flow Metab. 2017;37(2):671–82.
55. Meyer JS, Denny-Brown D. The cerebral collateral circulation. I. Factors influencing collateral blood flow. Neurology. 1957;7(7):447–58.
56. Mistri AK, Robinson TG, Potter JF. Pressor therapy in acute ischemic stroke: systematic review. Stroke. 2006;37(6):1565–71.
57. McCrary MR, Wang S, Wei L. Ischemic stroke mechanisms, prevention, and treatment: the anesthesiologist's perspective. J Anesth Perioper Med. 2017;4:76–86.
58. Mashour GA, et al. Perioperative care of patients at high risk for stroke during or after non-cardiac, non-neurologic surgery: consensus statement from the Society for Neuroscience in anesthesiology and critical care. J Neurosurg Anesthesiol. 2014;26(4):273–85.
59. Shin HK, et al. Mild induced hypertension improves blood flow and oxygen metabolism in transient focal cerebral ischemia. Stroke. 2008;39(5):1548–55.
60. Olavarria VV, et al. Head position and cerebral blood flow velocity in acute ischemic stroke: a systematic review and meta-analysis. Cerebrovasc Dis. 2014;37(6):401–8.
61. Munoz-Venturelli P, et al. Head Position in Stroke Trial (HeadPoST)—sitting-up vs lying-flat positioning of patients with acute stroke: study protocol for a cluster randomised controlled trial. Trials. 2015;16:256.
62. Liebeskind DS. Aortic occlusion for cerebral ischemia: from theory to practice. Curr Cardiol Rep. 2008;10(1):31–6.
63. Stokland O, et al. Cardiac effects of splanchnic and non-splanchnic blood volume redistribution during aortic occlusions in dogs. Acta Physiol Scand. 1981;113(2):139–46.
64. Stokland O, et al. Mechanism of hemodynamic responses to occlusion of the descending thoracic aorta. Am J Phys. 1980;238(4):H423–9.
65. Shuaib A, et al. Partial aortic occlusion for cerebral perfusion augmentation: safety and efficacy of NeuroFlo in Acute Ischemic Stroke trial. Stroke. 2011;42(6):1680–90.
66. Belayev L, et al. Albumin therapy of transient focal cerebral ischemia: in vivo analysis of dynamic microvascular responses. Stroke. 2002;33(4):1077–84.
67. DeFazio RA, et al. Albumin therapy enhances collateral perfusion after laser-induced middle cerebral artery branch occlusion: a laser speckle contrast flow study. J Cereb Blood Flow Metab. 2012;32(11):2012–22.

68. Chang TS, Jensen MB. Haemodilution for acute ischaemic stroke. Cochrane Database Syst Rev. 2014;(8):Cd000103.
69. Ginsberg MD, et al. High-dose albumin treatment for acute ischaemic stroke (ALIAS): Part 2. A randomised, double-blind, phase 3, placebo-controlled trial. Lancet Neurol. 2013;12(11):1049–58.
70. Seylaz J, et al. Effect of stimulation of the sphenopalatine ganglion on cortical blood flow in the rat. J Cereb Blood Flow Metab. 1988;8(6):875–8.
71. Levi H, et al. Stimulation of the sphenopalatine ganglion induces reperfusion and blood-brain barrier protection in the photothrombotic stroke model. PLoS One. 2012;7(6):e39636.
72. Khurana D, Kaul S, Bornstein NM. Implant for augmentation of cerebral blood flow trial 1: a pilot study evaluating the safety and effectiveness of the ischaemic stroke system for treatment of acute ischaemic stroke. Int J Stroke. 2009;4(6):480–5.
73. Willmot M, et al. A systematic review of nitric oxide donors and L-arginine in experimental stroke; effects on infarct size and cerebral blood flow. Nitric Oxide. 2005;12(3):141–9.
74. Terpolilli NA, et al. Inhalation of nitric oxide prevents ischemic brain damage in experimental stroke by selective dilatation of collateral arterioles. Circ Res. 2012;110(5):727–38.
75. Krishnan K, et al. Glyceryl trinitrate for acute intracerebral hemorrhage: results from the efficacy of nitric oxide in stroke (ENOS) trial, a subgroup analysis. Stroke. 2016;47(1):44–52.
76. Anderson CS, et al. Cluster-randomized, crossover trial of head positioning in acute stroke. N Engl J Med. 2017;376(25):2437–47.
77. Beretta S, et al. Cerebral collateral therapeutics in acute ischemic stroke: A randomized pre-clinical trial of four modulation strategies. J Cereb Blood Flow Metab. 2017. https://doi.org/10.1177/0271678X16688705.

Chapter 15
Controlled Reperfusion Against Ischemia Reperfusion Injury

Weijian Jiang, Jin Lv, Ying-Ying Zhang, and Kai Wang

Abstract Recanalization and reperfusion after an ischemic stroke sometimes could be harmful and deadly, and controlled perfusion may be a potential solution. This article summarized controlled perfusion literature, both animal and clinical, on heart, lung, kidney, and limbs and compared with brain controlled perfusion. Apparently more studies are needed to investigate if controlled perfusion may provide beneficial effects to stroke patients.

Keywords Recanalization · Stroke · Perfusion · Peripheral organs

1 Introduction

Timely blood flow restoration into ischemic tissue is the primary treatment of acute arterial occlusion diseases. However, the restoration sometimes may lead to serious ischemia reperfusion injury (IRI), an intractable problem or a dilemma characterized by a cascade of deleterious inflammatory responses and cell death in variety of conditions such as myocardial infarction, ischemic stroke and solid organ transplantation. With the understanding of IRI, controlled reperfusion concept was recently proposed as its protection against the IRI was demonstrated in the most of experimental studies. The controlled reperfusion includes perfusate composition-controlled one (i.e. ionic content, nutrients and acid-base balance) and physics condition-controlled one (i.e. pressure, flow and temperature). Both may be very promising reperfusion strategies because of easy implementation in clinical recanalization practices once their clinical net benefits were confirmed. The condition-controlled reperfusion was also termed by other names in several earlier studies, for example, gentle, gradual, ramped, progressive and staged reperfusion [1, 2]. It is worth mentioning that ischemic conditioning, including pre-, peri- and post-conditioning, is a well-known

W. Jiang (✉) · J. Lv · Y.-Y. Zhang · K. Wang
Vascular Neurosurgery Department, New Era Stroke Care and Research Institute,
The PLA Rocket Force General Hospital, Beijing, China
e-mail: cjr.jiangweijian@vip.163.com

© Springer International Publishing AG, part of Springer Nature 2018   231
W. Jiang et al. (eds.), *Cerebral Ischemic Reperfusion Injuries (CIRI)*, Springer Series
in Translational Stroke Research, https://doi.org/10.1007/978-3-319-90194-7_15

protection strategy for acute ischemia. However, there is a bit of difference in the status of target tissue aimed at by ischemic conditioning and controlled reperfusion. The former mainly aims at ischemia tissue in the territory of occluded artery to alleviate ischemia insult, and the latter at reperfusion tissue in the territory of recanalized artery to mitigate IRI [3, 4]. In this chapter, we will unify the terminology of controlled reperfusion and review its application in heart, lung, limb, kidney and brain.

2 Controlled Reperfusion of Heart/Myocardium

Timely blood flow restoration of ischemic myocardium is the primary treatment of acute myocardial infarction (AMI). However, a rapid restoration with uncontrolled flow rates and pressures sometimes may exacerbate myocardium damage, known as myocardial IRI [5].

In 1981, Follette et al. [6] firstly reported that modified reperfusate with enriched ionic content and hyperosmolarity significantly reduced the IRI after 1 h of ischemic arrest with canine left ventricular hypothermia. Furthermore, with the use of dog model undergoing left anterior descending (LAD) coronary occlusion, Allen and colleagues [7] demonstrated that substrate-enriched blood cardioplegic reperfusion alleviated reversible damage and recover cardiac contractile function even after up to 6 h of ischemia. Similar protective effects of composition-controlled reperfusion were also observed in porcine models of cardiopulmonary bypass [8] and LAD coronary occlusion [9].

Leukapheresis of normal blood is also a maneuver of composition-controlled reperfusion. Leukocyte depletion significantly inhibited coronary vascular resistance increase induced by reperfusion, and subsequently reduced myocardium damages and arrhythmias after 2 h of canine LAD coronary occlusion [10]. Perfusion of *ex vivo* hearts of pig with leukocyte-filtered blood also exhibited an effective recovery of cardiac function, as evidenced by reduction of creatine phosphokinase leakage and histopathologic injury score, as compared with hearts without ischemia or reperfused by normal blood [11]. Moreover, Okazaki et al. [12] demonstrated that reperfusion with leukocyte-depleted blood in rabbit model mitigated coronary endothelial damages and enhanced recovery of contractile function after 4 h of normothermic ischemia. In contrast, normal blood reperfusion caused significant endothelium damages and cardiac dysfunction.

Given the critical roles of reactive oxygen species (ROS) in reperfusion injury, concept of hypoxemic reperfusion, characterized by reperfusion with lower oxygen content or gradual increase of oxygen content, yielded favorable results by limiting ROS generation and oxidative damages. Massoudy et al. [13] observed that low PO_2 reperfusion of 300 mmHg for the first 5 min after 15 min global ischemia of isolated guinea pig heart caused a significant improvement of cardiac function. A similar study in Wistar rats also demonstrated that reperfusion with more physiological PO_2 (500 mmHg) resulted in significantly improved recovery of cardiac function, compared with high PO_2 (700 mmHg) [14]. Furthermore, Petrosillo and colleagues [15] showed that protective effect of hypoxic reperfusion was closely related to protection of mitochondrial function and subsequent reduction of ROS generation.

Pressure and flow-controlled reperfusion were initially studied in canine model of acute coronary occlusion. Reperfusion with constant low pressure/flow and gradual increase of pressure/flow was generally adopted. In 1986, Yamazaki et al. [16] showed that reperfusion in the initial stage at a constant rate of 20 mL/min with normal blood reduced arrhythmias and accelerated recovery of cardiac function. Almost at the same time, Okamoto and colleagues [17] observed the same protective effect in the similar experimental model at a constant reperfusion rate of 30 mL/min. The protective effect was also confirmed in studies of pig and rat models, showing that the constant low-flow reperfusion lessened end-diastolic wall thickness [18] and pressure [19], reduced calcium overload [20] and ultra-structural ischemic damages [21], and improved myocardial function and energetic recovery in post-ischemic myocardium [22]. Specifically, Ferrera [20] and Takeo et al. [19] further demonstrated that the cardioprotection of low-flow reperfusion might be contributed to reduction of calcium overload in myocardium, and there was a critical threshold of blood flow in improving both ventricular function and bioenergetics [22].

Gradual flow-controlled reperfusion had also been extensively investigated in various experimental animal models. An exponential augment of flow rate significantly reduced myocardial infarct size and endothelial dysfunction in ischemic dog and pig hearts after 60–90 min ischemia of LAD artery [23, 24]. Unexpectedly, the cardioprotective effect of gradual flow reperfusion in porcine models was found to be unrelated to reperfusion injury salvage kinase (RISK) pathway activation, which was generally considered to be critical for ischemic conditioning strategy [24]. Moreover, gradual reperfusion with stepwise increase of flow rate from 2 to 10 mL/min/g was also demonstrated to be cardioprotective in porcine models subjected to normothermic 25 min ischemia [25]. However, no beneficial effect of the gradual flow reperfusion was found on reducing myocardial infarct size in canine models of circumflex coronary artery occlusion [26].

Low-pressure reperfusion had a protective effect superior to constant flow reperfusion in canine models of LAD artery occlusion [27]. Bopassa and colleagues [28, 29] designed a series of exquisite experiments in isolated rats hearts exposed to 30 min of ischemia to determine an optimal pressure and duration of low-pressure reperfusion and the underlying mechanisms. Low-pressure reperfusion of 70 cmH$_2$O was found to provide a significant cardioprotection immediately after normothermic [30] and hypothermic [31] heart ischemia, by activating phosphatidylinositol 3-kinase (PI3K) signaling pathway to inhibit mitochondrial permeability transition pore (mPTP) opening [32]. Recently, they further demonstrated that there still had been the cardioprotective effect even after a delay of 10–20 min reperfusion [33].

Reperfusion with gradual pressure increase was also a strategy of pressure-controlled one. Ueno et al. [34] observed a significantly better systolic function and less interstitial edema after this kind of reperfusion in rabbit model. In hypothermic cardioplegic ischemia of rat hearts, a combination of the reperfusion and slowly rising temperature was also shown to have roles in improving energy preservation and reducing mitochondrial injury and myocyte edema [35, 36].

Although the cardioprotection of controlled reperfusion against IRI was well documented in animal models, the road from bench to bedside remains long to finish because of limited convincing clinical trials until now. The only randomized trial among 60 patients receiving pressure-controlled reperfusion with 50 and 75 mmHg demonstrated no believable benefits of the low-pressure reperfusion after coronary artery bypass grafting [37]. Further clinical studies are needed.

3   Controlled Reperfusion of Lung

Pulmonary IRI after lung transplantation is the main cause of graft dysfunction and early mortality after transplantation [38, 39]. In 1996, Hopkinson and Bhabra [40, 41] reported firstly application of two pressure-controlled reperfusion strategies in pulmonary injuries in an *ex vivo* rat lung experimental model. One was gradual increase of pressure every 15 min to physiologic level within 60 min [40]. Another one was constant pressure of 50% normal level for 5 or 10 min [41]. Both strategies significantly reduced pulmonary IRI and improved function of lung grafts, compared with normal pressure reperfusion. Moreover, their study showed that the pressure-controlled reperfusion reduced endothelial permeability, prevented pulmonary edema formation, and provided protection in early phase of lung graft [42]. Using a similar rat lung transplant model, Pierre et al. [43] studied flow-controlled reperfusion form initial flow rate of 0.4 mL/min up to 4 mL/min with gradual increment of 0.4 mL/min per minute. Their results showed an excellent gas exchange and improved oxygenation of ischemic lung with light and electron microscopy. The subsequent experiments in rabbit model of ischemic lung graft also demonstrated the favorable effect of pressure-controlled reperfusion as evidenced by prominently attenuated pulmonary injury in 4- to 24-h cold preserved lung graft [44, 45]. Moreover, a 30-min pressure-controlled reperfusion was determined as optimized duration length for 2-h warm ischemic lung grafts [46]. Specifically, a combination of gradual increase of flow rate of ventilation and reperfusion further provided the more favorable protection of lung function and against pulmonary injury [47].

The favorable effect had also been tested in porcine lung experimental model, where pressure-controlled reperfusion was found to result in reduction of IRI and improvement of pulmonary function as evidenced by improved pulmonary compliance, decreased pulmonary vascular resistance and reduced oxidative stress generation [48, 49].

Beneficial effect of composition-controlled reperfusion, such as modified reperfusate solution and leukocyte-depleting filter, was also reported in porcine lung model, as evidenced by reduction or complete avoidance of lung IRI [49–51]. Moreover, a further improvement of benefit was achieved by a combination of pressure-controlled reperfusion and additional pharmacological agents (such as pentoxifylline) or modified reperfusate solution [48, 49].

These promising experimental results accelerated clinical transformation in human lung transplantation. In 2000, Lick et al. [52] firstly reported that the controlled reperfusion achieved excellent functional results in all of five patients undergoing lung transplantation. Alvarado et al. [53] also provided striking data that the controlled reperfusion favored long-term outcomes post transplantation.

In recent years, the controlled reperfusion strategy had also been applied in cardiopulmonary bypass (CPB) surgery to alleviate pulmonary IRI. In Rong et al. study [54, 55], ischemic lung was protected by oxygen-controlled reperfusion against IRI in early stages through down-regulation of high mobility group box 1 (HMGB1) and receptor for advanced glycation end products (RAGE) expressions. Slottoscha et al. [56] investigated the protective effects of combination of composition-controlled and pressure-controlled reperfusion in porcine CPB model. The combination improved pulmonary mechanics and reduced markers of oxidative stress, but failed to avoid completely CPB-related lung injury.

4   Controlled Reperfusion of Limbs

Limb is another organ susceptible to severe IRI. Like ischemic injury in other organs, intracellular energy dysmetabolism is the main pathophysiological process of tissue ischemia, though permeability and filtration of capillary remain homeostasis for several hours after acute ischemia of limbs. Once reperfusion occurs suddenly, a series of catastrophic injuries will be triggered [57].

As early as the 1980s, many researchers considered that ischemic injury of limbs was not reversible if extensive and irreversible damages occurred. However, Beyersdorf and his colleagues [58] from John Wolfgang Goethe University observed some evidences that necrosis of skeletal muscle did not occur a few hours after warm ischemia. Until 1991, they demonstrated that condition and composition controlled reperfusion reduced IRI of limbs in a series of studies. Calcium, osmotic, amino acids, free-radical scavengers, PH, reperfusion pressure, temperature, and duration were all beneficial factors that could be controlled during reperfusion. Furthermore, an initial clinical trial in four patients demonstrated the deleterious consequence of reperfusion was minimized by the controlled reperfusion strategy.

Condition-controlled reperfusion was one of the most important strategies. In 2001, Unal et al. [59] demonstrated that gradual reperfusion reduced neutrophil accumulation, superoxide radical occurrence, and tissue infarction in the rat hind limb model. Their key step was gradual increase of blood flow rate from 25% to 100% of pre-ischemic baseline value by adjusting arterial clamp. For a while of 13 min, the blood-flow was controlled not to be over 1.5 times of the baseline value. In 2009, Dick et al. [60] reported another study with positive result. The initial reperfusion flow (0.3 mL/min) and reperfusate (15 °C, ad 1000 IU heparin and Ringer's lactate 1000 mL) were controlled for 20 min after 4 h of complete arterial ischemia of rodent limbs. They believed that IRI of limbs could be significantly alleviated by modifying initial perfusate and conditions. Single chemical composition such as recombinant bactericidal/permeability-increasing protein, plasma rich in growth factors, polyadenosine diphosphate-ribose polymerase had been shown to be beneficial in rat limb reperfusion [61–63].

Recent studies also demonstrated the protective effects of controlled reperfusion in Yorkshire swine. Reperfusion of one whole blood to one crystal solution at 60 mmHg pressure was used in ischemic limbs for 30 min, which yielded a significant reduction of oxidative stress and inflammatory response [64]. Another study on acute ischemic limbs of pig also showed that a small amount of reperfusate (hypertonic saline 6% dextran 70) significantly reversed ischemia-induced hemodynamics abnormality and tissue metabolic disorder [65].

In the past 30 years, most of animal studies had determined the protective effects of controlled reperfusion on ischemic limbs. However, there have been no enough evidences to support its application in routine clinical practices. Defraigne [66] in 1997 and Allen [67] in 1998 respectively reported one anecdotal case of successful salvage of ischemic limb with the use of perfusate and condition controlled reperfusion, respectively. A case series study of seven patients with severe acute ischemia of lower limb showed a good outcome of amputation-free survival. Six of them achieved complete recovery after surgical embolectomy and 30 min of controlled crystal reperfusion via a simplified perfusion system [68]. In one clinical trial of phase I, Walker et al. [69] observed a notably reduction of systemic complications in controlled reperfusion arm with the use of specific perfusion devices (pressure and flow controlled) and complex component solutions after lower limb revascularization (0/14 vs 5/21), but with no positive significance in leg edema (1/14 vs 5/21) and amputation (2/14 vs 1/21) as compared with control arm. In 2013, Heilmann et al. [70] reported results of a randomized, open-label and multicenter trial. The trial enrolled 174 patients from 14 centers, who were randomly assigned to perform thrombembolectomy plus normal blood reperfusion or plus controlled reperfusion (crystalline solution mixed with blood at a maximal perfusion pressure of 60 mmHg). Unfortunately, amputation-free survival rate was 82.6% in controlled reperfusion group and 82.4% in conventional treatment group ($p > 0.05$).

5 Controlled Reperfusion of Kidney

Renal IRI affects early and long-term survival of renal transplantation. Its potential mechanisms include ROS generation during reperfusion, triggering cascades of deleterious cellular responses that cause inflammation, cell death, and acute renal failure [71].

In 1996, Haab et al. [72] showed that pressure-controlled reperfusion with the use of initial perfusion pressure of renal artery of 60 mmHg for 20 min was beneficial for porcine renal tolerance to ischemia. They believed that it was reasonable that the renal protection during cellular damage repair was contributed to oxygen distribution, rather than sodium reabsorption.

Blood flow is also an essential factor for reperfusion condition. Durrani et al. [73] demonstrated that a gradual increase of blood flow by gradual release of renal artery microclamps after 45 min of renal ischemia mitigated histopathological changes of rat kidneys, as compared with a sudden increase of blood flow by immediate release of the microclamps. Furthermore, Mancina et al. [74] showed that pressure-controlled reperfusion, rather than flow-controlled reperfusion, maintained physiological and structural integrity of isolated porcine kidney model throughout 1-h reperfusion period. These studies suggested that renal ischemic tolerance may be induced by controlled reperfusion.

Following the promising animal studies, a series of clinical trials were performed on the protective effects of controlled reperfusion in renal graft. In Maathuis et al. study [75], reperfusion at 30/20 mmHg pressure significantly improved renal histopathological changes compared with 60/40 mmHg, with detection of less damage to proximal tubules, less release of ROS and inflammatory cytokines, and better cortical perfusion. Recently, Wszola et al. [76] found that renal reperfusion with a pressure-driven device during renal transplantation reduced renal resistance during reperfusion, shortened duration of delayed graft function (DGF) after renal transplantation, and alleviated incidence of interstial fibrosis and tubular atrophy (IFTA) after 1 year of transplantation. However, the protective mechanisms of pressure-controlled reperfusion have not yet been fully elucidated, and its clinical application should be carefully considered.

Several chemical substances, such as endothelin-A receptor antagonist, CO-releasing molecules 3 and erythropoietin, may play roles in the protection against renal IRI. In 2001, Knoll et al. [77] demonstrated that high doses of endothelin-A receptor antagonists accelerated recovery of acute renal failure of ischemia by improvement of renal reperfusion of rats. They found that endothelin-A receptor antagonists provided the protection only after reperfusion, involving mechanisms of vasoconstriction and free oxygen free radicals. In a series of preclinical studies, pretreatment with erythropoietin was showed to have protection against renal IRI. Two independent double-blind randomized trials of erythropoietin from Germany and the United States demonstrated the protective effects during kidney transplantation [78, 79]. Recently, a meta-analysis from China concluded that the pretreatment of high-dose erythropoietin protected renal ischemic injury without increasing the susceptibility to adverse events [80]. Mechanisms of the renal protection of erythropoietin remain uncertain, which might be achieved by reduction of pro-apoptotic Bax protein, induction of heat shock kinase B, and prevention of oxidative stress [81].

6 Controlled Reperfusion in Brain

Brain is more vulnerable to ischemic insult than any other organs. Despite the fact that restoration of blood flow as quickly as possible is the primary principle for saving ischemic penumbra in brain, paradoxical cerebral IRI during rapid reperfusion may further aggravate cerebral damages with severe clinical consequences [82]. The protective efficacy of controlled reperfusion had been extensively documented in a variety of experimental settings, such as acute myocardial infarction, ischemic limb and solid organ transplantation. However, there had been few studies on the controlled reperfusion of cerebral ischemia, and some conclusions were conflict.

Allen et al. [83] found that component-controlled reperfusion with leukodepleted blood significantly reduced cerebral IRI of porcine model subjected to 90 min of hypothermic circulatory arrest, as evidenced by less oxygen free radical formation and endothelin-1 release. In contrast, another study on 30-min global cerebral ischemia of porcine model failed to obtain any protective efficacy after controlled reperfusion with leukocyte-filtered blood [84]. Allen et al. [85, 86] developed a porcine global brain ischemia model by cross-clamping of the innominate artery and the left subclavian artery. They demonstrated that low-pressure controlled reperfusion (<50 mmHg), which was protective in other organs, exhibited no beneficial effects on the brain after 30 min of global cerebral ischemia. However, pulsatile-flow controlled reperfusion in the same model significantly improved neurological recovery and reduced infarct size [87]. In addition, Munakata et al. [88] tested protective efficacy of low-flow controlled reperfusion in global brain ischemia model of canine. They observed a significant reduction of brain edema and apoptosis in low-flow reperfusion animals. Gao and colleagues [89] also demonstrated the neuroprotective effects of flow-controlled reperfusion in local cerebral ischemia of rat with common carotid artery and distal middle cerebral artery (CCA/dMCA) occlusion, of which cerebral blood flow was allowed to recover gradually from 25% to 100% of the baseline through sequential release of bilateral CCA in 5 min.

7 Summary

Timely reperfusion to save a great quantity of salvageable ischemic tissue is the definitive treatment of acute arterial occlusive diseases. On the other hand, the reperfusion without control of physics reperfusion condition or perfusate component sometimes may cause a serious IRI. Protections of controlled reperfusion against the IRI had been demonstrated by the most of experimental studies in the various animals and organs, and in several clinical studies. However, the protection was not consistently confirmed, especially in clinical trials. It is a reasonable imaging that the controlled reperfusion will be a promising strategy but entangled with frustration for a long time in the future. We should make every endeavor to seek out an integrated solution to the IRI. The endeavor is especially important today in the era of reperfusion treatments for acute ischemic stroke.

References

1. Buckberg GD. Controlled reperfusion after ischemia may be the unifying recovery denominator. J Thorac Cardiovasc Surg. 2010;140:12–8. 18.e11–2

2. Vinten-Johansen J. Controlled reperfusion is a rose by any other name. J Thorac Cardiovasc Surg. 2015;150:1649–50.
3. Bopassa JC. Protection of the ischemic myocardium during the reperfusion: between hope and reality. Am J Cardiovasc Dis. 2012;2:223–36.
4. Vinten-Johansen J. Postconditioning and controlled reperfusion: the nerve of it all. Anesthesiology. 2009;111:1177–9.
5. Ibanez B, Heusch G, Ovize M, Van de Werf F. Evolving therapies for myocardial ischemia/reperfusion injury. J Am Coll Cardiol. 2015;65:1454–71.
6. Follette DM, Fey K, Buckberg GD, Helly JJ Jr, Steed DL, Foglia RP, Maloney JV Jr. Reducing postischemic damage by temporary modification of reperfusate calcium, potassium, ph, and osmolarity. J Thorac Cardiovasc Surg. 1981;82:221–38.
7. Allen BS, Okamoto F, Buckberg GD, Bugyi H, Young H, Leaf J, Beyersdorf F, Sjostrand F, Maloney JV Jr. Immediate functional recovery after six hours of regional ischemia by careful control of conditions of reperfusion and composition of reperfusate. J Thorac Cardiovasc Surg. 1986;92:621–35.
8. Holman WL, Skinner JL, Killingsworth CR, Rogers JM, Melnick S, Ideker RE, Digerness SB. Controlled postcardioplegia reperfusion: mechanism for attenuation of reperfusion injury. J Thorac Cardiovasc Surg. 2000;119:1093–101.
9. Davies JE, Digerness SB, Goldberg SP, Killingsworth CR, Katholi CR, Brookes PS, Holman WL. Intra-myocyte ion homeostasis during ischemia-reperfusion injury: effects of pharmacologic preconditioning and controlled reperfusion. Ann Thorac Surg. 2003;76:1252–8. discussion 1258
10. Kofsky ER, Julia PL, Buckberg GD, Quillen JE, Acar C. Studies of controlled reperfusion after ischemia. XXII. Reperfusate composition: effects of leukocyte depletion of blood and blood cardioplegic reperfusates after acute coronary occlusion. J Thorac Cardiovasc Surg. 1991;101:350–9.
11. Fedalen PA, Piacentino V 3rd, Jeevanandam V, Fisher C, Greene J, Margulies KB, Houser SR, Furukawa S, Singhal AK, Goldman BI. Pharmacologic pre-conditioning and controlled reperfusion prevent ischemia-reperfusion injury after 30 minutes of hypoxia/ischemia in porcine hearts. J Heart Lung Transplant. 2003;22:1234–44.
12. Okazaki Y, Cao ZL, Ohtsubo S, Hamada M, Naito K, Rikitake K, Natsuaki M, Itoh T. Leukocyte-depleted reperfusion after long cardioplegic arrest attenuates ischemia-reperfusion injury of the coronary endothelium and myocardium in rabbit hearts. Eur J Cardiothorac Surg. 2000;18:90–7.
13. Massoudy P, Mempel T, Raschke P, Becker BF. Reduction of oxygen delivery during post-ischemic reperfusion protects the isolated guinea pig heart. Basic Res Cardiol. 1999;94:231–7.
14. Kaneda T, Ku K, Inoue T, Onoe M, Oku H. Postischemic reperfusion injury can be attenuated by oxygen tension control. Jpn Circ J. 2001;65:213–8.
15. Petrosillo G, Di Venosa N, Ruggiero FM, Pistolese M, D'Agostino D, Tiravanti E, Fiore T, Paradies G. Mitochondrial dysfunction associated with cardiac ischemia/reperfusion can be attenuated by oxygen tension control. Role of oxygen-free radicals and cardiolipin. Biochim Biophys Acta. 2005;1710:78–86.
16. Yamazaki S, Fujibayashi Y, Rajagopalan RE, Meerbaum S, Corday E. Effects of staged versus sudden reperfusion after acute coronary occlusion in the dog. J Am Coll Cardiol. 1986;7:564–72.
17. Okamoto F, Allen BS, Buckberg GD, Bugyi H, Leaf J. Reperfusion conditions: importance of ensuring gentle versus sudden reperfusion during relief of coronary occlusion. J Thorac Cardiovasc Surg. 1986;92:613–20.
18. Peng CF, Murphy ML, Colwell K, Straub KD. Controlled versus hyperemic flow during reperfusion of jeopardized ischemic myocardium. Am Heart J. 1989;117:515–22.
19. Takeo S, Liu JX, Tanonaka K, Nasa Y, Yabe K, Tanahashi H, Sudo H. Reperfusion at reduced flow rates enhances postischemic contractile recovery of perfused heart. Am J Phys. 1995;268:H2384–95.

20. Ferrera R, Michel P. Protection of the ischemic heart during reperfusion: role of the low flow to avoid calcium overload into the myocardium in a pig model. Transplant Proc. 2002;34:3265–7.
21. Mrak RE, Carry MM, Murphy ML, Peng CF, Straub KD. Reperfusion injury in ischemic myocardium: protective effect of controlled reperfusion. Am J Cardiovasc Pathol. 1990;3:217–24.
22. Klawitter PF, Murray HN, Clanton TL, Palmer BS, Angelos MG. Low flow after global ischemia to improve postischemic myocardial function and bioenergetics. Crit Care Med. 2002;30:2542–7.
23. Sato H, Jordan JE, Zhao ZQ, Sarvotham SS, Vinten-Johansen J. Gradual reperfusion reduces infarct size and endothelial injury but augments neutrophil accumulation. Ann Thorac Surg. 1997;64:1099–107.
24. Musiolik J, van Caster P, Skyschally A, Boengler K, Gres P, Schulz R, Heusch G. Reduction of infarct size by gentle reperfusion without activation of reperfusion injury salvage kinases in pigs. Cardiovasc Res. 2010;85:110–7.
25. Pisarenko OI, Shulzhenko VS, Studneva IM, Kapelko VI. Effects of gradual reperfusion on postischemic metabolism and functional recovery of isolated guinea pig heart. Biochem Med Metab Biol. 1993;50:127–34.
26. Hattori R, Matsui H, Kitano M, Ichihara Y, Ogawa S, Hirai M, Hayashi H, Saito H. Staged reperfusion preserves the coronary flow reserve, especially in the regions not severely damaged by ischemic injury in the canine heart. Angiology. 1998;49:991–1004.
27. Acar C, Partington MT, Buckberg GD. Studies of controlled reperfusion after ischemia. XVII. Reperfusion conditions: controlled reperfusion through an internal mammary artery graft—a new technique emphasizing fixed pressure versus fixed flow. J Thorac Cardiovasc Surg. 1990;100:724–36.
28. Bopassa JC, Nemlin C, Sebbag L, Rodriguez C, Ovize M, Ferrera R. Optimal time duration for low-pressure controlled reperfusion to efficiently protect ischemic rat heart. Transplant Proc. 2007;39:2615–6.
29. Nemlin C, Benhabbouche S, Bopassa JC, Sebbag L, Ovize M, Ferrera R. Optimal pressure for low pressure controlled reperfusion to efficiently protect ischemic heart: an experimental study in rats. Transplant Proc. 2009;41:703–4.
30. Bopassa JC, Michel P, Gateau-Roesch O, Ovize M, Ferrera R. Low-pressure reperfusion alters mitochondrial permeability transition. Am J Physiol Heart Circ Physiol. 2005;288:H2750–5.
31. Bopassa JC, Vandroux D, Ovize M, Ferrera R. Controlled reperfusion after hypothermic heart preservation inhibits mitochondrial permeability transition-pore opening and enhances functional recovery. Am J Physiol Heart Circ Physiol. 2006;291:H2265–71.
32. Bopassa JC, Ferrera R, Gateau-Roesch O, Couture-Lepetit E, Ovize M. Pi 3-kinase regulates the mitochondrial transition pore in controlled reperfusion and postconditioning. Cardiovasc Res. 2006;69:178–85.
33. Ferrera R, Benhabbouche S, Da Silva CC, Alam MR, Ovize M. Delayed low pressure at reperfusion: a new approach for cardioprotection. J Thorac Cardiovasc Surg. 2015;150:1641–1648.e1642.
34. Ueno T, Yamada T, Yoshikai M, Natsuaki M, Itoh T. Effect of gradual reperfusion on ventricular function after 6-h preservation. Cardiovasc Surg. 1993;1:695–700.
35. Gunnes S, Ytrehus K, Sorlie D, Helgesen KG, Mjos OD. Improved energy preservation following gentle reperfusion after hypothermic, ischemic cardioplegia in infarcted rat hearts. Eur J Cardiothorac Surg. 1987;1:139–43.
36. Lindal S, Gunnes S, Lund I, Straume BK, Jorgensen L, Sorlie D. Ultrastructural changes in rat hearts following cold cardioplegic ischemia of differing duration and differing modes of reperfusion. Scand J Thorac Cardiovasc Surg. 1990;24:213–22.
37. Fontan F, Madonna F, Naftel DC, Kirklin JW, Blackstone EH, Digerness S. The effect of reperfusion pressure on early outcomes after coronary artery bypass grafting. A randomized trial. J Thorac Cardiovasc Surg. 1994;107:265–70.
38. Zenati M, Yousem SA, Dowling RD, Stein KL, Griffith BP. Primary graft failure following pulmonary transplantation. Transplantation. 1990;50:165–7.

39. de Perrot M, Liu M, Waddell TK, Keshavjee S. Ischemia-reperfusion-induced lung injury. Am J Respir Crit Care Med. 2003;167:490–511.
40. Hopkinson DN, Bhabra MS, Odom NJ, Bridgewater BJ, Van Doorn CA, Hooper TL. Controlled pressure reperfusion of rat pulmonary grafts yields improved function after twenty-four-hours' cold storage in university of Wisconsin solution. J Heart Lung Transplant. 1996;15:283–90.
41. Bhabra MS, Hopkinson DN, Shaw TE, Hooper TL. Critical importance of the first 10 minutes of lung graft reperfusion after hypothermic storage. Ann Thorac Surg. 1996;61:1631–5.
42. Bhabra MS, Hopkinson DN, Shaw TE, Onwu N, Hooper TL. Controlled reperfusion protects lung grafts during a transient early increase in permeability. Ann Thorac Surg. 1998;65:187–92.
43. Pierre AF, DeCampos KN, Liu M, Edwards V, Cutz E, Slutsky AS, Keshavjee SH. Rapid reperfusion causes stress failure in ischemic rat lungs. J Thorac Cardiovasc Surg. 1998;116:932–42.
44. Sakamoto T, Yamashita C, Okada M. Efficacy of initial controlled perfusion pressure for ischemia-reperfusion injury in a 24-hour preserved lung. Ann Thorac Cardiovasc Surg. 1999;5:21–6.
45. Fiser SM, Tribble CG, Kern JA, Long SM, Kaza AK, Kron IL. Controlled perfusion decreases lung transplant reperfusion injury in the setting of high flow reperfusion. J Heart Lung Transplant. 2001;20:183.
46. Guth S, Prufer D, Kramm T, Mayer E. Length of pressure-controlled reperfusion is critical for reducing ischaemia-reperfusion injury in an isolated rabbit lung model. J Cardiothorac Surg. 2007;2:54.
47. Singh RR, Laubach VE, Ellman PI, Reece TB, Unger E, Kron IL, Tribble CG. Attenuation of lung reperfusion injury by modified ventilation and reperfusion techniques. J Heart Lung Transplant. 2006;25:1467–73.
48. Clark SC, Sudarshan C, Khanna R, Roughan J, Flecknell PA, Dark JH. Controlled reperfusion and pentoxifylline modulate reperfusion injury after single lung transplantation. J Thorac Cardiovasc Surg. 1998;115:1335–41.
49. Halldorsson AO, Kronon MT, Allen BS, Rahman S, Wang T. Lowering reperfusion pressure reduces the injury after pulmonary ischemia. Ann Thorac Surg. 2000;69:198–203. discussion 204
50. Halldorsson AO, Kronon M, Allen BS, Rahman S, Wang T, Layland M, Sidle D. Controlled reperfusion prevents pulmonary injury after 24 hours of lung preservation. Ann Thorac Surg. 1998;66:877–84. discussion 884–5
51. Halldorsson A, Kronon M, Allen BS, Bolling KS, Wang T, Rahman S, Feinberg H. Controlled reperfusion after lung ischemia: implications for improved function after lung transplantation. J Thorac Cardiovasc Surg. 1998;115:415–24. discussion 424–5
52. Lick SD, Brown PS Jr, Kurusz M, Vertrees RA, McQuitty CK, Johnston WE. Technique of controlled reperfusion of the transplanted lung in humans. Ann Thorac Surg. 2000;69:910–2.
53. Alvarado CG, Poston R, Hattler BG, Keenan RJ, Dauber J, Griffith B, McCurry KR. Effect of controlled reperfusion techniques in human lung transplantation. J Heart Lung Transplant. 2001;20:183–4.
54. Rong J, Ye S, Wu ZK, Chen GX, Liang MY, Liu H, Zhang JX, Huang WM. Controlled oxygen reperfusion protects the lung against early ischemia-reperfusion injury in cardiopulmonary bypasses by downregulating high mobility group box 1. Exp Lung Res. 2012;38:183–91.
55. Rong J, Ye S, Liang MY, Chen GX, Liu H, Zhang JX, Wu ZK. Receptor for advanced glycation end products involved in lung ischemia reperfusion injury in cardiopulmonary bypass attenuated by controlled oxygen reperfusion in a canine model. ASAIO J. 2013;59:302–8.
56. Slottosch I, Liakopoulos O, Kuhn E, Deppe A, Lopez-Pastorini A, Schwarz D, Neef K, Choi YH, Sterner-Kock A, Jung K, Muhlfeld C, Wahlers T. Controlled lung reperfusion to reduce pulmonary ischaemia/reperfusion injury after cardiopulmonary bypass in a porcine model. Interact Cardiovasc Thorac Surg. 2014;19:962–70.
57. Fukuda I, Chiyoya M, Taniguchi S, Fukuda W. Acute limb ischemia: contemporary approach. Gen Thorac Cardiovasc Surg. 2015;63:540–8.

58. Beyersdorf F. Protection of the ischemic skeletal muscle. Thorac Cardiovasc Surg. 1991;39:19–28.
59. Unal S, Ozmen S, DemIr Y, Yavuzer R, LatIfoglu O, Atabay K, Oguz M. The effect of gradually increased blood flow on ischemia-reperfusion injury. Ann Plast Surg. 2001;47:412–6.
60. Dick F, Li J, Giraud MN, Kalka C, Schmidli J, Tevaearai H. Basic control of reperfusion effectively protects against reperfusion injury in a realistic rodent model of acute limb ischemia. Circulation. 2008;118:1920–8.
61. Harkin DW, Barros D'Sa AA, Yassin MM, Hoper M, Halliday MI, Parks TG, Campbell FC. Recombinant bactericidal/permeability-increasing protein attenuates the systemic inflammatory response syndrome in lower limb ischemia-reperfusion injury. J Vasc Surg. 2001;33:840–6.
62. Anitua E, Pelacho B, Prado R, Aguirre JJ, Sanchez M, Padilla S, Aranguren XL, Abizanda G, Collantes M, Hernandez M, Perez-Ruiz A, Penuelas I, Orive G, Prosper F. Infiltration of plasma rich in growth factors enhances in vivo angiogenesis and improves reperfusion and tissue remodeling after severe hind limb ischemia. J Control Release. 2015;202:31–9.
63. Hua HT, Albadawi H, Entabi F, Conrad M, Stoner MC, Meriam BT, Sroufe R, Houser S, Lamuraglia GM, Watkins MT. Polyadenosine diphosphate-ribose polymerase inhibition modulates skeletal muscle injury following ischemia reperfusion. Arch Surg. 2005;140:344–51. discussion 351–2
64. Kenyeres P, Sinay L, Jancso G, Rabai M, Toth A, Toth K, Arato E. Controlled reperfusion reduces hemorheological alterations in a porcine infrarenal aortic-clamping ischemia-reperfusion model. Clin Hemorheol Microcirc. 2016;63:235–43.
65. Waagstein LM, Jivegard L, Haljamae H. Hypertonic saline infusion with or without dextran 70 in the reperfusion phase of experimental acute limb ischaemia. Eur J Vasc Endovasc Surg. 1997;13:285–95.
66. Defraigne JO, Pincemail J, Laroche C, Blaffart F, Limet R. Successful controlled limb reperfusion after severe prolonged ischemia. J Vasc Surg. 1997;26:346–50.
67. Allen BS, Hartz RS, Buckberg GD, Schuler JJ. Prevention of ischemic damage using controlled limb reperfusion. J Card Surg. 1998;13:224–7.
68. Wilhelm MP, Schlensak C, Hoh A, Knipping L, Mangold G, Dallmeier Rojas D, Beyersdorf F. Controlled reperfusion using a simplified perfusion system preserves function after acute and persistent limb ischemia: a preliminary study. J Vasc Surg. 2005;42:690–4.
69. Walker PM, Romaschin AD, Davis S, Piovesan J. Lower limb ischemia: phase 1 results of salvage perfusion. J Surg Res. 1999;84:193–8.
70. Heilmann C, Schmoor C, Siepe M, Schlensak C, Hoh A, Fraedrich G, Beyersdorf F. Controlled reperfusion versus conventional treatment of the acutely ischemic limb: results of a randomized, open-label, multicenter trial. Circ Cardiovasc Interv. 2013;6:417–27.
71. Malek M, Nematbakhsh M. Renal ischemia/reperfusion injury; from pathophysiology to treatment. J Renal Inj Prev. 2015;4:20–7.
72. Haab F, Julia P, Nochy D, Cambillau M, Fabiani JN, Thibault P. Improvement of postischemic renal function by limitation of initial reperfusion pressure. J Urol. 1996;155:1089–93.
73. Durrani NK, Yavuzer R, Mittal V, Bradford MM, Lobocki C, Silberberg B. The effect of gradually increased blood flow on ischemia-reperfusion injury in rat kidney. Am J Surg. 2006;191:334–7.
74. Mancina E, Kalenski J, Paschenda P, Beckers C, Bleilevens C, Boor P, Doorschodt BM, Tolba RH. Determination of the preferred conditions for the isolated perfusion of porcine kidneys. Eur Surg Res. 2015;54:44–54.
75. Maathuis MH, Manekeller S, van der Plaats A, Leuvenink HG, t Hart NA, Lier AB, Rakhorst G, Ploeg RJ, Minor T. Improved kidney graft function after preservation using a novel hypothermic machine perfusion device. Ann Surg. 2007;246:982–8. discussion 989–91
76. Wszola M, Kwiatkowski A, Diuwe P, Domagala P, Gorski L, Kieszek R, Berman A, Perkowska-Ptasinska A, Durlik M, Paczek L, Chmura A. One-year results of a prospective, randomized

trial comparing two machine perfusion devices used for kidney preservation. Transpl Int. 2013;26:1088–96.

77. Knoll T, Schult S, Birck R, Braun C, Michel MS, Bross S, Juenemann KP, Kirchengast M, Rohmeiss P. Therapeutic administration of an endothelin-a receptor antagonist after acute ischemic renal failure dose-dependently improves recovery of renal function. J Cardiovasc Pharmacol. 2001;37:483–8.

78. Hafer C, Becker T, Kielstein JT, Bahlmann E, Schwarz A, Grinzoff N, Drzymala D, Bonnard I, Richter N, Lehner F, Klempnauer J, Haller H, Traeder J, Fliser D. High-dose erythropoietin has no effect on short- or long-term graft function following deceased donor kidney transplantation. Kidney Int. 2012;81:314–20.

79. Sureshkumar KK, Hussain SM, Ko TY, Thai NL, Marcus RJ. Effect of high-dose erythropoietin on graft function after kidney transplantation: a randomized, double-blind clinical trial. Clin J Am Soc Nephrol. 2012;7:1498–506.

80. Xin H, Ge YZ, Wu R, Yin Q, Zhou LH, Shen JW, Lu TZ, Hu ZK, Wang M, Zhou CC, Wu JP, Li WC, Zhu JG, Jia RP. Effect of high-dose erythropoietin on graft function after kidney transplantation: a meta-analysis of randomized controlled trials. Biomed Pharmacother. 2015;69:29–33.

81. Hussein Ael A, Shokeir AA, Sarhan ME, El-Menabawy FR, Abd-Elmoneim HA, El-Nashar EM, Barakat NM. Effects of combined erythropoietin and epidermal growth factor on renal ischaemia/reperfusion injury: a randomized experimental controlled study. BJU Int. 2011;107:323–8.

82. Zhang L, Zhang ZG, Chopp M. The neurovascular unit and combination treatment strategies for stroke. Trends Pharmacol Sci. 2012;33:415–22.

83. Allen BS, Castella M, Buckberg GD, Tan Z. Conditioned blood reperfusion markedly enhances neurologic recovery after prolonged cerebral ischemia. J Thorac Cardiovasc Surg. 2003;126:1851–8.

84. Lindblom RP, Tovedal T, Norlin B, Hillered L, Popova SN, Alafuzoff I, Thelin S. Mechanical reperfusion with leucocyte-filtered blood does not prevent injury following global cerebral ischaemia. Eur J Cardiothorac Surg. 2016;51:773–82.

85. Allen BS, Ko Y, Buckberg GD, Sakhai S, Tan Z. Studies of isolated global brain ischaemia: I. A new large animal model of global brain ischaemia and its baseline perfusion studies. Eur J Cardiothorac Surg. 2012;41:1138–46.

86. Allen BS, Ko Y, Buckberg GD, Tan Z. Studies of isolated global brain ischaemia: II. Controlled reperfusion provides complete neurologic recovery following 30 min of warm ischaemia – the importance of perfusion pressure. Eur J Cardiothorac Surg. 2012;41:1147–54.

87. Allen BS, Ko Y, Buckberg GD, Tan Z. Studies of isolated global brain ischaemia: III. Influence of pulsatile flow during cerebral perfusion and its link to consistent full neurological recovery with controlled reperfusion following 30 min of global brain ischaemia. Eur J Cardiothorac Surg. 2012;41:1155–63.

88. Munakata H, Okada K, Hasegawa T, Hino Y, Kano H, Matsumori M, Okita Y. Controlled low-flow reperfusion after warm brain ischemia reduces reperfusion injury in canine model. Perfusion. 2010;25:159–68.

89. Gao X, Ren C, Zhao H. Protective effects of ischemic postconditioning compared with gradual reperfusion or preconditioning. J Neurosci Res. 2008;86:2505–11.

Chapter 16
Therapeutic Window Beyond Cerebral Ischemic Reperfusion Injury

Wengui Yu and Liping Liu

Abstract Reperfusion therapy has been one of the major breakthroughs in clinical medicine. Intravenous thrombolysis (IVT) with recombinant tissue plasminogen activator (tPA) remains the only FDA approved therapy for acute ischemic stroke (AIS). It is effective within 4.5 h of symptom onset and better if given earlier. More recently, five randomized controlled trials (RCTs) demonstrated the efficacy of endovascular therapy (EVT) for AIS from large vessel occlusion (LVO) in the anterior circulation within 6–12 h of symptom onset. Reperfusion injury with symptomatic intracerebral hemorrhage (sICH) is the most feared complication of reperfusion therapy. However, the rate of sICH only increases slightly with longer delay to reperfusion. Numerous case reports and cohort studies have supported the expanding of therapeutic window up to 24 h in the anterior circulation and beyond in the posterior circulation. In patients with acute stroke from LVO in the posterior circulation, EVT up to 48 h after symptom onset was seldom futile in the absence of extensive baseline ischemia and longer treatment delay does not increase the risk of sICH. Such clinical studies have clearly pushed the boundary of our understanding about reperfusion injury and therapeutic time window. Laboratory and translational studies may better define collateral circulation, venous drainage, and the molecular mechanisms of contrast extravasation, hemorrhagic conversion, and sICH. A solid scientific foundation is needed for expanding therapeutic window way beyond reperfusion injury.

Keywords Acute ischemic stroke · Anterior circulation · Basilar artery occlusion · Posterior circulation · Reperfusion injury · Therapeutic window · Thrombectomy · And thrombolysis

W. Yu (✉)
Department of Neurology, University of California, Irvine, CA, USA
e-mail: wyu@uci.edu

L. Liu
Department of Neurology, Tiantan Hospital, Beijing Capital Medical University, Beijing, China

© Springer International Publishing AG, part of Springer Nature 2018   245
W. Jiang et al. (eds.), *Cerebral Ischemic Reperfusion Injuries (CIRI)*, Springer Series in Translational Stroke Research, https://doi.org/10.1007/978-3-319-90194-7_16

1   Introduction

Reperfusion therapies for acute ischemic stroke (AIS) may be the most exciting advances in clinical neuroscience in the last 2 decades. In 1995, intravenous thrombolysis (IVT) with recombinant tissue plasminogen activator (tPA) was shown to be effective for AIS within 3 h of symptom onset by a randomized controlled study (RCT) [1]. The United States Food and Agricultural Agency (FDA) approved iv tPA for AIS in 1996. A few RCTs were then conducted to examine the effect of intravenous (iv) tPA within 3–6 h of symptom onset [2–4]. In 2008, the European Cooperative Acute Stroke Study III (ECASS III) showed benefit of iv tPA for AIS within 4.5 h of symptom onset [5–8]. Analysis of pooled data revealed that time to treatment initiation is the single most important outcome predictor of IVT [9, 10].

Although there were reports of successful endovascular therapy (EVT) for large vessel occlusion (LVO) as early as 1999 [11–14], it was only in 2015 that five RCTs demonstrated the benefit of EVT for AIS from LVO in the anterior circulation [15–19]. The success was largely accredited to better study design and second-generation devices, primarily stent retrievers. Based on these recent landmark studies, the National guidelines and consensus statements in the United States, Europe, and Canada have all recommended EVT for AIS from LVO in the anterior circulation within 6 h of symptom onset. The Canadian guidelines have also recommended thrombectomy for selected patients up to 12 h of symptom onset according to evidence from the ESCAPE trial [17].

The therapeutic window has been increasing over the last few years to allow for maximal reversal of cerebral ischemia without worsening reperfusion injury.

2   Cerebral Ischemia Reperfusion Injury (CIRI)

In patients with acute ischemic stroke from LVO, the early pathophysiological responses include distal vasodilation to compensate for the poor circulation. The cerebrovascular reactivity or autoregulation is eventually lost. Recanalization of the occluded artery can therefore cause blood overflow into these dilated vessels, resulting in hyperperfusion, edema, capillary leak, or hemorrhage. Endothelial injury and breakdown of the blood-brain barrier (BBB) may be the underlying mechanism of CIRI [20]. Metabolic factors (such as lactic acid, adenosine, and free oxygen radical) and vasoactive neuropeptide also play a role. tPA promotes matrix degradation in the ischemic brain via activation of MMP-9 and increases both reperfusion damage and hemorrhage risk [21].

Hyperperfusion and reperfusion injury (RI) may cause headache, seizure, focal neurologic deficit, or mental status change. The diagnostic imaging criteria include an increase in cerebral blood flow (CBF) compared with baseline values, demonstration of hyperperfusion, intracerebral hemorrhage (ICH), or subarachnoid hemorrhage (SAH) on MRI or CT scans. Hyperperfusion syndrome or ICH may occur within hours or days of reperfusion therapy.

The major risk factors for reperfusion injury include age >75 years old, long-standing hypertension (HTN), severe chronic stenosis ≥90%, severe contralateral disease, poor collateral flow, diminished cerebrovascular reserve, postoperative HTN, and recent stroke [22]. The severity of reperfusion injury is dependent on the duration and intensity of cerebral hypoperfusion, grade of stenosis, and poor collateral flow.

Symptomatic ICH (sICH) is the most feared complication of reperfusion therapy. It was defined as intraparenchymal hematoma, SAH, or intraventricular hemorrhage associated with a worsening of the NIHSS score by ≥4 points within 24 h. The key imaging parameters associated with sICH are diffusion lesion volume [23], very low cerebral blood volume [24], and severely delayed blood flow [25]. These parameters indicate severe ischemia with poor collateral circulation.

Numerous preclinical studies and clinical trials have shown that reperfusion in the infarcted tissue without perfusion mismatch is ineffective and increases the risk of ICH [26]. Intensive control of HTN may reduce the incidence of reperfusion injury and ICH [27].

3   sICH from IVT in Randomized Controlled Studies (RCTs)

The rates of sICH from IVT trials and the control arms of recent thrombectomy studies are summarized in Table 16.1. In the original National Institute of Neurological Disease and Stroke (NINDS) study, IVT was shown to increase the risk of sICH by tenfold to 6.4% [1]. Patients with sICH had more severe deficits at base line (median NIHSS 20) than the whole study population (median NIHSS 14). Nine percent patients with ICH had CT evidence of cerebral edema at base line, as compared with 4% in the entire study population. Eleven deaths were attributed to ICH. Higher dose of tPA was associated with increased risk of ICH [2]. In the multivariate analysis of the pooled data, sICH was associated with rPA treatment and age, but not the onset to treatment time OTT or baseline NIHSS [4].

There appeared to have a trend of decreased sICH in recent years. The rate of sICH is 4.3% in the medical arm of the recent thrombectomy trials, as compared to 5.9% in those from pooled IVT data.

Of note, the EVT trials only enrolled patients with LVO in the anterior circulation [15–19].

Table 16.1 Symptomatic intracerebral hemorrhage (sICH) and outcomes in patients treated with IVT

Study	Patients (n)	Median NIHSS	IVT (h)	OTT time (min)	sICH (%)	Good outcome at 90 days (%)	Mortality (%)
NINDS	312	14	<3	–	6.4	39	17
ECASS[a]	247	11	<6	–	19.4	35.7	14.6
ECASS II	81	11	0–3	–	7		
	326	11	3–6	–	8.3		
ECASS III	418	9	3–4.5	–	7.9	52.4	6.1
ATLANTIS	272	10	3–5	–	7	34	11
MR CLEAN	267	18	<4.5	65–116	6.4	19.1	22
ESCAPE	150	16	<4.5	89–183	2.7	29.3	19
SWIFT PRIME	98	17	<4.5	80–155	3.1	35.5	12.4
ESTEND-IA	35	13	<4.5	105–180	6	40	20
RRVASCAT	103	17	<4.5	86–137	1.9	28.2	15.5

Abbreviation: *IVT* intravenous thrombolysis, *mRS* modified Rankin Score, *OTT* onset to treatment time, *sICH* symptomatic intracerebral hemorrhage
[a]iv tPA dose was higher at 1.1 mg/kg of body weight

4 sICH from IVT in Patients with Posterior Circulation Stroke (PCS)

The natural history of severe BAO was extremely poor. The Basilar Artery International Cooperation Study (BASICS) is the largest multicenter registry for acute basilar artery occlusion (BAO) [28].

Of the 27 BAO patients with either comatose (n = 26) or tetraplegic (n = 1) without acute therapy, 26 (96.3%) died within 1 month, and 1 (3.7%) patient survived with severe disability (mRS score 5) [28]. The grave prognosis of acute BAO makes it unethical to enroll such patients for RCTs.

As shown in Table 16.2, two of the largest case series demonstrated that IVT improves the outcome of acute BAO. Of the 72 patients with severe deficit from BASICS registry, 21% had good outcome with iv tPA therapy. The rate of sICH was 6% [28].

In the Helsinki series of BAO (n = 184) [29], 175 patients were treated with iv tPA, and 97% of them received concomitant full-dose heparin. Successful recanalization led to favorable outcome at 90 days in 50% of patients with no extensive brain infarction (pc-ASPECTS ≥ 8) and 5.9% of those with extensive brain infarction (pc-ASPECTS < 8; $P < 0.001$), irrespective of recanalization [29]. The rate of sICH was 11.5% in patients without extensive brain infarction (pc-ASECTS ≥ 8) and 25% in patients with extensive brain infarct (pc-ACPECTS < 8). Of note, only 40% these patients were treated with iv tPA within 6 h of symptom onset and 53.8% patients with extensive brain infarction (pc-ASECTS ≥ 8) received iv tPA beyond 12 h of symptom onset. The much higher rate of sICH was most likely the results of significantly delayed therapy for most of the patients and concomitant treatment with full-dose heparin.

In a prospective multi-center study with 883 consecutive patients treated with iv tPA within 4.5 h of symptom onset, patients with PCS (n = 95) had less sICH (0% versus 5%, $P < 0.05$) and more favorable outcome (66% versus 47%, $P < 0.001$) than patients with ACS [30]. Smaller infarct volumes and better collaterals may be the reason of lower sICH rates in PCS.

Taken together, the risk of sICH from IVT seems to be lower in PCS than in ASC. Onset to treatment time (OTT) and stroke severity predicts risk of sICH.

5   sICH from EVT for LVO in the Anterior Circulation

Five randomized controlled studies have demonstrated the benefit of second genera-tion endovascular devices (primarily, stent retriever) over medical therapy alone in patients with acute ischemic stroke due to LVO [15–19]. The rate of sICH ranged from 0% to 7.7% (Table 16.2). The aggregated risk of sICH was 4.3%, almost iden-tical to that of iv tPA control groups. The data suggest that thrombectomy with stent striever itself does not pose significant risk for sICH.

The MR CLEAN trial had a higher risk of sICH compared with other studies, likely due to less strict exclusion criteria. It did not use imaging to exclude par-ticipants with a large infarct core and poor collateral circulations. The probabil-ity of functional independence (mRS 0–2) at 3 months declined from 64.1% with symptom onset-to-reperfusion time of 180 min to 46.1% with symptom onset-to-reperfusion time of 480 min. However, rates of mortality, sICH, and major parenchymal hematoma did not significantly change with longer delay to reperfusion [31].

Table 16.2 sICH from IVT in patients with posterior circulation stroke (PCS)

Study	Patients (n)	Severity	Mean NIHSS	tPA within 6 h (%)	sICH (%)	Good outcome (%)	Mortality (%)
Schonewille (2009) [28]	49	Mild to moderate deficit*	–	84	6	53	16
	72	Severe deficit	–	81	6	21	46
Strbian (2013) [29]	132	pc-ACPECTS ≥ 8	16.5 ± 19	46	11.5	38	38.5
	52	pc-ACPECTS < 8	28 ± 12	25	25	5.9	66
Sarikaaya (2011) [30]	95	–	9.3 ± 7.9	100 (<4.5 h)	0	66	9

*Patients in a coma, with tetraplegia, or in a locked-in state were classified as having a severe stroke. Mild to moderate stroke was defined as any deficit that was less than severe.
sICH was defined by individual investigator [28] or NINDS criteria [29]. Good outcome is defined as mRS score 0–2
Abbreviation: *mRS* modified Rankin Score, *pc-ACPECTS* posterior circulation Acute Stroke Prognosis Early CT Score, *sICH* symptomatic intracerebral hemorrhage

6   sICH from EVT for LVO in the Posterior Circulation

There has been no randomized trial of EVT for AIS in the posterior circulation. Table 16.3 listed recent case series and multicenter registries of EVT in the posterior circulation. There were very low rates of sICH in most reports. However, a few case series had higher rate of sICH due to wire perforation, artery dissection, or use of intra-arterial (IA) urokinase [32–34]. Carneiro et al. described three cases of sICH from arterial dissection or perforation [33]. All of them received IVT prior to EVT and had right posterior cerebral artery injury by microwire and microcatheter manipulation (i.e. unrelated to the stent retriever). One patient had good recovery at 3 months but the other two patients had severe SAH and poor outcome.

In the largest multicenter registry of 148 patients with BAO, intracranial hemorrhage occurred in 9 (6%) patients (two SAH, five ICH, two combinations of SAH and ICH) [35]. Three of the ICHs were classified as parenchymal hematoma type II, and 2 as hemorrhagic infarction type II. One hemorrhagic complication was classified as asymptomatic and two SAHs were fatal. All hemorrhagic complications occurred in patients being treated with only one device (six cases with stent retrievers). Four of the nine hemorrhagic events (44%) occurred in patients receiving a combination of IVT, IA thrombolysis and mechanical thrombectomy.

Among the 705 patients from the pooled data, the rate of sICH is 4.5%. With improvement in EVT technology and reduction of guidewire or microcatheter-related injury, the rate of sICH should be much lower (Table 16.4).

Table 16.3 sICH from EVT for LVO in the anterior circulation

Study	Patients (n)	Median NIHSS	EVT (h)	Median OTT time (min)	Type of stent retriever	sICH (%)	Good outcome (%)	Mortality (%)
MR CLEAN	233	17	<6	210–313	Solitaire/Trevo	7.7	32.6	21
ESCAPE	165	16	<12	176–359	Solitaire/Trevo	3.6	53.0	10.4
SWIFT PRIME	98	17	<6	165–275	Solitaire	1.0	60.2	9.2
ESTEND-IA	35	17	<6	166–251	Solitaire	0	71.4	8.6
RRVASCAT	103	17	<8	201–340	Solitaire	1.9	43.7	18.4
Total	634					4.3	46.1	

Abbreviations: *OTT* onset to treatment time (groin puncture)

Table 16.4 sICH from EVT for LVO in the posterior circulation

Author and year	Patients (n)	Median NIHSS	iv tPA (%)	Median OTT (h)	Type of stent retriever	sICH (%)	Good outcome (%)	Mortality (%)
Espinosa de Rueda (2013) [53]	18	20.4	–	7.3	Solitaire, Trevo	0	50	22
Möhlenbruch (2013) [54]	24	24	87	5.5	Solitaire	4	33	22
Nagel (2013) [55]	36	–	64	–	Solitaire	3	31	29
Park (2013) [56]	16	12.2	25	7.4	Solitaire	0	56	36
Mourand (2013) [32]	31	38	61	8.5	Solitaire	16[a]	35	32
Raoult (2013) [57]	12	18.8	17	7.8	Solitaire	8	58	
Mordasini (2013) [58]	14	21 (5–36)	35.7	6.9 (4–24)	Solitaire	0	28.6	25
Andersson (2013) [59]	28	–	–	–	Solitaire	0	57	35.7
Lefevre (2013) [60]	25	20.3	–	–	Solitaire	4	56	21
Fesl (2014) [61]	21	–	52.4	–	Solitaire, Penumbra	0	28	33
Baek (2014) [62]	25	11	24	5.1	Solitaire	0	48	36
Gory (2015) [63]	22	13.6 (4–22)	36	8 (2.5–12.5)	Solitaire	14	27	12
Singer (2015) [35]	148	20	59	–	Stent retriever	6	34	35
Carneiro (2015) [33]	24	20 (12–36)	46	–	Trevo, Solitaire	12.5[b]	21	33
Yoon (2015) [46]	50	10.5 (7.7–16)	28	4.6 (3.4–5.7)	Trevo, Solitaire	0	54	33
Mokin (2016) [64]	100	19.2 ± 8.2	32	9.4 ± 7.8	Trevo, Solitaire Penumbra	5	35	12
Du (2016) [65]	21	25.57 ± 5.20	–	9.7 (6–16)	Solitaire	0	38.1	–
Van Houwelingen (2016) [34]	38	21 (15–32)	71	4.8 (3.6–6.3)	Trevo, Solitaire IA urokinase	5[c]	37	39

(continued)

Table 16.4 (continued)

Author and year	Patients (n)	Median NIHSS	iv tPA (%)	Median OTT (h)	Type of stent retriever	sICH (%)	Good outcome (%)	Mortality (%)
Alonso de Lecinana (2016) [66]	52	–	–	6.4 (5.3–9)	Trevo, Solitaire	2	40	33
Pooled data	**705**					**4.5**		

Abbreviations: *sICH* symptomatic intracranial hemorrhage, *OTT* onset to endovascular therapy time
sICH was defined as any parenchymal hematoma, subarachnoid hemorrhage or intraventricular hemorrhage associated with a worsening of the NIHSS score by ≥4 within 24 h. A modified Rankin score 0–2 was defined as good outcome
[a]Patient had higher NIHSS scores and five symptomatic intracranial hemorrhages were related to the procedure
[b]All the three complications were related to arterial well dissection/perforation
[c]Two sICH (5%) were seen in patients treated with IA urokinase

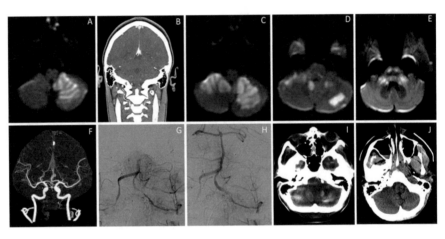

Fig. 16.1 Thrombectomy of BA occlusion in patient with recurrent stroke without reperfusion injury. Twenty-five year old woman with history of chiropractic neck manipulation presented with dizziness and unsteadiness. Image studies showed left cerebellar infarct (**a**) and bilateral vertebral artery dissection at C_1–C_2 level (**b**). She developed recurrent stroke on aspirin (**c**, **d**) and became locked-in from BA occlusion (**e**, **f**) on both aspirin and coumadin therapy within 2 weeks. Following successful thrombectomy (**g**, **h**), CT head showed contrast extravasation (**i**), which resolved overnight on repeat CT (**j**). Patient recovered well with only mild diplopia, hearing loss, and subtle left sided weakness

7   Contrast Extravasation Post EVT

Of note, it is very important to differentiate ICH from contrast extravasation. Contrast extravasation due to increased permeability of the BBB may be the most common hyperdense lesions on post-procedural CT scan [36–38]. Contrast staining usually lasts less than 24 h (Fig. 16.1). In contrast, post-procedural hemorrhage persists for days to weeks. Definite identification of contrast extravasation requires demonstration of contrast washout within 48 h.

8   Collateral Circulation and CIRI

The vasculature of the brainstem is different from that in the anterior circulation [39]. Brainstem receives collateral supply from anterior circulation via posterior communicating arteries (Pcoms). Acute BAO blocks small brainstem perforators and causes substantial ischemic risk to vital brainstem tissues. In patients with sufficient collateral retrograde flow from the Pcoms, EVT at any point in time would prevent significant infarctions. Collateral status was shown to be an independent predictor of outcome [35]. Increased OTT dose not increase rates of sICH in the posterior circulation.

Posterior circulation appears to have dynamic collaterals during BAO. Once the trunk of BA is occluded by a embolus, blood pressure at the junction of posterior cerebral arteries (PCAs) drops immediately and blood flow from the anterior circulation is partially diverted via Pcoms to fill the void in the distal BA, maintaining the patency of the superior cerebellar arteries (SCAs), the perforators, and the circumferential arteries. Effective collateral flow to the BA may prolong ischemic tolerance and lead to more favorable outcomes [35]. In a case series of acute BAO, five (83%) of the six patients with good distal BA collaterals had good functional outcome from intra-arterial urokinase [40]. Of the six patients without distal BA collaterals, only one (17%) had good functional outcome. Other studies have showed the significance of collaterals as a prognostic predictor in the anterior circulation [41, 42].

Brainstem also receives collateral supply from the vertebral arteries (VAs) and posterior inferior cerebellar arteries (PICAs). The anterior spinal arteries originate unilaterally or bilaterally from the distal VA, PICA, or the ascending cervical arteries [43, 44]. They can also provide reverse flow to PICAs or VAs. PICA provides collateral flow to the anterior inferior cerebellar artery (AICA) and SCA and thus to the brainstem perforating arterioles. Considerable amounts of artery-to-artery anastomoses exist between the superficial brainstem arteries. However, there is much less anastomoses between the internal brainstem arteries [44]. In patient with BAO, the collateral circulation may protect the brainstem for a while if the clot does not block the perforating arteries. Significant collaterals minimize the disruption of BBB and reduces the risk of sICH.

9   Expanding Therapeutic Window Beyond CIRI

In patients with LVO in the anterior circulation, ESCAPE trial has demonstrated that EVT is beneficial for selected patients up to 12 h of symptom onset [17]. Meta-analysis of the five landmark RCTs has demonstrated that the rate of sICH and major parenchymal hematoma increased slightly with longer delay to reperfusion (from 2.9% to 5.5%, and 2.5% to 8.1%, respectively) [31]. DAWN Trial is a clinical study designed to compare mechanical thrombectomy with the Trevo® Retriever plus medical therapy against medical therapy alone within 6–24 h after last known well time. On March 8, 2017, the independent Data Safety Monitoring Board (DSMB) recommended stopping study enrollment based on a pre-planned interim data review of the first 200 patients. The results were presented on May 16, 2017 at the European Stroke Organization Conference (ESOC). Treatment with the Trevo® Retriever significantly improved functional independence at 90 days compared to medical management alone (48.6% vs 13.1%). It is a 73% relative reduction in disability. The study showed that one in 2.8 patients treated with the Trevo Retriever within 24 h of a stroke is saved from severe disability. The Dawn trial showed eloquent data to support the expanding of therapeutic window up to 24 h in anterior circulation.

In patients with acute stroke from LVO in the posterior circulation, numerous case series and registries demonstrated that EVT of BAO up to 48 h after symptom onset was seldom futile in the absence of extensive baseline ischemia [28, 29, 35, 45, 46]. The observed efficacy was independent of time to treatment [28, 29, 45]. Longer treatment delays does not increase the risk of sICH [28, 29].

EVT has distinct advantages over IVT or intra-arterial thrombolysis. First, it may work rapidly and achieve recanalization within a few minutes rather than hours. Second, at experienced high volume centers, EVT is likely to be associated with lower risk of reperfusion injury due to comprehensive multidisciplinary care. Third, EVT is more effective in removing large embolus in proximal arteries. Lastly Multimodality imaging technology can be used to assess the benefit and risk of delayed EVT. While contrast extravasation is common in patients with large infarcts (Fig. 16.1), advanced age (Figs. 16.2 and 16.3), or matched perfusion deficit (Fig. 16.4) in the posterior circulation, sICH is actually very rare with tight periprocedural blood pressure control. Given significant efficacy and low risk of sICH, patients with BAO should be considered for treatment with EVT even beyond 48 h of symptom onset.

Highly-selected patients have been treated with EVT even at subacute or chronic phase well beyond days or weeks of symptom onset in the past [47–52, 67, 68]. However, such treatment has been controversial and considered experimental.

10   Implications for Translational Research

Numerous clinical studies and randomized trials have pushed the boundary of our understandings in acute ischemic stroke, recanalization and reperfusion injury. The molecular mechanisms of contrast extravasation, hemorrhagic conversion, and intracerebral hemorrhage remain unclear. Intensive blood pressure control may be important post EVT. However, we are not sure what will be the optimal blood pressure control post recanalization. It is also puzzling why pontine hemorrhage is

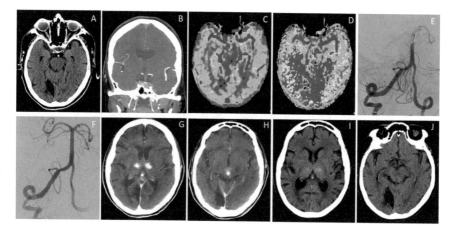

Fig. 16.2 Contrast extravasation from thrombectomy (OTT: 4 h) in elderly patient. Seventy-nine year old woman with history of A Fib on Coumadin presented with sudden onset of unresponsiveness at 1:30 pm. Image studies showed dense BA sign (**a**). Her INR was 1.57. After iv tPA at 4:20 pm, CTA and CT perfusion showed BA occlusion and CBV/MTT mismatch in posterior circulation (**b–d**). She underwent thrombectomy at 5:30 pm (**e, f**). Post-procedure CT showed contrast extravasation (**g, h**) in bilateral thalamus and ventral midbrain, which resolved on repeat CT next morning (**i, j**)

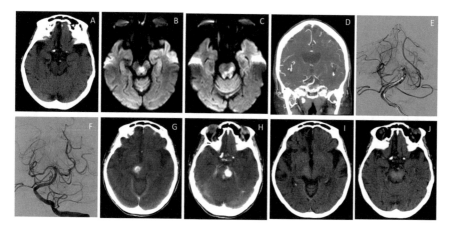

Fig. 16.3 Contrast extravasation from thrombectomy beyond 6 h window (OTT: 10 h and 51 min) in elderly patient. Seventy-five year old woman presented with drowsiness and speech difficulty at 3 am. Image studies showed dense BA sign (**a**), infarcts in midbrain tegmentum and left pons (**b, c**), and distal BA occlusion (**d**). She underwent thrombectomy at 1:51 pm (**e, f**). Post procedure CT showed contrast extravasation in right ventral midbrain and L pons (**g, h**), which resolved on repeat CT next day (**i, j**)

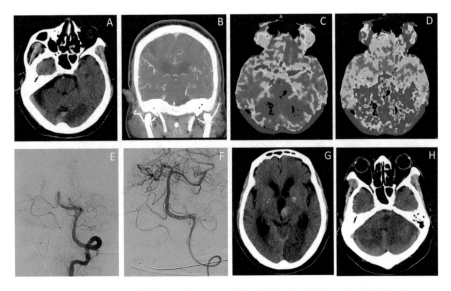

Fig. 16.4 Thrombectomy of BA occlusion beyond 6 h window in elderly patient. Eighty-five year old man with history of CAD and pacemaker was found unresponsiveness at 3:00 am. Last known well was 10 pm the prior night. She was GCS 4 on arrival. Image studies snowed dense BA sign and acute L occipital infarct (**a**, **b**). CTA and CT perfusion showed BA occlusion and CBV/MTT match (**c**, **d**). She underwent thrombectomy at 5:30 pm (**e**, **f**). Post procedure CT showed small left thalamic hemorrhage, left pontine and occipital infarcts (**g**, **h**)

common in hypertensive emergency while CIRI is rare in the brainstem after effective EVT. Better understanding of the collateral circulation and venous drainage in the posterior circulation will also be instrumental in formulating precision reperfusion therapy. We cannot keep "try and fail" in clinical practice for too long as in the past. Better laboratory and translational studies may hold the keys to some of the above unsolved mysteries.

References

1. The National Institute of Neurological Disorders and Stroke rt-PA Stroke Study Group. Tissue plasminogen activator for acute ischemic stroke. N Engl J Med. 1995;333:1581–7.
2. Hacke W, Kaste M, Fieschi C, Toni D, Lesaffre E, von Kummer R, et al. Intravenous thrombolysis with recombinant tissue plasminogen activator for acute hemispheric stroke. The European Cooperative Acute Stroke Study (ECASS). JAMA. 1995;274(13):1017–25.
3. Hacke W, Kaste M, Fieschi C, von Kummer R, Davalos A, Meier D, et al. Randomised double-blind placebo-controlled trial of thrombolytic therapy with intravenous alteplase in acute ischaemic stroke (ECASS II). Second European-Australasian Acute Stroke Study Investigators. Lancet. 1998;352(9136):1245–51.
4. Hacke W, Donnan G, Fieschi C, Kaste M, von Kummer R, Broderick JP, et al. Association of outcome with early stroke treatment: pooled analysis of ATLANTIS, ECASS, and NINDS rt-PA stroke trials. Lancet. 2004;363(9411):768–74.

5. Hacke W, Kaste M, Bluhmki E, Brozman M, Dávalos A, Guidetti D, et al. Thrombolysis with alteplase 3 to 4.5 hours after acute ischemic stroke. N Engl J Med. 2008;359(13):1317–29.
6. Wahlgren N, Ahmed N, Dávalos A, Hacke W, Millán M, Muir K, et al. Thrombolysis with alteplase 3–4.5 h after acute ischaemic stroke (SITS-ISTR): an observational study. Lancet. 2008;372(9646):1303–9.
7. Lansberg MG, Bluhmki E, Thijs VN. Efficacy and safety of tissue plasminogen activator 3 to 4.5 hours after acute ischemic stroke: a metaanalysis. Stroke. 2009;40(7):2438–41.
8. Saver JL, Gornbein J, Grotta J, et al. Number needed to treat to benefit and to harm for intravenous tissue plasminogen activator therapy in the 3- to 4.5-hour window. Stroke. 2009;40(7):2433–7.
9. Lees KR, Bluhmki E, von Kummer R, Brott TG, Toni D, Grotta JC, et al. Time to treatment with intravenous alteplase and outcome in stroke: an updated pooled analysis of ECASS, ATLANTIS, NINDS, and EPITHET trials. Lancet. 2010;375:1695–703.
10. Yeo LL, Paliwal P, Teoh HL, Seet RC, Chan BP, Liang S, et al. Timing of recanalization after intravenous thrombolysis and functional outcomes after acute ischemic stroke. JAMA Neurol. 2013;70:353–8.
11. Furlan A, Higashida R, Wechsler L, et al. Intra-arterial prourokinase for acute ischemic stroke. The PROACT II study: a randomized controlled trial. Prolyse in Acute Cerebral Thromboembolism. JAMA. 1999;282:2003–11. https://doi.org/10.1001/jama.282.21.2003.
12. Chopko BK, Kerber C, Wong W, Georgy B. Transcatheter snare removal of acute middle cerebral artery thromboembolism: technical case report. Neurosurgery. 2000;46:1529–31.
13. Mayer TE, Hamann GF, Brueckmann H. Treatment of basilar artery embolism with mechanical extraction device: necessity of flow reversal. Stroke. 2002;33:2232–5.
14. Yu W, Binder D, Foster-Barber A, Malek R, Smith WS, Higashida RT. Endovascular embolectomy of acute basilar artery occlusion. Neurology. 2003;61:1421–3.
15. Berkhemer OA, Fransen PS, Beumer D, van den Berg LA, Lingsma HF, Yoo AJ, et al. A randomized trial of intraarterial treatment for acute ischemic stroke. N Engl J Med. 2015;372:11–20.
16. Campbell BC, Mitchell PJ, Kleinig TJ, Dewey HM, Churilov L, Yassi N, et al. Endovascular therapy for ischemic stroke with perfusion-imaging selection. N Engl J Med. 2015;372:1009–18.
17. Goyal M, Demchuk AM, Menon BK, Eesa M, Rempel JL, Thornton J, et al. Randomized assessment of rapid endovascular treatment of ischemic stroke. N Engl J Med. 2015;372:1019–30.
18. Saver JL, Goyal M, Bonafe A, Diener HC, Levy EI, Pereira VM, et al. Stent-retriever thrombectomy after intravenous t-PA vs. t-PA alone in stroke. N Engl J Med. 2015;372:2285–95.
19. Jovin TG, Chamorro A, Cobo E, de Miquel MA, Molina CA, Rovira A, et al. Thrombectomy within 8 hours after symptom onset in ischemic stroke. N Engl J Med. 2015;372:2296–306.
20. Khatri P, McKinney AM, Swenson B, et al. Blood- brain barrier, reperfusion injury, and hemorrhagic transformation in acute ischemic stroke. Neurology. 2012;79(suppl 1):S52–7.
21. Cheng T, Petraglia AL, Li Z, Thiyagarajan M, Zhong Z, Wu Z, Liu D, Maggirwar SB, Deane R, Fernandez JA, LaRue B, Griffin JH, Chopp M, Zlokovic BV. Activated protein C inhibits tissue plasminogen activator-induced brain hemorrhage. Nat Med. 2006;12:1278–85.
22. Abou-Chebl A, Yadav JS, Reginelli JP, et al. Intracranial hemorrhage and hyperperfusion syndrome following carotid artery stenting: risk factors, prevention, and treatment. J Am Coll Cardiol. 2004;43:1596–601.
23. Singer OC, Humpich MC, Fiehler J, Albers GW, Lansberg MG, Kastrup A, Rovira A, Liebeskind DS, Gass A, Rosso C, Derex L, Kim JS, Neumann-Haefelin T. Risk for symptomatic intracerebral hemorrhage after thrombolysis assessed by diffusion-weighted magnetic resonance imaging. Ann Neurol. 2008;63:52–60.
24. Albers GW, Thiji VN, Wechsler L, et al. Magnetic resonance imaging profiles predict clinical response to early reperfusion: the diffusion and perfusion imaging evaluation for understanding stroke evolution (DEFUSE) study. Ann Neurol. 2006;60:508–17.
25. Campbell BC, Christensen S, Parsons MW, et al. Advanced imaging improves prediction of hemorrhage after stroke thrombolysis. Ann Neurol. 2013;73:510–9.
26. Gupta R, Vora NA, Horowitz MB, et al. Multimodal reperfusion therapy for acute ischemic stroke: factors predicting vessel recanalization. Stroke. 2006;37:986–90.

27. Abou-Chebl A, Reginelli J, Bajzer CT, Yadav JS. Intensive treatment of hypertension decreases the risk of hyperperfusion and intracerebral hemorrhage following carotid artery stenting. Catheter Cardiovasc Interv. 2007;69:690–6.
28. Schonewille WJ, Wijman CA, Michel P, et al. Treatment and outcomes of acute basilar artery occlusion in the Basilar Artery International Cooperation Study (BASICS): a prospective registry study. Lancet Neurol. 2009;8:724–30.
29. Strbian D, Sairanen T, Silvennoinen H, et al. Thrombolysis of basilar artery occlusion: impact of baseline ischemia and time. Ann Neurol. 2013;73:688–94.
30. Sarikaya H, Arnold M, Engelter ST, Lyrer PA, Mattle HP, Georgiadis D, et al. Outcomes of intravenous thrombolysis in posterior versus anterior circulation stroke. Stroke. 2011;42:2498–502.
31. Saver JL, Goyal M, Lugt A, et al. Time to treatment with endovascular thrombectomy and outcomes from ischemic stroke: a meta-analysis. JAMA. 2016;316:1279–88.
32. Mourand I, Machi P, Milhaud D, et al. Mechanical thrombectomy with the solitaire device in acute basilar artery occlusion. J Neurointerv Surg. 2014;6:200–4.
33. Carneiro AA, Rodrigues JT, Pereira JP, Alves JV, Xavier JA. Mechanical thrombectomy in patients with acute basilar occlusion using stent retrievers. Interv Neuroradiol. 2015;21(6):710–4. https://doi.org/10.1177/1591019915609781.
34. van Houwelingen RC, Luijckx GJ, Mazuri A, et al. Safety and outcome of intra-arterial treatment for basilar artery occlusion. JAMA Neurol. 2016;73:1225–30.
35. Singer OC, Berkefeld J, Nolte CH, et al. Mechanical recanalization in basilar artery occlusion: the ENDOSTROKE study. Ann Neurol. 2015;77(3):415–24.
36. Smith WS. Safety of mechanical thrombectomy and intravenous tissue plasminogen activator in acute ischemic stroke. Results of the multi Mechanical Embolus Removal in Cerebral Ischemia (MERCI) trial, part I. AJNR Am J Neuroradiol. 2006;27:1177–82.
37. Nikoubashman O, Reich A, Gindullis M, et al. Clinical significance of post-interventional cerebral hyperdensities after endovascular mechanical thrombectomy in acute ischaemic stroke. Neuroradiology. 2014;56:41–50.
38. Parrilla G, García-Villalba B, Espinosa de Rueda M, et al. Hemorrhage/contrast staining areas after mechanical intra-arterial thrombectomy in acute ischemic stroke: imaging findings and clinical significance. AJNR Am J Neuroradiol. 2012;33:1791–6.
39. Lindsberg PJ, Pekkola J, Strbian D, Sairanen T, Mattle HP, Schroth G. Time window for recanalization in basilar artery occlusion: speculative synthesis. Neurology. 2015;85:1806–15. https://doi.org/10.1212/WNL.0000000000002129.
40. Cross DT, Moran CJ, Akins PT, Angtuaco EE, Derdeyn CP, Diringer MN. Collateral circulation and outcome after basilar artery thrombolysis. Am J Neuroradiol. 1998;19:1557–63.
41. Jung S, Mono ML, Fischer U, et al. Three-month and long-term outcomes and their predictors in acute basilar artery occlusion treated with intra-arterial thrombolysis. Stroke. 2011;42:1946–51.
42. Jung S, Gilgen M, Slotboom J, et al. Factors that determine penumbral tissue loss in acute ischaemic stroke. Brain. 2013;136:3554–60.
43. Kang HS, Han MH, Kim SH, Kwon OK, Roh HG, Koh YC. Anterior spinal artery as a collateral channel in cases of bilateral vertebral arterial steno-occlusive diseases. Am J Neuroradiol. 2007;28:222–5.
44. Lescher S, Samaan T, Berkefeld J. Evaluation of the pontine perforators of the basilar artery using digital subtraction angiography in high resolution and 3D rotation technique. Am J Neuroradiol. 2014;35:1942–7.
45. Vergouwen MD, Algra A, Pfefferkorn T, Weimar C, Rueckert CM, Thijs V, Kappelle LJ, Schonewille WJ. Time is brain(stem) in basilar artery occlusion. Stroke. 2012;43:3003–6.
46. Yoon W, Kim SK, Heo TW, Baek BH, Lee YY, Kang HK. Predictors of good outcome after stent-retriever thrombectomy in acute basilar artery occlusion. Stroke. 2015;46:2972–5.
47. Lai TT, Lavian M, Al-Khoury L, Suzuki S, Yu W. Case report of endovascular therapy of acute basilar artery occlusion: beyond the restricted time window. Neurol Cases. 2016;3(2):1–4.
48. Yu W, Kostanian V, Fisher M. Endovascular recanalization of basilar artery occlusion 80 days after symptom onset. Stroke. 2007;38(4):1387–9.

49. Grigoriadis S, Gomori JM, Grigoriadis N, Cohen JE. Clinically successful late recanalization of basilar artery occlusion in childhood: what are the odds? Case report and review of the literature. J Neurol Sci. 2007;260(1–2):256–60.
50. Kostanian, Yu W. Successful recanalization of chronic basilar artery occlusion. Neurol Cases. 2014;1(1):1–3.
51. Lin R, Aleu A, Jankowitz B, et al. Endovascular revascularization of chronic symptomatic vertebrobasilar occlusion. J Neuroimaging. 2012;22(1):74–9.
52. Aghaebrahim A, Jovin T, Jadhav AP, Noorian A, Gupta R, Nogueira RG. Endovascular recanalization of complete subacute to chronic atherosclerotic occlusions of intracranial arteries. J Neurointerv Surg. 2014;6:645–8.
53. Espinosa de Rueda M, Parrilla G, Zamarro J, et al. Treatment of acute vertebrobasilar occlusion using thrombectomy with stent retrievers: initial experience with 18 patients. AJNR Am J Neuroradiol. 2013;34:1044–8.
54. Möhlenbruch M, Stampfl S, Behrens L, et al. Mechanical thrombectomy with stent retrievers in acute basilar artery occlusion. AJNR Am J Neuroradiol. 2014;35:959–64.
55. Nagel S, Kellert L, Möhlenbruch M, et al. Improved clinical outcome after acute basilar artery occlusion since the introduction of endovascular thrombectomy devices. Cerebrovasc Dis. 2013;36:394–400.
56. Park BS, Kang CW, Kwon HJ, et al. Endovascular mechanical thrombectomy in basilar artery occlusion: initial experience. J Cerebrovasc Endovasc Neurosurg. 2013;15:137–44.
57. Raoult H, Eugène F, Ferré JC, et al. Prognostic factors for outcomes after mechanical thrombectomy with solitaire stent. J Neuroradiol. 2013;40:252–9.
58. Mordasini P, Brekenfeld C, Byrne JV, et al. Technical feasibility and application of mechanical thrombectomy with the solitaire FR revascularization device in acute basilar artery occlusion. AJNR Am J Neuroradiol. 2013;34:159–63.
59. Andersson T, Kuntze Söderqvist Å, Söderman M. Mechanical thrombectomy as the primary treatment for acute basilar artery occlusion: experience from 5 years of practice. J Neurointerv Surg. 2013;5:221–5.
60. Lefevre PH, Lainay C, Thouant P, et al. Solitaire FR as a first-line device in acute intracerebral occlusion: a single-centre retrospective analysis. J Neuroradiol. 2014;41:80–6.
61. Fesl G, Holtmannspoetter M, Patzig M, et al. Mechanical thrombectomy in basilar artery thrombosis: technical advances and safety in a 10-year experience. Cardiovasc Intervent Radiol. 2014;37:355–61.
62. Baek J, Yoon W, Kim S. Acute basilar artery occlusion: outcome of mechanical thrombectomy with solitaire stent within 8 hours of stroke onset. Am J Neuroradiol. 2014;35:989–93.
63. Gory B, Eldesouky I, Sivan-Hoffmann R, et al. Outcomes of stent retriever thrombectomy in basilar artery occlusion: an observational study and systematic review. J Neurol Neurosurg Psychiatry. 2016;87(5):520–5. https://doi.org/10.1136/jnnp-2014-310250.
64. Mokin M, Sonig A, Sivakanthan S, et al. Clinical and procedural predictors of outcomes from the endovascular treatment of posterior circulation strokes. Stroke. 2016;47:782–8.
65. Du S, Mao G, Li D, Qiu M, Nie Q, Zhu H, Yang Y, Zhang Y, Li Y, Wu Z. Mechanical thrombectomy with the Solitaire AB stent for treatment of acute basilar arteryocclusion: a single-center experience. J Clin Neurosci. 2016;32:67–71.
66. Alonso de Leciñana M, Kawiorski MM, Ximénez-Carrillo Á, Cruz-Culebras A, García Pastor A, Martínez-Sánchez P, et al. Madrid Stroke Network. Mechanical thrombectomy for basilar artery thrombosis: a comparison of outcomes with anterior circulation occlusions. J Neurointerv Surg. 2016. https://doi.org/10.1136/neurintsurg-2016-012797.
67. Dashti SR, Park MS, Stiefel MF, McDougall CG, Albuquerque FC. Endovascular recanalization of the subacute to chronically occluded basilar artery: initial experience and technical considerations. Neurosurgery. 2010;66(4):825–32.
68. Abou-Chebl A. Endovascular treatment of acute ischemic stroke may be safely performed with no time window limit in appropriately selected patients. Stroke. 2010;41:1996–2000.

Printed in the United States
By Bookmasters